WIDE-GAP LUMINESCENT MATERIALS:
Theory and Applications

THE KLUWER INTERNATIONAL SERIES IN:

ELECTRONIC MATERIALS: Science and Technology

Series Editor
Harry L. Tuller
Massachusetts Institute of Technology

Other books in the series:

DIAMOND: *Electronic Properties and Applications*
Lawrence S. Pan and Don R. Kania, Editors
ION IMPLANTATION: *Basics to Device Fabrication*
Emanucle Rimini, Author
SOLID STATE BATTERIES: *Materials Design and Optimization*
C. Julien and G. Nazri, Authors
SOL-GEL OPTICS: Processing and Applications
L.C. Klein, Editor
PHOTOREFRACTIVE EFFECTS AND MATERIALS
David Nolte, Editor
MATERIALS FOR OPTOELECTRONICS
Maurice Quillec, Editor
PIEZOELECTRIC ACTUATORS AND ULTRASONIC MOTORS
Kenji Uchino, Author

The Series **ELECTRONIC MATERIALS: Science and Technology** will address the following goals:

* Bridge the gap between theory and application
* Foster and facilitate communication among materials scientists, electrical engineers, physicists and chemists.
* Provide publication with an interdisciplinary approach in the following topic areas:

- Sensors and Actuators
- Electrically Active Ceramics & Polymers
- Structure-Property-Processing-Performance Correlations in Electronic Materials
- Electronically Active Interfaces
- High Tc Superconducting Materials
- Optoelectronic Materials
- Composite Materials
- Defect Engineering
- Solid State Ionics
- Electronic Materials in Energy Conversion-Solar Cells, High Energy Density Microbatteries, Solid State Fuel Cells, etc.

WIDE-GAP LUMINESCENT MATERIALS: Theory and Applications

edited by

Stanley R. Rotman

Ben Gurion University of the Negev

SPRINGER SCIENCE+BUSINESS MEDIA, LLC

ISBN 978-0-7923-9837-0 ISBN 978-1-4615-4100-4 (eBook)
DOI 10.1007/978-1-4615-4100-4

Library of Congress Cataloging-in-Publication Data

A C.I.P. Catalogue record for this book is available
from the Library of Congress.

Printed on acid-free paper.

TABLE of CONTENTS

Authors' Biographies

YEHOSHUA KALISKY

Yehoshua Kalisky received the Ph.D. degree in physical chemistry in 1980 from the Hebrew University of Jerusalem, Israel, where he engaged in research on the laser spectroscopy and energy transfer in solid state laser materials and laser system analysis.

During the period of 1981 - 1983, he was a postdoctoral employee at Xerox Corporation (Webster Research Center), investigating the photophysics of organic/polymeric materials for non-linear optical applications.

Since 1984 he has been with the Laser Department of the Nuclear Research Centre - Negev, Beer-Sheva, Israel. His research interests have primarily concentrated on the spectroscopic studies of rare earth and transition metal-doped crystals. This research led to new insights into the multisite nature of garnets and energy transfer processes among doped ions situated at different sites. This research also yielded practical applications such as the development of a surgical CW holmium laser at 2.1 μm and the pulsed version of that laser. Further studies of various garnets doped with Cr^{4+} led to a novel configuration of passively Q-switched Nd:YAG laser.

While on sabbatical at Innovative Lasers (Tucson, AZ), Dr. Kalisky designed and operated successfully the first Cr^{4+}:$Lu_3Al_5O_{12}$ tunable solid state laser. This laser was integrated into an analytical detection laser tool used by the microelectronics industry.

Dr. Kalisky is the author and co-author of more than 100 papers including conference presentations and invited talks. He also holds two US patents in laser technology.

Dr. Kalisky is a member of the Optical Society of America, SPIE - The International Society for Optical Engineering, and the Israel Physical Society.

E. V. ZHARIKOV

Evgenii V. Zharikov received the Ph.D. degree in electrical engineering from the P. N. Lebedev Physical Institute in 1975 and the Doctor of Sciences degree in electrical engineering from the General Physics Institute, Russian Academy of Sciences, in 1990.

In 1975 after receiving his Ph.D., he joined the P. N. Lebedev Physical Institute, Moscow, as a research scientist and worked there in the field of growth and study of nonlinear and laser crystals.

After the formation of the General Physics Institute of Academy of Sciences, Moscow, in 1983, he joined it as head of laboratory for crystal growth, and has been working there until now.

Since 1990, he has been working also as a professor at the Mendeleyev University of Chemical Technology of Russia in Moscow. Since 1995, he is the head of the Department for Materials of Quantum Electronics in Mendeleyev University. His main interests are material science, crystal growth, and the study of the physical properties of crystals for solid state lasers.

V. A. SMIRNOV

Valery A. Smirnov received his Ph.D. degree in physics from Moscow State University in 1970, and had been working there until 1981. In 1981, he joined the P. N. Lebedev Physical Institute, Moscow, as a research scientist.

After the formation of the General Physics Institute of Academy of Sciences, Moscow, in 1983, he joined it. He received his Doctor of Sciences degree in physics from the General Physics Institute, Moscow, in 1988, and the title of Professor in 1991.

Now he is working at GPI, Moscow, as the head of the laboratory for the spectroscopy of crystals. His main interests are the spectroscopy of laser crystals and energy transfer in solids.

STANLEY R. ROTMAN

Stanley R. Rotman received the BS, MS and PhD degrees from the Electrical Engineering and Computer Science Department of the Massachusetts Institute of Technology in 1979, 1980, and 1985, respectively. His doctoral thesis was entitled "Defect structure of luminescent garnets." In 1985 he joined the staff of the Science and Technology Division of the Institute for Defense Analyses, where he did research on solid-state lasers, IR target acquisition, and electronic materials characterization. In 1987 he was appointed senior lecturer in the Department of Electrical and Computer Engineering at Ben-Gurion University of the Negev, where he is now associate professr. His present research interests are the characterization of optical materials for solid-state lasers, modeling IR target acquisition, and digital image processing. He is a member of OSA, IEEE, and SPIE.

ALEXANDER LEMPICKI

Research Professor, Boston University, Boston, MA

Alexander Lempicki is a Research Professor in the Chemistry and Physics Departments of Boston University, Boston, MA. He earned his Ph.D. at the University of London, London, UK. He came to BU in 1983 from GTE Laboratories, where he was a Research Manager in Quantum Electronics, Electro-Optics and Luminescence. At GTEL he poineered the development of the first laser in the liquid state, and successfully extended the principles into the realm of ultrahigh power applications. His interests also included II-VI compounds, rare earth ions in crystals and liquids, nonlinear light scattering from liquids, high-resolution laser spectroscopy, development of new solid state laser materials, energy transfer in various material systems, optical properties of magnetic semiconductors, optically induced magnetization, luminescent solar energy collectors and glass ceramics, and the optical properties of transition metals in glasses and crystals. At Boston University he has applied this expertise to the phenomenon of radiation-induced scintillation and has substantially advanced both the state-of-the-art of materials development and the fundamental understanding of the governing processes. He has been a prime contributor in the development of two promising new scintillator materials, cerium-doped lutetium orthophosphate and orthoaluminate. Prof. Lempicki has written some 140 scientific publications and has received 18 patents.

ANDRJEZ J. WOJTOWICZ

Professor, N. Copernicus University, Torun, Poland
Research Professor, Boston Univesity, Boston, MA

Andrjez J. Wojtowicz is a Professor in the Physics Department of N. Copernicus University, Torun, Poland, where he directs a research group in the Laboratory of Solid State Optoelectronics. He also is a Research Professor in the Chemistry Department of Boston University, Boston, MA, to which he came some ten years ago as a postdoctoral fellow. His research interests include the optical, electronic, and magnetic properties of semiconducting materials, solid state lasers, new detector materials for high-energy radiation, low- and high-energy phosphors, and other materials for optoelectronic applications. He has been a major contributor in advancing the understanding of the interactions between ionizing radiation and matter, on both theoretical and practical levels. Prof. Wojtowicz is Principal or Co-Principal Investigator of several projects funded by the Polish Government, the European Community, the US Dept. of Energy, and the US National Institutes of Health. He has authored or coauthored some 20 scientific papers.

CHARLES BRECHER

Research Professor, Boston University, Boston, MA

Charles Brecher is a Research Professor in the Chemistry Department of Boston University, Boston, MA. HE joined BU in 1993 after some thirty years at GTE Laboratories in Waltham, MA and Bayside, NY. His career has focused on various aspects of the interaction of light with matter and their implications with regard to material structure and kinetic processes. He earned his Ph.D. at Columbia University, where he performed doctoral and postdoctoral research on the infrared spectroscopy of molecular crystals. At GTEL he conducted optical investigations on liquid laser materials, coordination chemistry, crystal structure, photoflash combustion kinetics, magnetic bubble memory materials, and photoluminescent and cathodoluminescent phosphors. He was one of the pioneers in the use of laser-induced fluroescence line-narrowing to probe the structure of glasses, and of time-resolved spectroscopy to characterize ablative processes in electric arcs. He developed new processes for the fabrication and processing of metal and ceramic powders by means of thermal plasmas, and has made numerous contributions in the field of optical ceramics. At Boston University, he is currently exploring the potential of activated optical ceramics as scintillator materials, with particular attention to the effect of grain size on the kinetic processes that govern the scintillation phenomenon. Prof. Becher is author or coauthor of some sixty scientific papers and has received thirteen patents.

BERNARD JACQUIER

Prof. Bernard Jacquier is a full professor at the University of Lyon 1, Lyon, France. He was recently the director of the Laboratoire de Physico-Chimie des Materiaux Luminescents. His present research is in the development of fibers for optical communications.

PREFACE

Electro-optic devices based on doped wide-band materials are present in industrial uses, in military applications and in everyday life. Whether one engages in laser surgery with a neodymium-YAG laser or one communicates overseas using optical fibers, the development of these materials is both scientifically and commercially of great interest.

Much of the most innovative work has been done in the last 15 years in this area. A minor revolution in optical fiber communications has occurred with the development of erbium-doped fiber amplifiers. Solid-state laser development shifted into high-gear with the theoretical and experimental study of doubly-doped garnet lasers. Recent developments on semiconductor laser arrays are making diode-pumped solid-state lasers commercially feasible.

The purpose of this book is to detail these developments and to point out that many of the same underlying physical processes control advances in several diverse applications. For example, the basic science of energy transfer will be discussed by Zharikov et al. and Rotman for energy transfer and dopant-defect interactions, respectively; it will also be crucial in understanding cerium-doped scintillators, neodymium-chromium lasers, and up-conversion fiber lasers. As another example, phonon-induced non-radiative relaxation will appear in every chapter in this book.

Conceptually, one can divide this book into two sections. The first three chapters are primarily involved with the basic physical processes involved. Chapters 1 and 2 (Kalisky and Zharikov et al.) contain the basic science involved in the doped wide-bandgap oxides and fluorides. The main hosts and the most popular dopants which are used are covered. Chapter 3 (my contribution) looks at the effects that defects can have on these luminescent materials, both through competition for energy and by shifts in the energy levels of the dopants.

The last three chapters are the applications, in which experts in each field examine the latest developments in solid-state lasers (Kalisky), scintillators (Lempicki et al.), and fiber optics (Jacquier). These fields are certainly not the only ones which are possible to include for applications; they are definitely areas in which major advances have occurred recently. In these chapters, certain basic scientific principles, e.g. f-shell and d-shell electrons, energy transfer and upconversion, repeat themselves several times, with different implications and specific uses.

The manner of presentation of the subject matter was left to the discretion and personal appreciation of the chapter author(s). We believe this book is useful as reference for hosts and dopants for doped wide band-gap materials, a starting text for researchers entering the field and, hopefully, a stimulus for all to consider new viewpoints in our common research.

HOSTS FOR SOLID-STATE LUMINESCENT SYSTEMS

Yehoshua Kalisky

Laser Department, Nuclear Research Centre-Negev
P.O. Box 9001, Beer-Sheva 84190, Israel

Abstract

Solid-state luminescent sources are based on luminescent dopants located in wide-bandgap non-luminescent hosts. Often, the particular host chosen can be as important as the dopant used. In this chapter, we briefly review the properties of the hosts which will influence the luminescent properties of our source.

HOSTS FOR SOLID-STATE LUMINESCENT SYSTEMS

Yehoshua Kalisky

Contents

Hosts for Solid-State Luminescent Systems

Yehoshua Kalisky

Laser Department, Nuclear Research Centre-Negev
P.O. Box 9001, Beer-Sheva 84190, Israel

1. Introduction

The success of a laser or phosphorescent system depends on the correct choice of the emitting ion or ions and of the host lattice. These interact in complicated ways which often are not apparent until experimentally tested. Nevertheless, certain parameters of the crystalline laser host, both which directly affect the dopants (e.g. crystal fields, site symmetries) and those which only indirectly affect the dopants (crystal hardness, thermal conductivity), can be analyzed ahead of time such that system performance can be predicted and improved. In this chapter, we will discuss the assumptions leading to a proper choice of a laser host material, which eventually lead to an efficient lasing system. (Note that while we emphasize lasers, the conclusions can be extrapolated to the luminescent systems.)

We will illustrate our conclusions by using the recent development of mid-infrared (Mid-IR) laser systems as our primary example. The main common hosts used for mid IR lasers are oxides such as yttrium aluminium garnet -$Y_3Al_5O_{12}$ (YAG), and other kinds of garnets, as well as oxide hosts based on aluminates and silicates. Another family of crystals widely used are based on fluorides like yttrium lithium fluoride-$YLiF_4$ (YLF), or barium yttrium or ytterbium fluoride - BaY_2F_8 or $BaYb_2F_8$, respectively. The family of disordered materials includes various types of glasses (especially oxides, phosphates and fluorides type glasses) and glass-ceramics, which combine properties of both glasses and crystals. The properties of most common laser hosts are discussed in detail by Kaminskii [1] and by Koechner [2].

2. Crystal Structure [3]

2.1 INTRODUCTION

A solid state material can be in an ordered or disordered state. An ordered structure is the basis for an ideal crystal, while a disordered state can be either in a glassy or amorphous state. An ideal crystalline state is constructed by an infinite repetition of the same structure; this structure by itself consists of a single atom or group of atoms or molecules.

The lattice is a group of space points arranged periodically. These points are a mathematical representation of the crystal structure. When atoms or molecules, defined as a ***basis of atoms,*** are coupled to these periodic space points in a one-to-one correspondence, the crystal structure is defined. The lattice is defined by three translation vectors a_1, a_2, a_3, and three corresponding arbitrary integers u_1, u_2, u_3 , whereas the smallest unit that defines lattice structure is the ***lattice unit cell.***

2.2 TYPES OF CRYSTALS

All three dimensional lattices are grouped into seven types of cells as follows: triclinic, monoclinic, orthorhombic, tetragonal, cubic, trigonal and hexagonal. As an example, for a cubic cell, its cell axes are $a_1=a_2=a_3$ and the angles between the axes are:$\alpha=\beta=\gamma$.

The orientation of a crystal plane can be specified by indices (hkl) which are the indices of the plane. The index that is defined by a plane that cuts an axis on the negative side has a negative sign and can be symbolized by a minus sign above the index (h k l).

2.3 FORCES BINDING THE CRYSTAL LATTICE

The solid state phase is characterized by binding forces that are stronger than those in liquid or gas phase. There are several binding forces in the solid state phase:

2.3.1 Van der Waals Interaction:

Van der Waals interaction results from local distortion of charge distribution around inert gas atoms at a distance R that is larger than their ionic radii. This distortion leads to local redistribution of the electrostatic charges around the atoms and to the formation of local induced dipole moments in the atoms of the crystal. The dipole moment interaction is a source for attractive forces between the crystal atoms at short distances. It varies as $-C/R^6$ where R is the separation between the two atoms and C is a constant proportional to the electronic polarizability, α, which is defined as: $P=\alpha \cdot E_{loc}$. In this definition P is the polarization defined as the dipole moment per unit volume, and E_{loc} is the local electric field at the atom.

2.3.2 Repulsive Forces:

Repulsive forces result from two main sources: The first source is the mutual overlap between the charge distribution wavefunctions around the atoms and is explained as follows: when atoms are brought together their charge distribution begins to overlap and the total energy of the atomic system is changed. At a close separation of the two atoms, the overlap of the charge distribution will increase to a point where electrons from one atom will occupy states of the neighboring atom thus violating the ***Pauli exclusion principle***. To avoid the violation of Pauli principle, electrons will have to occupy high energy states, thus increasing the total energy of the system. A second source for the repulsive interaction is the repulsive forces of the two nuclei. These repulsive forces increase the total energy of the system.

2.3.3 Ionic Bonding:

In the case of ionic crystals, the bonding between the crystal ions is of Coulombic nature, which results from electrostatic interaction between the charged ions. The electronic configuration of ionic crystals are closed electronic shells, like inert atoms. However, unlike the case of inert gases, the formation of a closed-shell configuration leads to a charged atom. As an example, if we take sodium chloride, the electronic configuration of neutral Na is: $1S^2\ 2S^2\ 3S^1$ while that of neutral Cl atom is: $1S^2\ 2S^2\ 3S^2\ 3P^5$. The configuration of the ionic state with filled shells will be:

Na^+: $1S^2\ 2S^2$ 3S, and for Cl^- it will be:

Cl^-: $1S^2\ 2S^2\ 3S^2\ 3P^6$

The electrical charge distribution around the ions in the ionic crystal is approximately spherically symmetric but is slightly distorted at short distances, close to the contact point. For ions with a charge $\pm q$, separated by an interionic distance r, the electrostatic field is given by $\pm q/r^2$, where the forces are attractive for charges of opposite signs, and repulsive for charges of the same signs. Although there is some charge polarization that may contribute to Van der Waals interaction in ionic crystals, most of the interaction results from electrostatic interaction.

2.3.4 Covalent Bonding:

Covalent or homopolar bonds are a strong bonding formed by two electrons contributed by the two neutral atoms that form the molecular structure. Electrons that form the bond are partly localized in the contact region of the atoms. The binding energy depends on the spin orientation so that the strongest binding corresponds to two antiparallel spins. This spin - dependent energy is called **exchange interaction.** If the electronic shells are not filled, the repulsive forces will become weaker and the bond length will be shorter. Electrons can overlap efficiently to form a low energy state (bonding state) or a high energy state (antibonding state).

3. Materials Hosting Luminescent Ions

There are generally three groups of hosts for luminescent ions:
1. Ordered materials such as crystals.
2. Disordered media such as glasses.
3. Materials that incorporate the properties of both glasses and crystals - glass ceramics.

Generally, a candidate for a host laser material should exhibit good optical, mechanical and thermal properties as will be discussed in the following section.

3.1 OPTICAL PROPERTIES

Optical properties of a host material include transmission of light through the crystal, scattering properties, absorption in the UV (which is extremely important in the case of flash lamp pumped lasers) and resistance of the laser material to color center formation under UV excitation. We also have to consider the changes of the refraction index with temperature, polarization properties of the host as a result of birefringence and the effect of the host on radiative and non-radiative processes. Upon doping the crystal with the active ion, optical and spectroscopic properties of the lasing ion should also considered. As a general rule, it is essential to minimize the processes that increase the losses in the crystal. Therefore absorption of light through the crystal and scattering of the transmitted light in the crystal will be minimized. In a doped crystal, it is also desirable to minimize the non-radiative rates, excited state absorption, or multiphoton excitations, and utilize efficiently the pumping light. Table 1 gives the optical properties of several standard laser hosts.

3.2 MECHANICAL AND THERMAL PROPERTIES

Mechanical properties of a laser host include hardness, strain and elastic modulus. The mechanical properties are important in two aspects:

1. Good mechanical properties will enable the crystal to sustain the extreme conditions of pumping and laser energy flux.
2. Good mechanical properties simplify the fabrication technique during drilling the crystal out of the boule, and later in the following steps of cutting and polishing. These advantages make the product more cost effective.

Table 1. A summary of the optical properties of YAG, GSGG and YLF [4]

Property	YLF ($YLiF_4$)	YAG ($Y_3Al_5O_{12}$)	GSGG ($Gd_3Sc_2Ga_3O_{12}$)
Index of Refraction, n	n_0=1.443 n_e=1.464 @ 628 nm n_0=1.448 n_e=1.470 @ 1.06mm	n=1.8347 @ 628 nm 1.82 @ 1.06 mm	n=1.97 @ 628 nm n=1.925 @1.06 mm
dn/dT, C^{-1}	$-0.91x10^{-6}$ ordinary $-2.86x10^{-6}$ extraordinary @578 nm	$8.9x10^{-6}$ @ 1060 nm $7.3x10^{-6}$	10.1x10-6 @1060nm
UV absorption	below 0.2 mm	@380 nm	color centers
Crystalline Structure	Tetragonal	Cubic 12.01 A lattice constant	Cubic 12.57 A lattice constant

Thermal properties which include: thermal conductivity, thermal shock parameter, and thermal expansion coefficient of the crystal are important parameters in the design of a laser crystal. Temperature gradients in the laser crystal will lead to gradients in refraction index, and hence to poor beam quality. Table 2 summarizes the main thermal properties of several host crystals.

Table 2: A summary of the main thermal and mechanical properties of YLF, YAG and GSGG at 300°K. [4]

Property	YLF ($YLiF_4$)	YAG ($Y_3Al_5O_{12}$)	GSGG ($Gd_3Sc_2Ga_3O_{12}$)
Thermal conductivity, k(W/cm-°K)	0.06	0.13-0.12	0.058-0.07
Thermal expansion coeff., α, (°C^{-1})	a axis: 13x 10^{-6} c axis:8x10^{-6}	[100] 8.2x10^{-6} [110] 7.7x10^{-6} [111] 7.8x10^{-6}	7.8x10^{-6}
Thermal shock resistance, (Watt/m)	140	1100-800	660
Growth Temperature (°C)	820	1950-1970	1860
Density, g*cm^{-3}	3.99- undoped	4.46±0.04	6.46-doped 6.439-undoped
Hardness (Moh)	4-5	8.5	7.5-8
Elastic modulus (Dynes/cm^2)	7.5x10^{11}	33x10^{11}	21x10^{11}
Specific heat, $Jg^{-1}K^{-1}$	0.79	0.59	0.488

3.3 CHEMICAL PROPERTIES

The chemical properties that are important for a laser material are chemical inertness, crystallographic matching between the size of the host unit cell and the ionic radius of the dopant, and charge compensation to maintain the electrical neutrality of the host.

In addition, the phonon energies of the network forming molecules of the crystal, the symmetric properties of the host and the crystal field within the ionic site are important parameters. All affect the luminescent properties of the dopant ions. Table 3 shows how the different host crystals, with their different lattice constants affect the distribution constant of the ions (i.e., the fraction of the ions in the crystal relative to the fraction in the melt), for the doping of Nd^{3+} and Cr^{3+}. Table 4 shows how the energy gap between the $^4T_2$ and ^2E levels of Cr^{+3} is affected by the host.

Table 3. Distribution coefficient for Cr^{3+} and Nd^{3+} and lattice constant parameters for several garnets

Host	Lattice Constant, Å	Distribution Coefficient, Nd^{3+}	Distribution Coefficient, Cr^{3+}
YAG	12.01	0.18 [5]	
GGG	12.3844	0.63	3.3
Ca,Zr-sub GGG	12.4535*	N/A	2.8
GSGG	12.561	0.6-0.75 [5]	~1

* Average value as a result of a multisite structure in this host.

Table 4. Energy gap between $^4T_2$ and ^2E levels of Cr^{3+} in several garnets

Energy gap ΔE, cm-1	Laser host
350	YSGG
300-335	GGG
600	YGG
50	GSGG
1043	YAG
-1000	LLGG
10	(CA,ZR)GGG*
481	YIGG**

*$(Gd,Ca)_3(Ga,Zr,Gd)_2Ga3O_{12}$ [6]

**$Y_3(In,Ga)_2Ga_3O_{12}$ [7]

4. Conclusions

In this brief chapter, we have tried to emphasize the characteristics of laser hosts which can affect the performance of a laser or a phosphor. In the next chapter, we will consider how the dopant is influenced by our hosts and how the dopant can be chosen to produce optimum results.

References

[1] Kaminskii, A., 1981: "Laser Crystals: Their Physics and Properties", Springer-Verlag, Berlin, Heidelberg.

[2] Koechner, W., 1996: *Solid State Laser Engineering,,* Springer-Verlag, Berlin, Heidelberg, New York.

[3] Kittel, C., 1976: "Introduction to Solid State Physics", 5th Edition, Chapter 3, John Wiley & Sons, New York, 1976.

[4] Most of the data for YLF and GSGG in Tables 1 and 2 are taken from product data sheet of Litton Airtron. Most of the data for YAG was taken from data sheet of Union Carbide Crystal Products.

[5] Denker, B. I., Osiko, V. V., Starikov, B. P., Timoshenkin, M. I., Shcherbakokv, I. A. and Yablonski, A. Ya., 1976: "Spectroscopic Properties of Neodymium-Activated Scandium Garnets," *Sov. J. Quant. Electron.* **6**, pp. 334-336, and references therein.

[6] Brenier, A., Boulon, G., Pedrini, C. and Madej, K., 1992: "Effects of Ca^{2+}, Zr^{4+} Ion Pairs on Spectroscopic Properties of Ca^{3+}-Doped $Gd_3Ga_5O_{12}$ Garnets", *J. Appl. Phys.* **71**, pp. 6062-6068.

[7] Yun-kui-Li, Zhang, B., Yu-xing Xia. In: *Conference on Lasers and Electro-Optics (Optical Society of America, 1993)*, p. 312, Paper CWJ37.

LUMINESCENT DOPANTS

E.V.Zharikov, V.A.Smirnov

General Physics Institute of Russian Academy of Sciences,

38 Vavilov Street, Moscow, 117942, Russia

Abstract

This chapter is a short review of the spectroscopy and laser operation of luminescent dopants in solids, focusing on the most relevant problems. The transition metal ions of the most practical interest from the iron, rare-earths, and actinides groups will be described. The processes involved in the conversion of the initial pump excitation to electron excitation, in particular intracenter transitions, interionic interactions, quenching and sensitization of luminescence, and up-conversion processes, will be discussed. The corresponding models and schemes of energy transfer will be given for the cases of isolated ions and a group of particles.

LUMINESCENT DOPANTS

E. V. Zharikov, V. A. Smirnov

Contents

Luminescent Dopants

E.V.Zharikov, V.A.Smirnov

General Physics Institute of Russian Academy of Sciences,

38 Vavilov Street, Moscow, 117942, Russia

1. Introduction

This chapter is devoted to the description of the spectroscopy and laser operation of luminescent dopants in solids. Luminescent dopants in crystals and glasses are usually transition group elements. A specific feature of these ions is the presence of internal partially filled electron shells, giving a magnetic moment to the ions and causing various electron transitions in the UV, visual, and IR regions of spectrum. The ions of transition metals possess a rich system of energy levels, which make them an excellent source of dopants for efficient luminophors emitting in various spectral ranges.

Both radiative and non-radiative transitions, non-radiative energy transfer, and up-conversion processes in luminescent dopants greatly influence the transformation of the pumping energy into luminescence.

The discovery of the nature of these phenomena and their contribution to the appearance of luminescence are of great importance, as is the development of the theory and corresponding quantitative models.

A great number of papers have been devoted to the study of the luminescent properties of transition metal ions, and numerous solid state lasers with various wavelengths and regimes of operation have been developed using these ions and active particles. Nevertheless, the potential of transition metal ions is still not exhausted.

2. Transition Metal Ions as Luminescent Dopants

The periodic system of elements contains five transition groups of elements: the iron group *(3d-elements),* the palladium group *(4d-elements),* the lanthanide (or rare earth elements) group *(4f-elements),* the platinum group *(5d-elements),* and the actinide group *(5f-elements).* In the elements belonging to these groups, the strict order of electron shells filling according to the atomic number of the element in the periodic table is broken. The filling of the *d*-shell competes with the filling of the *s*-shell having the next quantum number , i.e. the *3d*-shell competes with the *4s*-shell, *4d* with *5s*, and *5d* with *6s*. The filling of f-shells is even more complicated. Here, the *4f*-shell competes with *5d*- and *6s*-ones, and the *5f*-shell competes with *6d* and *7s*.

Incorporated into the crystal or glass lattice, luminescent dopants form "activator-host" systems, which can operate as the active media of solid state lasers, luminophors, scintillators, etc. With rare exception, paramagnetic ions used for these applications belong to *3d*- or *4f*-elements. That is why we will concentrate our attention on these elements. A short description of transition metal ions will be given below; it will focus on the ions with unfilled *3d*- and *4f*-shells which are the most widely used in solid state lasers. More detailed information can be found in reviews [1-7].

2.1. 3d-TRANSITION METALS

The transition metal group is also called the iron group, i.e. *3d*-elements with incomplete shells. This group includes elements from Sc

(21) to Zn (30). In the process of ion formation, a transition metal atom loses external *4s*-electrons and some *3d*-electrons and its electron configuration looks like $1s^2 2s^2 2p^6 3s^2 3p^6 3d^n$. Due to the lack of shielding *3d* electrons are strongly influenced by the external crystal field and, therefore, by lattice vibrations. As a result, vibronic structure arises in the spectrum of electron transitions of the ion. The transitions become electron-vibronic, and are accompanied by a change of both the electron energy of the dopant and the energy of the vibronic modes of the crystal.

The Tanabe-Sugano diagrams [8,9] give an indication of the behavior of transition metal ionic energy levels depending on the crystal field strength for different electron configurations and can be found in various reviews (see, for example, [1,5,6]). Vibronic-coupled broad bands, which comform to the vibronic structure of the spectrum of the impurity center states and are caused by electron-phonon coupling, arise in luminescence spectra alongside narrow lines comforming to pure electron transitions. The interaction between the electron levels of the dopant and the vibrational modes of the lattice causes the typical four-level structure to have widely tunable emission bands and large Stokes shifts. Tunability results from the strong vibronic coupling which has its own penalties, such as non-radiative relaxation and broadband ESA. The latter is a typical problem for $3d^2$-$3d^8$ configurations since they have a number of possible excited states [6].

The chemical aspect is also important, since it links the incorporation of transition metal ions to the crystal lattice in the desired valent state. To incorporate titanum ions into the Al_2O_3 crystal as Ti^{3+}, special procedures are required during the crystal growth process since titanum tends to the 4+ state (see Table 1).

Table 1

Chemical stability of valency states of transition metal ions (adapted from [11]).

Electron configuration	Ions	Stability against	
		oxidation	reduction
$3d^0$	Ti^{4+}	high	moderate
$3d^1$	Ti^{3+}	low	high
	V^{4+}	moderate	low
$3d^2$	Ti^{2+}	low	high
	V^{3+}	moderate	moderate
	Cr^{4+}	moderate	low
	Mn^{5+}	high	low
$3d^3$	V^{2+}	low	high
	Cr^{3+}	high	high
	Mn^{4+}	high	low
$3d^4$	Cr^{2+}	low	high
	Mn^{3+}	moderate	low
$3d^5$	Mn^{2+}	high	high
	Fe^{3+}	high	low
$3d^6$	Fe^{2+}	low	high
	Co^{3+}	high	low
$3d^7$	Co^{2+}	high	high
	Ni^{3+}	high	moderate
$3d^8$	Ni^{2+}	low	high
$3d^9$	Cu^{2+}	high	low

The nature of the host ions has a large influence on the valence state of the doped transition metal ions. For example, when manganese ions are inserted in sapphire $Al^{3+}{}_2O_3$, they demonstrate, in general, a 3+ valent state; in rutile, $Ti^{4+}O_2$, they possess a 4+ valent state; and in fluorides of divalent metals MeF_2, manganese ions usually keep a 2+ valency [1]. Additional incorporation of divalent ions Ca^{2+} or Mg^{2+} during crystal growth in Cr^{3+}-containing garnet host (containing only trivalent host ions) causes a compensational transition of the trivalent chromium to the 4+ state accompanied by a transfer from the octahedral site to a tetrahedral one [10].

In crystals, 3d ions usually reside in octahedral or tetrahedral coordination. Octahedral sites demonstrate higher crystal field splittings. This results in two important consequences. First, increasing the splitting reduces the non-radiative decay of the excited states and, thus, leads to an enhancement of the lifetime. On the other hand, octahedral sites have inverse symmetry and are characterized by a weaker transition strength, which results in a reduction of the emission cross-section. It should be emphasized that usually we deal with transition metal ions in slightly distorted octahedral sites, which cause the enlargement of the emission cross-section, accompanied with a corresponding reduction of lifetime. Since the optical transitions within the $3d^n$ configuration are parity-forbidden, the lifetime of the excited state is rather high (usually around 100 μs), while the emission cross-section is low, providing high energy storage, but requiring low-loss materials to obtain reasonable output laser efficiency [11].

On the other hand, the incorporation of 3d-ions into low symmetry positions (for example, tetrahedral sites) causes considerably higher transition oscillator strength. In this case, enlargement of emission cross-section is accompanied by a reduction of the radiative lifetime of the metastable states, which makes energy storage more difficult.

Trivalent Titanum

The Ti^{3+} ion is the simplest among the transition metal ions. Its $3d^1$ electron configuration has only two electron states in the octahedral site: the upper ^2E and low $^2T_2$ states (see Figure 1). The trigonal component of the field splits the low manifold into two levels. The lowest of them is split in turn by spin-orbital coupling into two components. Due to the Jahn-Teller dynamic effect, the ^2E manifold is split into two levels.

The main advantage of the $3d^1$ configuration is the absence of any upper *d*-electron states, which prevents undesired excited state absorption (ESA). On the other hand, in [12] it was noted that the possibility of ESA to uppper non-*d*-electron states, such as *p*-states or charge-transfer states, exists, but usually the states are too high in energy to have any effect. The authors of [13] have demonstrated, with the example of $Ti{:}Al_2O_3$, that the Ti-systems can exhibit strong ESA into the charge transition band which occurs in the Ti^{3+}-Ti^{4+} complex centers. In the case of $Ti{:}YAlO_3$ this leads to a nearly total suppression of lasing.

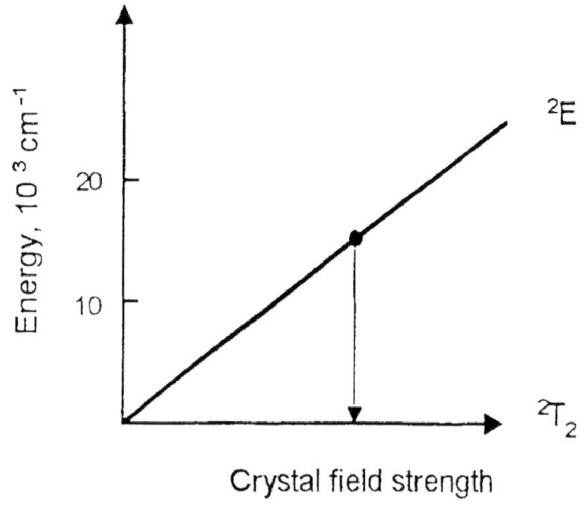

Fig. 1 Energy level diagram for Ti^{3+} ion ($3d^1$) in octahedral coordination

Laser action of Ti^{3+} ions was demonstrated in Al_2O_3 crystals [14] and many further efforts have been made to enhance the performance of this laser. The lifetime of Ti^{3+} in Al_2O_3 is τ=3.2 μs at room temperature and the emission cross-section σ_{em}~$3 \cdot 10^{-19}cm^{-2}$; this system enables tunability from ~0.7 to 1.1 μm [14-16].

It was mentioned above that it is difficult to keep titamum in the trivalent state. Although Ti^{4+} ion has a $3d^0$ electron configuration and is optically inactive in the NIR region, some parasitic absorption in the range of the Ti^{3+} emission assigned to Ti^{3+}-Ti^{4+} pairs is observed [7]. The laser quality of the Ti:Al_2O_3 crystals is characterized by a so called figure of merit (FOM) determined by the ratio of the peak absorption at 490 nm to the parasitic absorption at 800 nm (the peak wavelength of emission of Ti^{3+} ion). Material with FOM>100 (for π-polarized light) is commercially available [16].

Further laser action of Ti^{3+} ions was obtained in other hosts (in particular, in $BeAl_2O_4$ [17] and $YAlO_3$ [18]), but Ti^{3+}:Al_2O_3 remains the best titanum laser. The reasons for this are both the unique thermomechanical properties of the Al_2O_3 host and the spectroscopic advantages of the Ti^{3+}:Al_2O_3 system.

Divalent vanadium

The electron configuration of the V^{2+} ion is Ar($3d^3$), the same as the electron configuration of the Cr^{3+} ion. Although V^{2+}-systems are similar to Cr^{3+}-systems in many respects, V^{2+} usually occurs in octahedral sites characterized by a weaker crystal field, which causes a shift of its absorption and emission bands to the long wavelength spectral region. In the case of fluoride perovskites like $KMgF_3$ this shift is about 200 nm, and the usual emission spectral range (the $^4T_2 \rightarrow {}^4A_2$ transition) is from 0.9 to 1.1 μm [19].

V^{2+} ions possess low oxidation resistance unlike Cr^{3+} ions (see Table 1). Thus, V^{3+} (and V^{4+} ions) can be detected in crystals containing V^{2+}. V^{2+}-

systems demonstrate higher probabilities of non-radiative decay in comparison to Cr^{3+}-systems; hence, V^{2+}-lasers operate only at low temperatures. ESA produced losses at the emitting wavelength have a notably negative contribution to the laser operation of V^{2+}-systems and restrict the spectral range.

Laser generation from V^{2+} ions was obtained for the first time in the MgF_2 crystal [16]. Although the luminescence of the V^{2+} ion in MgF_2 covers the 1.0-1.4 μm region, ESA restricts the achieved tuning range to the 1.07-1.15 μm range at T=80 K [5].

Divalent chromium

Recently the pulse generation of Cr^{2+} ions in ZnS and ZnSe crystals was observed in the 2.3-2.5 μm spectral range at room temperature [21]. This fact has attracted great attention to divalent 3d-element ions (TM^{2+}) in the II-VI hosts.

In principle, the absorption and emission properties of Cr^{2+} and other TM^{2+} in the II-VI compounds ions are well known [22]. A new study carried out in the Lawrence Livermore National Laboratory [23] showed that these ions are very promising for solid state lasers. Among the investigated dopants, Co^{2+}, Ni^{2+}, and Cr^{2+} in ZnS, ZnSe, and ZnTe, the last ion, Cr^{2+} ($3d^4$ electron configuration) is the most promising laser activator.

In zinc chalcogenides, the TM^{2+} ions substitute for Zn^{2+} ions in tetrahedral crystallographic sites instead of octahedral sites. The tuning range of Cr^{2+}-doped II-VI compounds covers at least the 2100-3000 nm region. In the case of ZnSe:Cr^{2+} the luminescence band ($^5E \rightarrow ^5T_2$ transition) goes from 1800 to 3000 nm with a decay time of τ=9 μs at room temperature.

The laser action of ZnSe:Cr^{2+} pumped by a 40 μs-pulse emission of MgF_2:Co^{2+} laser (1860-1895 nm), was observed in the 2280-2530 nm range

[24]. ZnSe: Cr^{2+} has a broad intensive absorption band with a peak of around 1800 nm, which allows the use of InGaAsP laser diode pumping.

Trivalent chromium

The Cr^{3+} ion is the most widely used activator among the iron group ions. This fact is due to the unique properties of the Cr^{3+} ion, mostly caused by its electron configuration. These properties are summurized below [11,25,26]:

1. The stability of the Cr^{3+} valent state with respect to oxidizing and reducing.

2. The presence of three very strong absorption bands in the UV and visible region (spin-allowed transitions), which provides highly efficient flash-lamp pumping.

3. The strong tendency of the Cr^{3+} ions to occupy one-type sites (with an octahedral configuration).

4. The possibility to utilize the fast effective non-radiative energy transfer from Cr^{3+} to RE^{3+} ions, which makes possible the use of Cr^{3+} as an efficient sensitizer.

5. The relative ease of growth of Cr^{3+}-doped crystals due to the close to unity distribution coefficients and the chemical stability of the ions.

The Cr^{3+} ion has three electrons in the outer shell. A simplified scheme of the low-lying energy levels for the $3d^3$ electron configuration in an octahedral crystal field as a function of crystal field strength is shown in Figure 2. The scheme does not show energy level splitting due to lower crystal fields and spin-orbit coupling, and is convenient for describing the Cr^{3+} luminescence process which includes the $^4T_2$, ^2E, and ground $^4A_2$ levels.

In known Cr^{3+}-doped crystals and glasses, the ratio D_q/B varies in the 2.0-2.8 range, and the energy gap ΔE between the $^4T_2$ and ^2E states varies from -1200 to +2300 cm^{-1}. Let us note that, since the position of the ^2E level is practically independent of the crystal field strength (see Figure 2), a change of the ΔE value is due only to the change of the position of the $^4T_2$ level. Therefore, the absolute change of the energy position of the $^4T_2$ level responsible for the electron-vibronic luminescence of the chromium ions is ~3500 cm^{-1}. This is the reason for the strong shift of the luminescence spectra of Cr^{3+}-doped crystals. Actually, in various hosts, the position of the Cr^{3+} emission band varies from ~700 to 1100 nm.

Referring to Figure 2, the value D_q/B~2.2 is the cross-over point of the ^2E and $^4T_2$ energy levels (i.e., at this value, the energy gap ΔE between them is 0). The area to the right of this point corresponds to higher values of the D_q/B parameter; here, Cr^{3+} ions are in crystallographic sites with a strong crystal field, and the energy difference $\Delta E>0$. The area to the left of the point corresponds to lower values of the D_q/B parameter; here, Cr^{3+} ions are in crystallographic sites with a low crystal field, and $\Delta E<0$. With a strong crystal field, the ^2E state is the lower excited state (see Figure 2). This state has a rather long lifetime of radiative decay (on the order of milliseconds) [6] since the $^2E \rightarrow {}^4T_2$ transition is spin-forbidden and luminescence occurs at one or two zero-phonon lines (R-lines) with a wavelength around 690 nm. A typical material with Cr^{3+} ions in a strong crystal field is ruby $Cr^{3+}:Al_2O_3$, where ΔE ($^4T_2$ - ^2E)=2300 cm^{-1}.

In the alexandrite crystal, $Cr^{3+}:BeAl_2O_4$ [27], trivalent chromium is in an intermediate crystal field (ΔE=800 cm^{-1}). The character of the Cr^{3+} luminescence in the crystal differs from the above. At liquid nitrogen temperature, sharp R-lines are dominant. At room temperature, the $^4T_2$ state is partially populated, i.e., 6% of excited ions are in this state. However, since the probability of the $^4T_2 \rightarrow {}^4A_2$ transition is two orders of magnitude higher than the probability of the $^2E \rightarrow {}^4A_2$ transition [6], at room temperature the broadband $^4T_2$ emission with an effective lifetime of 260 μs is dominant. Note that in crystals with ΔE>0, the lifetime of an excited state is determined by the lifetime of the metastable ^2E level and the extent of the population of the $^4T_2$ upper level. In the case of alexandrite, the inherent lifetime of the ^2E level is about 2 ms and, of the $^4T_2$ level, about 10 μs [27].

The structure of the Cr^{3+} energy levels corresponding to the cross-over point of the ^2E and $^4T_2$ levels is seen in gadolinium-scandium-gallium garnet GSGG:Cr^{3+}, where the energy gap ΔE is 0-50 cm^{-1} [25,28]. In the crystal an effective thermalization of ^2E and $^4T_2$ states takes place, causing the appearance of an intensive electron-vibronic fluorescence band, which persists even to liquid helium temperatures [29,30].

In low crystal field media, the $^4T_2$ state is the lower energy level. That is why broadband luminescence completely determines the emission spectrum of the crystals even at cryogenic temperatures. The typical example of such a material is lanthanum-lutetium-gallium garnet LLGG:Cr^{3+} where ΔE= -1000 cm^{-1} [31].

Among the materials with ΔE<0, $LiCaAlF_6$:Cr^{3+} and $LiSrAlF_6$:Cr^{3+} crystals [32,33] demonstrating very high efficiency [34] should be noted. The crystals are now among the most used materials for Cr^{3+} tunable lasers. In comparison to other Cr^{3+}-tunable laser materials, these crystals demonstrate almost the highest values of *"intrinsic efficiency"* [32], i.e., the hypothetical value of slope efficiency taking place in the absence of any passive loss. As was demonstrated in [35], the excited state absorption is the main source of the reduction of Cr^{3+}-doped materials intrinsic efficiency. Note that studies by many researchers (see, for instance, [13,35,36]) have shown that Cr^{3+} ions in different materials are strongly influenced by ESA, which considerably reduces the number of usable Cr^{3+}-doped materials.

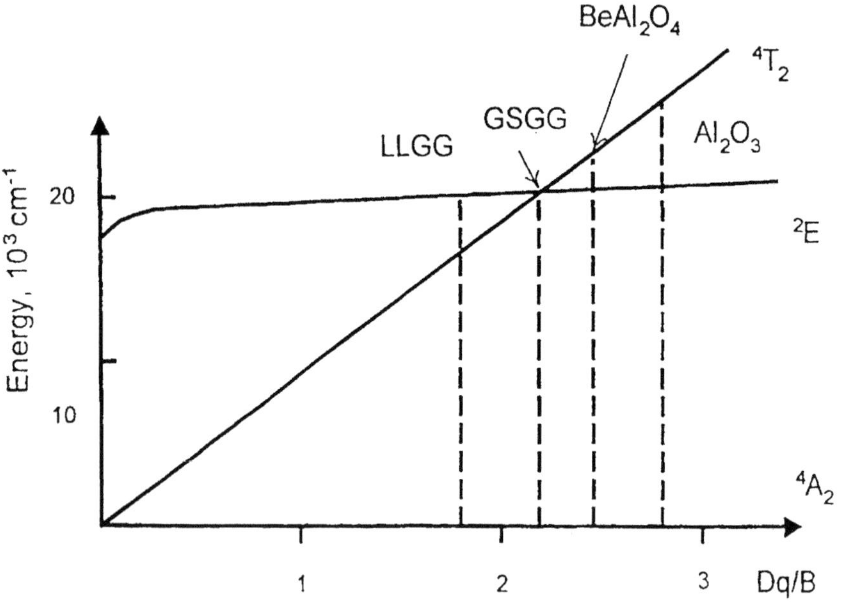

Fig. 2 Diagram of low-lying energy levels of Cr^{3+} ions ($3d^3$) in octahedral coordination

As was shown in [29], there is a relation between the intensive broadband luminescence of Cr^{3+} ions and effective Cr^{3+}-sensitization of RE^{3+} luminescence. Crystallochemical design [26,37] has allowed the development of a number of garnet hosts, including GSGG, with an efficient Cr^{3+}-sensitization of Nd^{3+}, Er^{3+}, Ho^{3+}, Tm^{3+} luminescence [25,38].

During the last few years the development of visible laser diodes has caused new interest in the Cr^{3+} ion as a target of diode pumping. Cr^{3+} and Cr^{3+} - Nd^{3+} systems attract now the most research attention [39,40]. Cr^{3+}-sensitized Nd-lasers with diode pumping via chromium at 670 nm provide higher overall system efficiency than in the case of direct pumping of Nd^{3+} ions at 808 nm. The necessary condition for this is effective non-radiative energy transfer, like that taking place in scandium garnets, where efficiency is close to 100% [38,40]. In diode pumped Cr^{3+}-lasers, active temperature control of the diode is not necessary due to a considerably shorter diode emission wavelength as well as the broad absorption band of the Cr^{3+} ions. This allows the elimination of the power consumption of the diode coolers and, therefore, considerably increases the overall efficiency of the lasers.

Tetravalent chromium

The unusual oxidizing state of the chromium ion very rarely occurs in nature and, up to most recent time, has been barely studied. Tetravalent chromium came to the attention of researchers in the middle 1980s, when, concurrently in several laboratories, the unusual 1-μm absorption of Cr-doped GSGG crystals was discovered (see [10]). The authors of [41] ascribed the absorption to the formation of tetrahedrally coordinated Cr^{4+} ions. Soon after that, the tunable lasing of Cr^{4+} ions in YAG [42,43] and forsterite Mg_2SiO_4 (MSO) [44,45] crystals was reported, although the physical interpretation of the phenomenon was not quite correct. Many Cr^{4+}-doped crystals have been suggested (see reveiw [46]), but up to now YAG and MSO crystals attract the most attention and exhibit the best results. Yttrium orthosilicate Y_2SiO_5 [47,48] should also be noted among other Cr^{4+}-doped crystals, as well as the aluminium and gallium melilites [49], and, as recently suggested, the wurtzite-like $LiAlO_2$ and $LiGaO_2$ crystals [50,51].

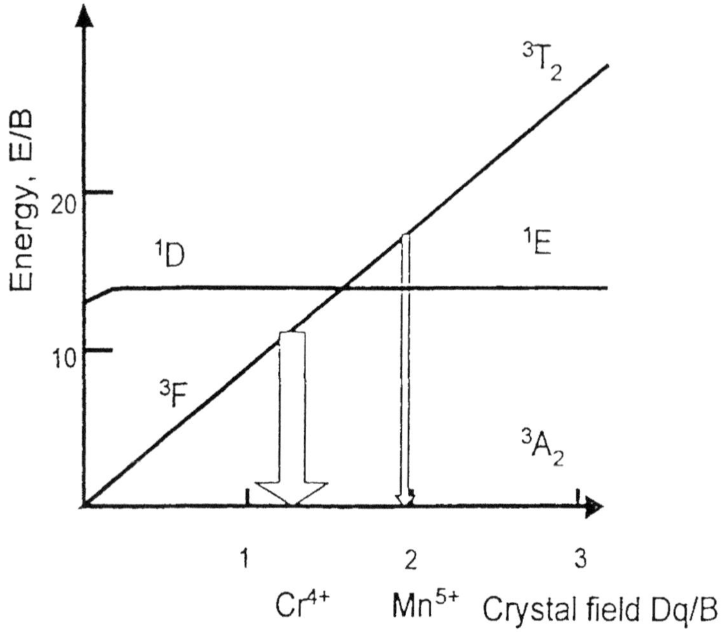

Fig. 3 Diagram of low-lying energy levels of a $3d^2$ ion in tetrahedral coordination

The electronic configuration of the Cr^{4+} ion is $3d^2$, and, due to its small ionic radius, the ion prefers to occupy tetrahedral sites in crystal structure. The simplified Tanabe-Sygano diagram is shown in Figure 3. It is the same as for the Ni^{2+} ion ($3d^8$) in octahedral symmetry. The ground state is the spin triplet $^3A_2$ (e^2-electron configuration). The Cr^{4+} luminescence spectra in various hosts exhibit a weak crystal field. In [46], it was shown that in various crystal hosts, the crystal field strength for tetrahedral Cr^{4+} varies in the range 1.2-1.6 Dq/B. In this case, the first excited state is the $^3T_2$ triplet (t_2e-electron configuration), which results in broad absorption and emission bands. According to the calculations in [46], in Cr^{4+}:YAG and Cr^{4+}:MSO, the ^1E level is above the $^3T_2$ one by $\Delta E = E(^1E) - E(^3T_2) = 1440$ and 3090 cm^{-1}, correspondingly.

In these crystals, Cr^{4+} ions are usually accompanied by Cr^{3+} and, occasionally, by Cr^{2+} ones (as in Cr:MSO [52]). In garnet crystals having both octahedral and tetrahedral ones sites, a typical Cr^{4+} concentration is a few percent of that of the Cr^{3+} ions occupying octahedral sites [10]. The incorporation of Cr^{4+} ions into the structure during the growth is usually very difficult. In Y_2SiO_5, an example of a purely Cr^{4+}-system [48], the distribution coefficient of Cr^{4+} ions is only 0.027 [53].

An oxidizing ambient atmosphere during crystal growth assists the formation of Cr^{4+} ions. This fact is confirmed by comparing both absorption and luminescence spectra. According to [54], the concentration of Cr^{4+} ions in single-crystalline Cr:MSO fibres grown by the laser-heated pedestal method in air and pure O_2 is greater by a factor of 10 and 20, respectively, in comparison to that in fibers grown in a pure N_2 atmosphere.

However, after the growth process, the transformation $Cr^{3+} \rightarrow Cr^{4+}$ proceeds with a considerably lower yield. In [52], a Cr:MSO crystal grown by the Czochralski method in a slightly oxidizing atmosphere (N_2 +1 vol.% O_2) underwent annealing in pure oxygen at 1750°C. Even after 100 hours of annealing, the absorption corresponding to tetravalent chromium increased by only a small extent.

On the other hand, as was shown in [49], the introduction of optically inactive impurities influencing the redistribution of tetravalent chromium in the structure into the crystal could be a very efficient way of enhancing the spectral and luminescence properties of Cr^{4+}-doped crystals.

A typical example for the Cr^{4+} ion absorption bands in the visible and near IR spectral regions are those in the YAG crystal with peaks at 465 nm, 610 nm, and around 1 μm [55]. In forsterite, which possesses notable pleochroism, the peaks are situated around 400 nm, 700 nm, and 1000 nm [52,56]. Cr^{4+} ions occupying tetrahedral crystallographic sites have large absorption cross-sections due to the absence of inverse symmetry. For example, in the case of YAG, σ_{abs} (1 μm)=$3.6 \cdot 10^{-19}$ cm^2, and σ_{abs}(650 nm)=$7.4 \cdot 10^{-19}$ cm^2, twenty times larger than for the octahedral Cr^{3+} ion [57]. Due to this fact, Cr^{4+}-doped materials have a high absorbance capability despite the low tetravalent chromium concentration in the crystals (usually 10^{18}-10^{19} cm^3). The absorption of Cr^{4+}-doped YAG in the visible range is so high, it is sometimes called "black YAG".

Cr^{4+} luminescence is characterized by a broad band emission in 1000-1850 nm range, with a maximum situated in the 1150-1520 nm range (see review [46]). Spin-allowed broadband luminescence in Cr^{4+}-systems has short decay times. At room temperature, the luminescence lifetime in Cr^{4+}-crystals is usually 2-5 μs, varying from ~0.1 μs in $LiNbGeO_5$ [58] to ~30 μs in $LiAlO_2$ [50,51].

In YAG, the lifetime of Cr^{4+} at room temperature is 4 μs, with a peak cross-section of stimulated emission $4 \cdot 10^{-19}$ cm^2 [43]; in forsterite, the values are 2.7 μs and $1.94 \cdot 10^{-19}$ cm^2, respectively [56]. The tunablity range for Cr:MSO is 1.15-1.34 μm [45,59], and for Cr^{4+}:YAG is 1.31-1.59 μm [60].

The short decay times of the excited state notably limit the possibility of energy storage in Cr^{4+}-systems and prevent laser performance with high average power. On the other hand, intensive absorption of tetravalent chromium at the wavelength of Nd-laser emission allows Cr^{4+}-systems to be considered as effective converters of 1.06 μm emission into tunable near IR emission.

The authors of [61] have proposed the use of Cr^{4+} ions as sensitizers of luminescence of some RE^{3+} ions whose absorption bands overlap well with the broadband emission of tetravalent chromium. During the process of sensitization, direct non-radiative energy transfer from the Cr^{4+} excited ion to the upper laser level of RE^{3+} ion takes place (see Figure 4). Here, the first excited states of Er^{3+} ($^4I_{13/2}$), Tm^{3+} ($^3F_4$), Ho^{3+} ($^5I_7$), and Dy^{3+} ($^6H_{13/2}$) act as upper laser levels. The inversion of the population of the upper laser level is of special interest for achieving a 1.5 μm crystalline erbium laser operating at the $^4I_{13/2} \rightarrow {}^4I_{15/2}$ transition.

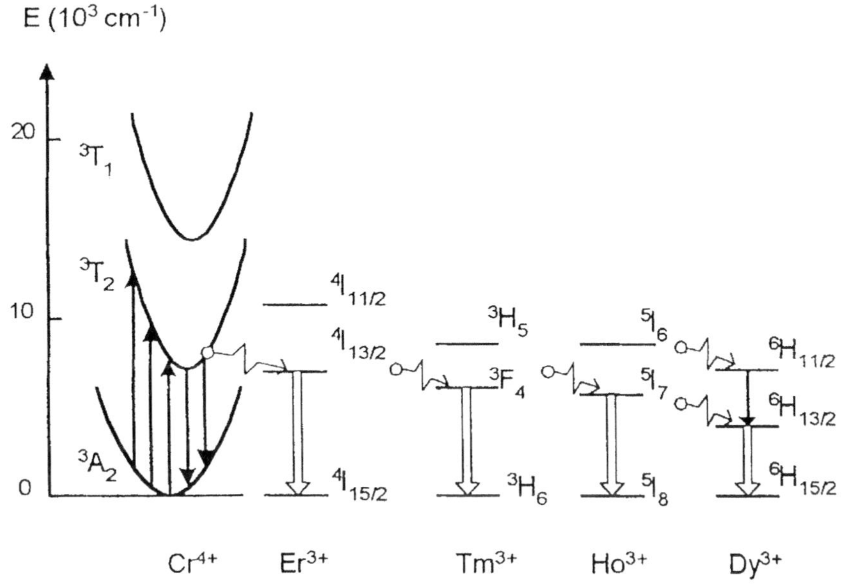

Fig. 4 Schematic diagram of energy transfer from the Cr^{4+} ion to some RE^{3+} ions

In [49,61], the sensitization of the luminescence of $Cr^{4+} \rightarrow Er^{3+}$ and $Cr^{4+} \rightarrow Tm^{3+}$ with quantum efficiencies of up to 50-65% was demonstrated in YAG and melilite $CaYAl_3O_7$ and $SrGdGa_3O_7$ crystals. In Y_2SiO_5 the quantum efficiency of $Cr^{4+} \rightarrow Er^{3+}$ reaches 19% [62]. Laser action at the $^3H_4 \rightarrow {}^3H_6$ transition of Tm^{3+} ion assisted by energy transfer from tetravalent chromium was reported in Cr^{4+},Tm^{3+}:YAG [63].

Along with activator and sensitizer functions, tetravalent chromium can also be used for other purposes. First of all, due to saturated one-micron absorption, Cr^{4+}-systems are used as passive Q-switches for Nd-lasers [53,64-66]. The shift of absorption bands in Mg_2SiO_4 and Y_2SiO_5 crystals to the short-wavelength spectral region make possible the passive Q-switched operation of the ruby laser (λ=694 nm) using Cr^{4+}-doped silicates as an intracavity saturable absorbers [97].

Cr^{4+}-centers are also used for Nd-lasers mode-locking [68]. Co-doping by Cr^{4+} and Nd^{3+} ions causes a switch of regimes [69,70]. Phototropic centers absorbing at the main transition of Nd^{3+} ions (λ=1.06 μm)

suppress the parasitic escape of inversion due to amplified spontaneous emission [71] and can provide efficient Cr,Nd:GSGG laser operation at the 1.32 μm wavelength [70].

ESA plays a substantially negative role in Cr^{4+}-systems. However, in lutetium-aluminium garnet LuAG:Cr^{4+}, reduced ESA values were obtained. For LuAG:Cr^{4+}, the cross-section of excited state absorption is $\sigma_{ESA}(1.06\ \mu m)<3 \cdot 10^{-20}$ [72], which is one order of magnitude lower than for isomorphous YAG:Cr^{4+}. This encourages the study of Cr^{4+}-systems for applications as an active media as well as passive Q-switches.

Divalent manganese

Mn^{2+} is the most abundant natural luminophor; this is due to its high resistance to both oxidation and reduction (see Table 1). The electron configuration of the Mn^{2+} ion (r_{ion}=0.66 Å) is $3d^5$ (see Figure 5), and the ion can occupy both octahedral and tetrahedral crystallographic sites. In comparison to the octahedral configuration; in the tetrahedral one, the energy of all transitions increases due to the weaker crystal field strength (D_q(tetr)=4/9 D_q(oct)). Therefore, the change of the structural coordination causes a considerable change of position of the only broad luminescence band of Mn^{2+}, which determines the color of luminescence [73]. Thus, for octahedral Mn,$^{2+}$ the red-orange emission in calcite ($^4T_{1g} \rightarrow {}^6A_{1g}$ transition) is typical, while, for the tetrahedral ion, green luminescence in willemite ($^4T_1 \rightarrow {}^6A_1$ transition) occurs. Along with the coordination change, the difference in crystal field strength due to covalency, interatomic distances, etc., should be taken into account. The position of the luminescence band maximum is influenced by the concentration of the Mn^{2+} ion [73]. For an unaltered Mn^{2+} coordination in silicate glasses the luminescence color changes from green at 0.05-0.1 wt % of MnO to mixed red-green (two peaks in the red and green spectral region) at 1-3 wt % of MnO, to red, if the concentration is more than 3 wt %. Green luminescence is assigned to single Mn^{2+} centers, and red luminescence to pairs and clusters of Mn^{2+} ions [74].

The energy levels scheme of Mn^{2+} does not contain broad absorption bands, and the excitation is usually produced by ultraviolet irradiation. The intensity of X-ray luminescence and cathodo-luminescence are quite high. Mn^{2+} ions are also the most abundant thermoluminescent centers. A comprehensive analysis of halophosphates was recently performed in [75] due to the importance of the system for the development of phosphors for luminescent lamps.

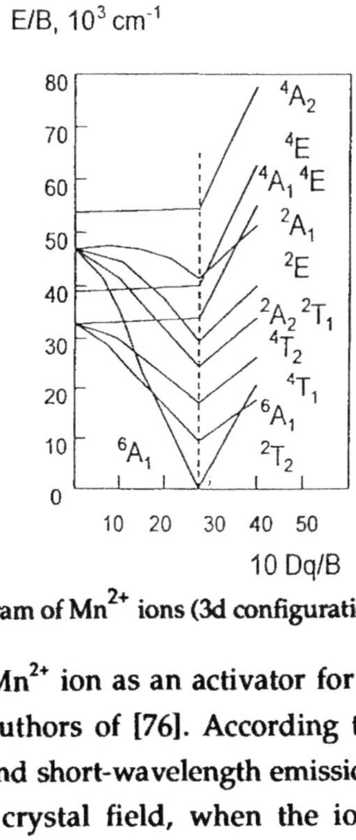

Fig. 5 Energy level diagram of Mn^{2+} ions (3d configuration)

The attempt to use Mn^{2+} ion as an activator for solid state lasers was undertaken by the authors of [76]. According to the Tanabe-Sugano diagram [8], broadband short-wavelength emission of $3d^5$ ions could be possible in a weak crystal field, when the ion is in a tetrahedral coordination.

Due to the lack of inverse symmetry, the Laporte selection rule for the parity forbidden *d-d* transitions is lifted. Unfortunately, all transitions to the ground state are spin forbidden, resulting in a low transition probabilities. This causes low absorption and emission cross-sections, and therefore requires high dopant concentrations for compensation.

In the spinel crystals studied in [76], the measured peak Mn^{2+} absorption cross-section at λ=497 nm was $\sigma_{abs}=2.6\cdot10^{-21}$ cm^2. Broadband luminescence in the 500-600 nm range of the $^4T_1\rightarrow{}^6A_1$ transition had a relatively high lifetime: τ=6.2 ms. The luminescence peak shifted from approximately 515 nm to 535 nm as the Mn^{2+} concentration increased from 1 at.% to 18 at.%. The emission cross-section was $\sigma_{em}=2.3\cdot10^{-22}$ cm^2, forty times smaller than for Cr^{3+}:GSGG.

To compensate for low absorption and emission cross-sections, high Mn^{2+} concentrations up to 20 at.% were used. However, strong excited state absorption with a cross-section one order of magnitude larger than that for excitation from the ground state and a spectrum overlapping the luminescence spectrum prevented both lasing and gain.

Hence, the Mn^{2+} ion is not a very promising prospective laser ion, though in principle, it could be possible to obtain Mn^{2+} lasing in a host with less ESA (obtainable in a host with an extremely low crystal field, if such a host can be found).

Pentavalent manganese

The electron configuration of the Mn^{5+} ion is the same as that of the Cr^{4+} ion - $Ar(3d^2)$; Mn^{5+} ions occupy tetrahedral crystallographic sites. However, in contrast to tetravalent chromium, Mn^{5+} centers correspond to the strong crystal field situation (Dq/B=1.9-2.2 [46]), and Mn^{5+}-doped compounds occupy the right side with respect to the cross-over point of the ^1E and $^3T_2$ levels (see Figure 3). The first excited state of the Mn^{5+} ion is the ^1E singlet (the e^2 electron configuration). Therefore, despite the similar visible and near-infrared absorption spectra of the Mn^{5+}-system are similar to the Cr^{4+}-doped compounds, their emission properties are notably distinct.

While Cr^{4+}-systems demonstrate broadband luminescence at room temperature, emission spectra of Mn^{5+}-doped materials are strongly structured and have one or two predominant sharp peaks. Narrowband luminescence of Mn^{5+}-systems at the spin-forbidden $^1E \rightarrow {}^3A_2$ transition is essentially longlived. At room temperature, the lifetimes of Mn^{5+} centers are 1-2 order of magnitude longer than the lifetimes of Cr^{4+}-doped compounds. Usually they vary from several tens to several hundreds of microseconds [46], which is favorable for high power laser operation. In contrast to Cr^{4+}-doped materials demonstrating four-level schemes of laser action, for Mn^{5+}-systems, a three-level scheme with a rather high threshold and low efficiency is typical. The laser action of the Mn^{5+} ion at λ=1181 nm was reported for the first time in $Mn{:}Ba_3(VO_4)_2$ pumped by a dye-laser at room temperature [77]. A dreadful drawback of the Mn-system is the presence of notable amounts of Mn^{4+} alongside Mn^{5+} ions in the same tetrahedral crystallographic sites. According to the EPR-data [78] in $Mn{:}Ba_3(VO_4)_2$, the amount of Mn^{4+} is about 1/5 of the total amount of Mn^{5+} ions.

Nevertheless, Mn^{5+}-systems possess a number of important advantages. They have very strong absorption bands, which make the materials suitable for pumping by different sources, including AlGaAs diodes. This, as well as the long lifetimes and high cross-sections of stimulated emission [79,80], still attract researchers to the Mn^{5+}-systems.

Divalent cobalt

The electronic configuration of Co^{2+} is $Ar(3d^7)$. Co^{2+} ions can occupy both octahedral and tetrahedral sites, but systems with octahedral Co^{2+} are more common. The Tanabe-Sugano diagram for the $3d^7$ configuration in the octahedral crystal field calculated for C/B=4.5 [9] is shown in Figure 6. It can be seen that the increase in crystal field strength causes a change in the Co^{2+} ground state. At low crystal field (Dq/B<2.1), $^4T_1$ is the ground state for the $3d^7$ configuration, while for Dq/B>2.1, the ^2E state becomes the ground state. In a majority of known hosts (MgF_2,

MgO, $KZnF_3$, etc.), Co^{2+} ions are located in a low crystal field (the Dq/B value does not exceed 1.2).

The luminescence of the $^4T_2 \rightarrow ^4T_1$ transition of $Co^{2+}:MgF_2$ at low temperatures lies in the 1.5-2.3 μm region and contains sharp zero-phonon lines and broad sidebands. This broadband emission is of considerable interest due to the possibility of extremely wide tuning at such a long-wavelength spectral region.

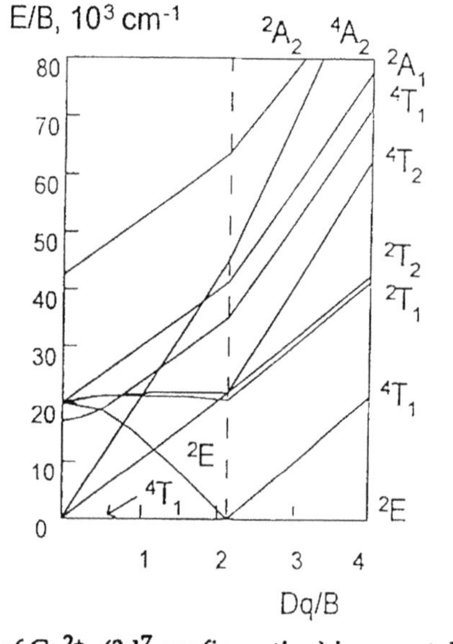

Fig. 6 Energy levels of Co^{2+} ($3d^7$ configuration) in an octahedrally symmetrical crystal field

The laser action of Co^{2+} was demonstrated for the first time in a MgF_2 crystal by Johnson et. al [81] with flashlamp pumping. At low temperature (T=80 K) laser operation of $Co^{2+}:MgF_2$ is very efficient by selectively pumping with the 1.32 μm emission of the Nd:YAG laser [82]. At room temperature, the lasing parameters notably deteriorate due to multiphonon non-radiative relaxation processes. Temperature quenching of the Co^{2+} luminescence in MgF_2 is extremely high: at 80° K, the lifetime of the $^4T_2$ level is 1.05 ms, while at room temperature, the value reduces to only 36.5 μs [83]. It should be noted that in contradiction to Ni^{2+}, Cr^{3+}, V^{2+} and other systems with 3d-transition metal ions, Co^{2+}-lasers are practically not influenced by ESA both at the pumping wavelength and the emission wavelength [30,82].

Co^{2+} ions can also occupy tetrahedral sites, in ZnO, $LiGa_5O_8$, etc. The energy structure of Co^{2+} in tetrahedral sites is similar to the structure of Cr^{3+} in octahedral sites, excluding the Dq value, which is notably less in the case of tetrahedral sites. In $LiGa_5O_8$ the lifetime of metastable $^4T_1$ level of Co^{2+} is only 200 ns due to lack of inverse symmetry in the tetrahedral sites [84].

In last few years, the Co^{2+} ion as an activator for solid state lasers is of less interest since it works efficiently only at cryogenic temperatures. On the other hand, the authors of recent papers [85,86] have shown that Co^{2+}-containing garnets YAG and YSGG are promising saturable absorber Q-switches for 1.5 μm laser emission.

Divalent nickel

Phonon-terminated laser action was demonstrated for the first time in 1963 in the MgF_2:Ni^{2+} crystal at the $^3T_2 \rightarrow {}^3A_2$ transition of Ni^{2+} at liqiud nitrogen temperature [87]. Later the laser action of Ni^{2+} ions has been demonstrated in many hosts, including oxides and fluorides.

The electronic configuration of Ni^{2+} is Ar($3d^8$). According to [6], Ni^{2+} can occupy both octahedral and tetrahedral positions. The energy level diagram for Ni^{2+} ($3d^8$) in an octahedral crystal field is the same as for Cr^{4+} ($3d^2$) in tetrahedral surroundings. According to the luminescence spectra of Ni^{2+}-containing compounds, Ni^{2+} ions are usually in low crystal fields of octehedral configuration. In the case of Ni^{2+}:MgO the value is $Dq/B \approx 1$, which is near the crossover point of the $^3T_1$ and ^2E levels (see Figure 3). In fluorides, the crystal field strength for Ni^{2+} is even less [84].

The Ni^{2+}-system possesses two metastable levels, $^1T_2$ and $^3T_2$. In octahedral crystallographic positions, the luminescence of Ni^{2+} occurs in the near IR region for the $^3T_2 \rightarrow {}^3A_2$ transition, and in the visual spectral region from the $^1T_2$ level to the T_2 level ($^1T_2 \rightarrow {}^3T_2$ transition, red emission) and $^3A_2$ level ($^1T_2 \rightarrow {}^3A_2$ transition, green emission). Let us note that oscillator strengths for absorption and emission transitions in Ni^{2+} occupying an octahedral position are 2-3 orders of magnitude lower than in the case of the Cr^{4+} ion in tetrahedral symmetry with analogous energy structure [30].

In fluoride crystals at low temperatures, the lifetime τ of the $^1T_2$ state is about 1 ms, and the lifetime of the excited $^3T_2$ state is near 10 ms. In Ni^{2+}:MgF_2, the lifetime τ ($^3T_2$) is 12.8 ms at T=77°K. The decay times of the $^1T_2$ and $^3T_2$ excited states decline, when the temperature increases, due to the increase of multiphonon non-radiative transitions.

Lasing of Ni^{2+} in the near IR region has been obtained in a number of crystal hosts: MgF_2, MgO, $KZnF_3$, $MgAl_2O_4$, etc. In Ni^{2+}:MgF2 crystals, both pulsed and continuous wave laser action were obtained at low temperatures. The tuning range at T=80°K was 1.6-1.74 μm [5].

Laser action of Ni^{2+} takes place only at low temperatures. In the case of Ni^{2+}:MgO, the observed decay time of the excited state remains constant up to 400 K, evidence of a high quantum yield. However, even in this case, the laser action takes place only below 200 K [6]. This behavior of the Ni^{2+}-systems is assigned to the combined influence of thermo-induced ground state absorption and, especially, excited state absorption. Non-radiative decay and ESA restrict considerably the lasing of Ni^{2+}, which reduces the number of studies of Ni^{2+}-systems in recent years.

Other ions

V^{3+}

The V^{3+} ion has the same electron configuration as the Cr^{4+} ion - Ar($3d^2$). V^{3+} ions in octahedral sites in Al_2O_3 demonstrate, at room temperature, broadband luminescence at the ${}^3T_{2g}\rightarrow{}^3T_{1g}$ transition in the 0.765-1.1 μm spectral range, with a lifetime τ=4 μs (the lifetime is constant in the 4.2-300°K temperature range) and an emision cross-section $\sigma_{em}=2.2\cdot10^{-19}$ cm^2 [88].

Broadband luminescence of V^{3+} in forsterite Mg_2SiO_4 was observed in the 1.1-1.7 μm spectral region at room temperature [89]. In [90], tetrahedral V^{3+} in YAG was shown to be an efficient saturable absorber for the passive mode-locking and Q-switching of solid state lasers in the far red and near infrared spectral range (from 0.747 to 1.34 μm), including Nd-lasers and Ti:Al_2O_3.

The lifetime of the ${}^3T_2$ excited state of V^{3+} ions in YAG is 5 ns. Peak absorption cross-sections at ${}^3A_2\rightarrow{}^3T_2$ transition are $(8.2\pm2.5)\cdot10^{-18}$ cm^2 at 1.08 μm and $(1.7\pm0.5)\cdot10^{-17}$ cm^2 at 1.33 μm [90]. The latter value is higher than the peak absorption cross-sections of Cr^{4+} at the analogous transition in garnets and forsterite.

V^{4+}

V^{4+} ion is isoelectronic with the Ti^{3+} ion and has the same electron configuration - Ar($3d^1$). In [88], the luminescence of V^{4+}:Al_2O_3 at the ${}^2E\rightarrow{}^2T_{2g}$ transition was observed in the 540-850 nm spectral range with a lifetime τ=2.4 μs and an emission cross-section $\sigma_{em}=2.4\cdot10^{-19}$ cm^2. Incorporation of tetravalent vanadium into the Al_2O_3 crystal is accompanied by the compensation of excess charge of the V^{4+} ions by intristic defects (for example, cation vacancies) and by impurities. Usually the crystals contain V^{2+} and V^{3+} ions along with V^{4+}. The laser action of V^{4+}:Al_2O_3 in the 590-620 nm spectral range at T=300°K under the pumping by second harmonic of Q-switched ruby laser was obtained in [88].

Cr^{5+} and Mn^{6+}

The electron configuration of Cr^{5+} and Mn^{6+} ions is Ar($3d^1$), the same as for Ti^{3+}. The ions occupy tetrahedral positions in different oxygen compounds. Cr^{5+} ions incorporate into different phosphate(V), vanadate (V), and arsenate (V) hosts, and Mn^{6+} ions incorporate into sulfate (VI), chromate (VI), selenate (VI), and molybdate (VI) hosts.

Broad band luminescence of Cr^{5+} and Mn^{6+} at the $^2T_2 \rightarrow {}^2E$ transition in all compounds investigated so far is very similar and covers the 900-1400 nm spectral range at T=10°K [91]. In this paper Mn^{6+}:$BaSO_4$ was offered as a promising laser material, posessing a lifetime τ=2.75 μs, and peak emission cross-section $\sigma_{em}=9.4 \cdot 10^{-19}$ cm^2.

Cr^{5+} and Mn^{6+}-systems demonstrate considerable thermal quenching. The luminescence intensity of Cr^{5+}:Li_3PO_4 reduces approximately by a factor of 4 when the temperature increases from 10°K to 300 K. In the case of Mn^{6+}:K_2CrO_4 the intensity reduces quicker, and, at room temperature, it comprises only 6% of the initial value.

Fe^{3+}

The electron configuration of the Fe^{3+} ion is Ar($3d^5$), similar to Mn^{2+} (see Figure 5). Fe^{3+} ions can occupy both octahedral and tetrahedral crystallographic sites. In garnet crystals, the absorption of Fe^{3+} ions due to intraconfiguration 3d optical transitions covers the visible region up to the UV and has several sharp intensive lines around 400 nm [92]. In addition, the Fe^{3+} ion has a broad absorption band between 0.9 and 1.0 μm, which can be used for InGaAs laser diode pumping [93].

Fe^{3+} ions in tetrahedral sites in garnets demonstrate intensive broadband luminescence at the $^4T_1(^4G) \rightarrow {}^6A_1$ transition around 800 nm.

The emission can be excited by pumping into any of the 3d intraconfiguration bands or into the charge transition band (λ_{max}=255 nm) [92,94]. Fe^{3+} ions have been repeatedly used as sensitizers in garnet hosts doped with Ho^{3+} [93,95], Nd^{3+} [94], Yb^{3+} [94], and Tm^{3+} [92].

2.2. RARE-EARTHS

The electronic configuration of the rare earth (RE) ions is: $1s^2 2s^2 2p^6 3s^2 3p^6 3d^{10} 4s^2 4p^6 4d^{10} 4f^n 5s^2 5p^6$. The filling of the $4f$-shell shielded from the external crystalline field by the $5s^2 5p^6$ shells proceeds with the increasing of the atomic number of the RE element.

Generally, RE luminescent dopants exist in crystals and glasses as di- or trivalent ions. If a lanthanide atom loses the two outer 6s-electrons, it transforms to a divalent ion. To produce a trivalent ion, the atom additionally loses one $5d$-electron (as in case of Gd or Lu) or one $4f$-electron.

Since the valent electrons of lanthanide atom are usually $5d^1 6s^2$ ones (naturally occuring when the ion is formed from the atom), the stable oxidizing state of RE ions is, as a rule, (+3). However, elements adjoining lanthanum ($4f^0$), gadolinium ($4f^7$), and lutetium ($4f^{14}$) have variable oxidizing states (see Table 2).

A considerable disadvantage of divalent RE ions is the difficulty of their incorporation into the crystalline host without changing their valent state, since this state is not very stable. To increase the concentration of divalent RE ions at the cost of non-desirable trivalent ions, the latter are transformed to the divalent state by intensive γ-radiation [96]. Divalent RE ions produced by additive coloring [97,98] demonstrate more stability under thermal and optical bleaching compared to radiatively produced ones. It is worth noting that some crystalline host, for example SrB_4O_7, possess unique properties allowing the stabilization of the divalent state of RE ions [99].

Table 2.

Typical oxidizing states of RE elements [147]

La	Ce	Pr	Nd	Pm	Sm	Eu	Gd
+3	+3, +4	+3, +4, (+5)	+3	+3	+3, +2	+3, +2	+3
Tb	Dy	Ho	Er	Tm	Yb	Lu	
+3, +4	+3, (+4)	+3	+3	+3, (+2)	+3, +2	+3	

Deep shielding of the $4f$-shell and hence the weak influence of the external crystalline host explains the similarity of the chemical properties of the lanthanides and of their narrow band spectra. The incompletely filled 4f-shell is responsible for the presence of unpopulated energy levels in this shell. Therefore, the absorption and luminescence spectra of RE ions contain a number of narrow bands due to electron transitions.

The $4f$-$4f$ transitions are parity-forbidden in the electric-dipole approximation. The exclusion can be lifted in case of non-centrosymmetrical crystal fields. The characteristic oscillator strength of these transitions f is 10^{-7}-10^{-5}; the radiative lifetime τ is usually from 50 to 10 μs. The energy level scheme of trivalent RE ions was presented in the classic work by Dieke and Crosswhite [100] and then was reproduced in many reveiws. Figure 7 shows the energy levels and laser transitions of trivalent rare-earth ions.

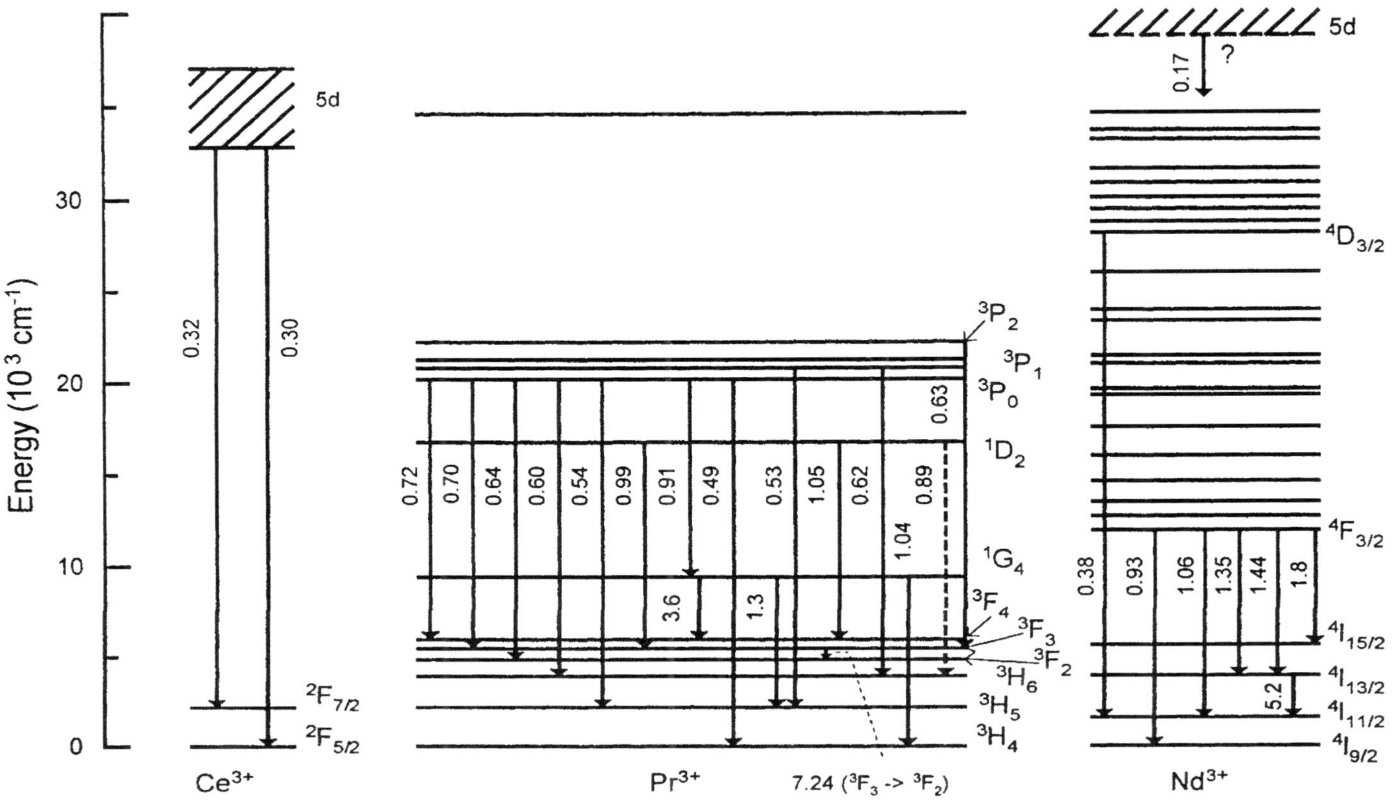

Fig.7 (a) Laser transitions in trivalent rare-earth ions (adapted from [7])

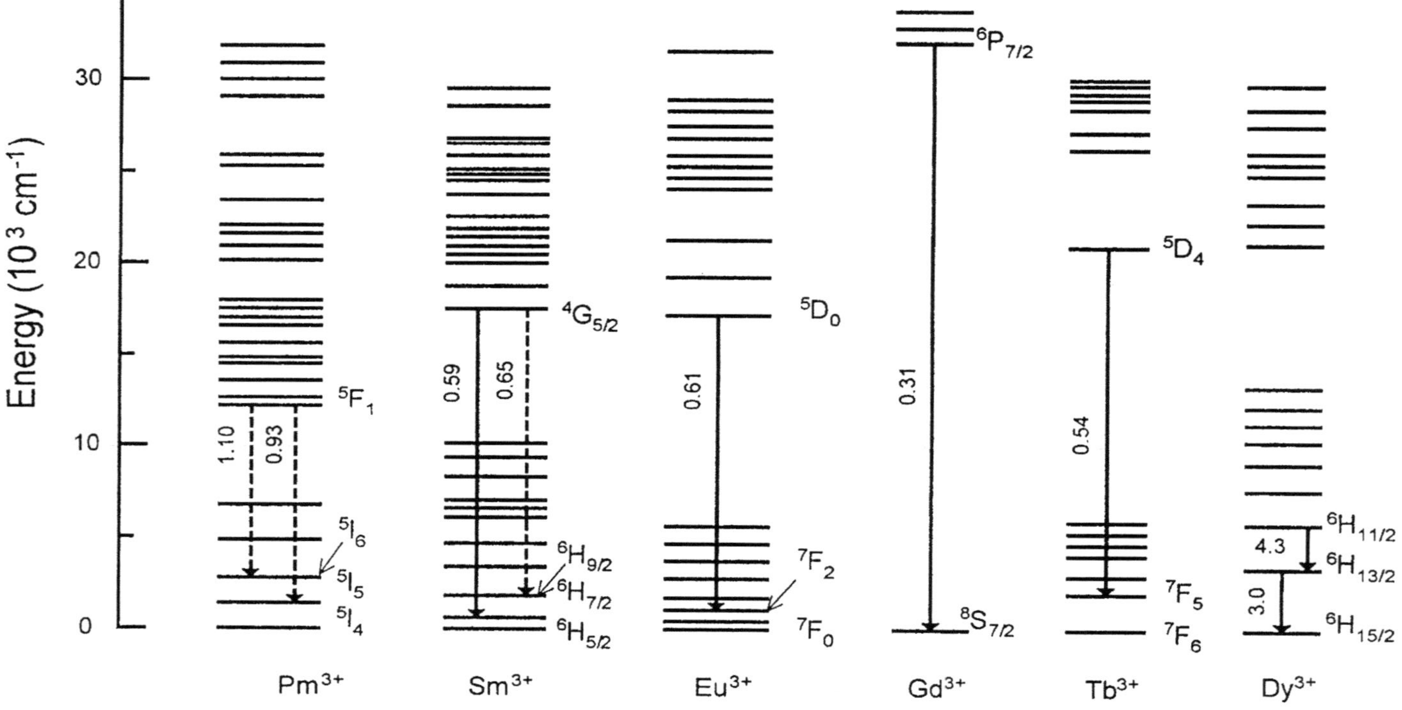

Fig.7 (b) Laser transitions in trivalent rare-earth ions (adapted from [7])

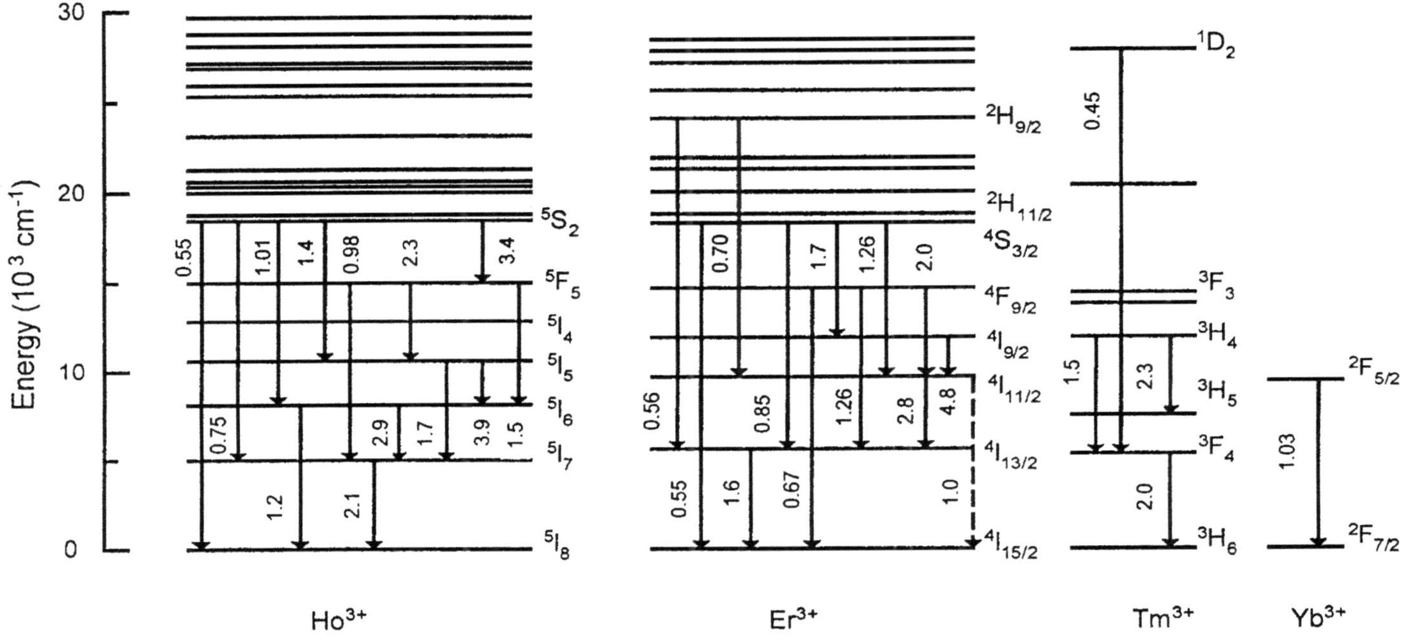

Fig.7 (c) Laser transitions in trivalent rare-earth ions (adapted from [7])

Transitions from the $4f^n$ to the $4f^{n-1}5d$ configuration (or, simply, $4f \rightarrow 5d$ transitions) have been studied for both di- and trivalent RE ions. $4f \rightarrow 5d$ transitions are allowed electric-dipole transitions having close to unity oscillator strength, and, thus they are much more intensive than forced electric-dipole $4f \rightarrow 4f$ transitions. Due to the strong interaction of the 5d state with the lattice, $4f \rightarrow 5d$ absorption has a large linewidth.

The luminescence of $4f \rightarrow 5d$ transitions has been obtained on many trivalent RE ions in the vacuum UV range. Divalent RE ions have an additional $4f$-electron, which decreases the energy of the $5d$-state. Thus, $4f \rightarrow 5d$ absorption bands are located in the visual spectral range.

The lifetime of the $4f^{n-1}5d \rightarrow 4f^n$ transitions for the lanthanides belonging to the beginning of the row is about 20-50 ns (Ce^{3+} and Pr^{3+}), whereas for ions from the middle of the row (Sm^{2+} and Eu^{2+}) this value is around 1 μs, and, for the lanthanides from the end of the row (Tm^{2+} and Yb^{2+}) ions, it is around several tens of μs [101].

Except for La^{3+} and Lu^{3+}, with totally unfilled ($4f^0$) and filled ($4f^{14}$) $4f$-shells, respectively, laser generation was obtained for all trivalent lanthanides. In crystals and glasses doped with RE ions, stimulated emission was obtained on more then fifty $4f$-$4f$ and several $5d$-$4f$ laser transitions in the spectral range from the vacuum UV to the medium IR.

Up to now, the shortest wavelength emission of a solid state laser was obtained at 0.172 μm at the$5d$-$4f$ transition of Nd^{3+} in LaF3 pumped by Kr_2^* eximer emission [102,103]. The longest wavelength stimulated emission was recently obtained at 7.24 μm at the $^3F_3 \rightarrow {}^3F_2$ transition of Pr^{3+} in $LaCl_3$ pumped by the two-micron Tm:YAG laser at 148 K [104].

Cerium

Ce^{3+} ion is one of the most important luminescent dopants used in stintillators (see the Chapter "Inorganic Scintillators" in the present book) emitting at 0.35-0.6 μm. The Ce^{3+} ion is of interest as an activator for the production of stimulated UV emission at the *5d-4f* transition. In several Ce^{3+}-doped fluorides, laser action was obtained. The vibrationally broadened *5d-4f* transition allows tunable laser emission of Ce^{3+} ions in the 0.3 μm region pumped by the fourth harmonic of a Nd:YAG laser. In Ce:LiCAF [105] and Ce:LiSAF [106], crystals tuning in the ~280-320 nm range was obtained. Ce:YLF and Ce:$LuLiF_4$ crystals pumped by the 248 nm emission of a KrF-eximer laser operate in the 305-335 nm range [107,108]. Up to now, continuous wave emission of Ce^{3+} ions was not demonstrated.

Recently Ce^{3+} ions have been also used for the sensitizing of luminescence of some RE ions, in particular Er^{3+} ions [109].

Praseodymium

The Pr^{3+} ion demonstrates the greatest number of *4f-4f* laser transitions among trivalent RE ions. Pr^{3+} has a number of transitions (from the $^3P_0$ or $^3P_1$ levels) in the visible range, which distinguishes the Pr^{3+} ion from the other RE ions.

The visual spectral range is important for a number of applications in the field of chemistry, biology and medicine. However, in this region there is no tunable vibronic lasers; even the second harmonic of NIR lasers produces only a partial overlapping of the region. An alternative for dye-lasers is the Pr:YLF laser, demonstrating CW operation at room temperature at a number of wavelengths in the visual and NIR spectral region from 552 to 904 nm [110].

The strong absorption bands of Pr^{3+} ion in the 420-480 nm range make possible efficient flash-lamp pumping.

Recently, great interest has been expressed in the production of tunable UV lasing at the *5d-4f* transition of the Pr^{3+} ion (analogous to Nd^{3+} and Ce^{3+} ions). The lifetime values of Pr^{3+} at the *5d-4f* transition in different fluoride hosts are very similar and close to 25-29 ns [111]. Although the emission of the Pr^{3+} ion at the *5d-4f* transition has been demonstrated in various hosts, excited state absorption (ESA) from the *5d* state to the conduction band prevents laser action [111].

Neodymium

Nd^{3+} ion is the most popular activator for solid state lasers since its energy level structure is very suitable for excitation and laser action, and it demonstrates intensive absorption and luminescence transitions. Widespread use of Nd ions in solid state lasers is also due to the high values of emission cross-sections of the Nd^{3+} main laser transitions ($^4F_{3/2} \rightarrow {}^4I_{11/2}$), which leads to low generation thresholds of Nd-lasers operating both in the CW and pulse regimes.

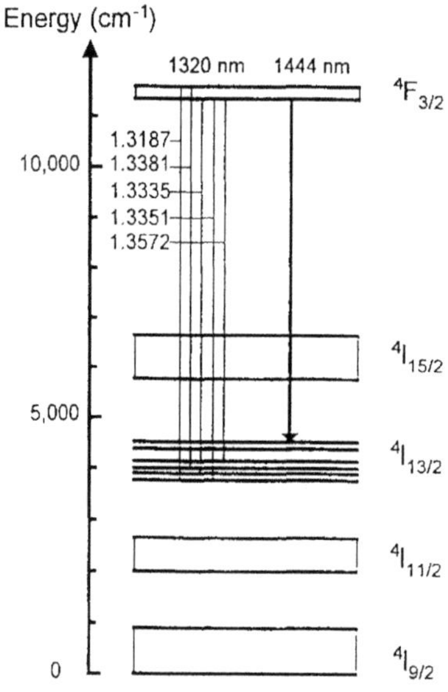

Fig. 8 Fragment of energy level diagram of Nd^{3+} in YAG with laser transitions

Stimulated emission of Nd^{3+} ions was obtained in more than 170 various hosts [2]. The most important among them is Nd:YAG operating at the $^4F_{3/2} \rightarrow {}^4I_{11/2}$ transition with λ_{em}=1.064 μm, possessing low threshold and high lasing efficiency at room temperature. More than 2/3 of solid-state lasers produced in the world are Nd:YAG lasers. Another important host for Nd^{3+} ions is glass; powerful glass lasers operating at $^4F_{3/2} \rightarrow {}^4I_{11/2}$ transition have been developed.

Table 3 contains the values of the emission cross-sections and lifetimes of Nd^{3+} at the $^4F_{3/2} \rightarrow {}^4I_{11/2}$ transition in different hosts derived from numerous papers.

Table 3. Emission cross-sections and lifetimes of Nd^{3+} ion in different hosts.

Host	σ_{em} (10^{-19} cm^2)	τ (μs)	$\sigma_{em}\cdot\tau$ (10^{-23} cm^2s)
YAG ($Y_3Al_5O_{12}$)	3.0 (4.0)	230 (260)	7.8-9.2
GSGG ($Gd_3Sc_2Ga_3O_{12}$)	3.1 (1.4)	280	3.9-8.7
GGG ($Gd_3Ga_5O_{12}$)	2.0 (2.3)	240 (280)	4.8-5.9
GGG:Ca,Mg,Zr	0.49	260	1.3
YVO_4	10.0 (8.7; 10.7)	98 (115)	8.5-12.3
$GdVO_4$	6 (7.6)	94 (100)	5.6-7.6
BEL ($La_2Be_2O_5$)	3.3 (0.85)	120 (150)	1.3-4.0
YLF ($YLiF_4$)	3.0 (2.2; 3.7)	480 (570)	11.4-15.6
$YAlO_3$	3.7	210	7.8
Y_2SiO_5	1.0	214	2.1
LNA, LMA ($LaMgAl_{11}O_{19}$)	0.3 (0.4; 0.57)	260 (375)	1.0-1.5
ASN ($SrMgAl_{12}O_{19}$)	0.48	230	1.1
phosphate glass	0.40-0.42	290-330	1.1-1.4
silicate glass	0.27-0.29	340	0.9-1.0
NdP_5O_{14}	2	115	2.3
$NdAl_3(BO_3)_4$	10	19	1.9
SGGM ($SrGdGa_3O_7$)	0.66	245	1.6
LGG ($La_3Ga_5GeO_{14}$)	0.8	220	1.8
FAP ($Ca_5(PO_4)_3F$)	5	230	11.5
SFAP ($Sr_5(PO_4)_3F$)	6.5 (5.4)	285 (298)	16.2-18.4
SVAP ($Sr_5(VO_4)_3F$)	5	215	10.8

Stimulated emission of neodymium ions was obtained at seven different transitions (see, for instance, [7]). The most important among them are: the above mentioned $^4F_{3/2}\rightarrow{}^4I_{11/2}$ transition (λ~1.06 μm), $^4F_{3/2}\rightarrow{}^4I_{9/2}$ transition (λ~0.9μm), and the $^4F_{3/2}\rightarrow{}^4I_{13/2}$ transition (λ~1.3 μm). Recent investigations have rekindled the interest in the last transition as a way of producing eye-safe lasing at 1.444 μm in the Nd:YAG crystal on the other Stark sublevels (see Figure 8) [112,113].

Holmium

The Ho^{3+} ion is intensively studied as an activator for solid state lasers. The list of laser crystals activated by Ho^{3+} presently contains more than 70 entries and it is second in length only to the Nd^{3+}-doped laser crystals list [2]. More than a dosen laser transitions have been found in Ho-activated crystals, but most papers deal with obtaining and optimizing laser action at the $^5I_7\rightarrow{}^5I_8$ transition, demonstrating eye-safe emission with a wavelength of 2.1 μm. This is due to the fact that, at 2.1 μm, the atmosphere has a high transmission, which provides the possibility of using Ho^{3+}-lasers in remote sensing and atmosphere monitoring. Special interest in the $^5I_7\rightarrow{}^5I_8$ transition is explained by the rather high lifetime of the upper laser level $^5I_7$ (τ_{em}~10 ms), which provides for high energy storage and allows efficient energy transfer even at relatively low transfer rates. The emission cross-sections of Ho^{3+} at the $^5I_7\rightarrow{}^5I_8$ transition in some oxide and fluoride crystals are given in Table 4.

Table 4
Spectroscopic parameters of Er^{3+} ($^4I_{13/2} \rightarrow {}^4I_{15/2}$ transition); Tm^{3+} ($^3F_4 \rightarrow {}^3H_6$ transition); and Ho^{3+} ($^5I_7 \rightarrow {}^5I_8$ transition) ions in different crystals (from [114]).

Host	Er^{3+}				Tm^{3+}				Ho^{3+}				CN
	λ_{em}	σ_{em}	τ_{em}	ΔE	λ_{em}	σ_{em}	τ_{em}	ΔE	λ_{em}	σ_{em}	τ_{em}	ΔE	
$Y_3Al_5O_{12}$	1644	0.45	6.9	470	2011	0.22	10.5	580	2098	0.98	7.8	460	8
$LiYF_4$	1615	0.42	11.9	350	1902	0.33	15.6	360	2050	1.20	16.1	280	8
BaY_2F_8	1600	0.27	15.4	280	1858	0.11	18.3	310	2057	1.07	17.9	310	8
LaF_3	1568	0.37	15.7	220	1850	0.25	15.4	210	2050	0.33	21.0	320	~9
$KCaF_3$	1612	0.12	20.1	260	1855	0.15	28.3	250	2085	0.28	22.2	300	6
$YAlO_3$	1625	0.31	7.2	450	1936	0.50	5.0	460	2118	0.82	8.1	460	12
$La_2Be_2O_5$	1588	0.23	4.8	250	1880	0.23	0.43	450	2088	0.42	1.04	340	~10

Parameter abbreviations: λ_{em}, *emission wavelength (in nm)*; σ_{em}, *effective emission cross-section (in* 10^{-20} cm^2*)*; τ_{em}, *measured emission lifetime (in ms)*; ΔE, *energy gap between terminal laser level and ground state (in* cm^{-1}*)*; *CN, coordination number of activator ion*

It should be noted, however, that lasing at the $^5I_7 \rightarrow {}^5I_8$ transition happens by a truly three-level scheme. The lower laser level $^5I_8$ is the ground state of the Ho^{3+} ion; it demonstrates a significant population at room temperature, causing a high lasing threshold and relatively low efficiency. This is why cryogenic temperatures are required for Ho^{3+} two-micron lasing. The three-level scheme necessitates the use of low concentrations of Ho^{3+} in active medium (usually less than 1 at.%).

To increase pumping efficiency, holmium laser materials are usually codoped by sensitizers. The suggestion to put high concentrations of Er^{3+} and Tm^{3+} ions in Ho^{3+}-media has had the most success [115]. Such multiply doped laser materials obtained by this method, for instance, the so called "alphabetic" YAG (Ho^{3+}:αβYAG) and "alphabetic" YLF (Ho^{3+}:αβYLF), demonstrate highly efficient operation under flash-lamp pumping, but only at cryogenic temperatures.

The use of Cr^{3+} ions instead of Er^{3+} [116] in such multiply doped materials has allowed efficient two-micron laser action at room temperature under flash-lamp pumping. This scheme uses the efficient absorption of pump light by Cr^{3+} ions with further energy transfer to Tm^{3+} ions. The cross-relaxation process in Tm^{3+} from the $^3H_4$ state with a quantum yield of 2 transfers the excitation to the $^3F_4$ level of Tm^{3+}, followed by energy transfer to the upper laser level $^5I_7$ Ho^{3+}. The processes will be considered in further detail. As an alternative, Tm^{3+}-Ho^{3+} pumping by laser diodes is very efficient. In this case, Cr^{3+} ions are not needed, and the pumping proceeds directly to the $^3H_4$ state Tm^{3+} by means of laser diode emission with λ_{em}~0.8μm. Further energy transfer in the system proceeds in a similar way to the above process. Highly efficient sensitization of Ho^{3+} ion using a cross-relaxation "two for one" pumping process is why the two-micron Ho-laser is one of the most efficient and used in actual practice.

Another laser transition of Ho^{3+} ions of interest is the $^5I_6 \rightarrow {}^5I_7$ transition with an emission wavelength of about 2.9 μm. Laser action at this transition has been demonstrated in various oxide and fluoride crystals

[2]. Obtaining tunable three-micron Ho^{3+} emission at $^5I_6 \rightarrow ^5I_7$ transition is of practical interest. In [117], laser action in Ho:$YAlO_3$ at seven distinct lines in the 2.844-3.017 μm range under pumping by the 1.08 μm emission of a Nd:$YAlO_3$ laser is reported.

Note that the energy level structure of the Ho^{3+} ion, as in the case of Er^{3+} and Tm^{3+} ions, is the basis of the development of numerous cascade and cross-cascade laser schemes [118].

A Ho^{3+}-doped crystal (Ho:YLF) has been demonstrated [119] to be a good solid-state saturable absorber for the passive Q-switching of 2-μm lasers.

Erbium

Rather large energy differences between multiplets in Er^{3+} ions are favourable for getting stumulated emission at various wavelengths. Among numerous laser transitions discovered in Er^{3+} ions, the following two are of the most interest: the $^4I_{11/2} \rightarrow ^4I_{13/2}$ transition (λ=2.8 μm) and the $^4I_{13/2} \rightarrow ^4I_{15/2}$ transition (λ=1.6 μm). The energy level structure of the Er^{3+} ion is favorable for the realization of cascade generation [2].

The intensive absorption bands of Er^{3+} ion situated near 0.8 μm and 1.0 μm allow the use of AlGaAs laser diodes (λ_{em}=800 nm) or InGaAlAs laser diodes (λ_{em}=970 nm) as pump sources.

Erbium lasers operating at $^4I_{11/2} \rightarrow ^4I_{13/2}$ transition are of significant medical interest since 3-μm emission is absorbed by water. In practice, lasers based on two garnet crystals, YAG [120,121] and YSGG [122], are used. Lasers operating at the $^4I_{11/2} \rightarrow ^4I_{13/2}$ transition usually have good energy characteristics despite the fact that the transition is self-terminated (i.e. the lifetime of the lower level $^4I_{13/2}$ is considerably higher than the lifetime of the upper level $^4I_{11/2}$). Er^{3+}-systems demonstrate strong up-conversion processes greatly influencing the populations of the $^4I_{13/2}$ and $^4I_{11/2}$ levels. These processes are:

$$^4I_{13/2} + {}^4I_{13/2} \rightarrow {}^4I_{9/2} + {}^4I_{15/2} \qquad (w_1)$$

$$^4I_{11/2} + {}^4I_{11/2} \rightarrow {}^4F_{7/2} + {}^4I_{15/2} \qquad (w_2)$$

The w_1 process reduces the lifetime of the $^4I_{13/2}$ level and populates the upper laser level of the 3-μm transition, while the w_2 process increases the losses from the $^4I_{11/2}$ level. Up-conversion processes in the Er^{3+}-system have been studied and will be considered in detail later. Note that the up-conversion processes allow lasing in the visible spectral range, i.e., green generation at λ=551 nm at the $^4S_{3/2} \rightarrow {}^4I_{15/2}$ transition using diode pumping [123].

To obtain lasing at the $^4I_{13/2} \rightarrow {}^4I_{15/2}$ transition (λ=1.6 μm) is of great interest too, despite the three-level lasing scheme and the rather high thresholds. This interest is related to the fact that the erbium laser at the $^4I_{13/2} \rightarrow {}^4I_{15/2}$ transition emits in the eye-safe spectral range and has a long lifetime for the upper laser level, which makes possible high energy storage for pulse generation. However, despite numerous attempts to develop an efficient crystalline erbium laser at the $^4I_{13/2} \rightarrow {}^4I_{15/2}$ transition using sensitization and selective pumping [124-126], in practice, the majority of one-and-a-half micron emission sources are the erbium glasses. This is mainly due to the much faster intracenter relaxation $^4I_{11/2} \rightarrow {}^4I_{13/2}$ in glasses compared to that in crystals. The values of the emission cross-section at the $^4I_{13/2} \rightarrow {}^4I_{15/2}$ transition in a number of oxides and fluorides are given in Table 4.

Crystals containing erbium are of interest as saturable absorbers for the passive Q-switching of the 1.5 μm emission of Er lasers (see, for example [127]).

Thulium

The Tm^{3+} ion has only a few absorption bands in the visual spectral range. That is why, until recently, the efficient pumping of Tm-systems could only proceed using Er^{3+}, Yb^{3+}, and Cr^{3+} ions as sensitizers [3]. Diode laser pumping radically revised the situation. Tm^{3+} ions are particulary attractive for pumping by AlGaAs diodes due to the strong absorption at the $^3H_4 \rightarrow {}^3H_6$ transition (λ~0.8 μm). In the case of YAG doped with Tm^{3+} ions, the pump wavelength is 785 nm.

The energy levels scheme of the Tm^{3+} ion is relatively simple and has large energy differences between *J*-states. Three laser transitions occuring in the eye-safe spectral range are of the most interest presently: $^3H_4 \rightarrow {}^3F_4$ (λ_{em}=1.5 μm), $^3H_4 \rightarrow {}^3H_5$ (λ_{em}=2.3 μm), and $^3F_4 \rightarrow {}^3H_6$ (λ_{em}=2 μm).

The last transition is the most important. The values of the emission cross-section at this transition are given in Table 4. In principle, it proceeds according to the typical three-level scheme. However, the cross-relaxational "two for one" pumping process makes it a quasi-four-level transition possessing unique characteristics.

At high Tm concentrations, cross-relaxation process couples the transition from the $^3H_4$ excited state of one Tm^{3+} ion with the transition from the $^3H_6$ ground state of another one. Thus, after the absorption of one pump photon, the first Tm^{3+} ion relaxes from the $^3H_4$ state to the $^3F_4$, and simultaneously, another Tm^{3+} ion excites from the $^3H_6$ ground state to the $^3F_4$ state as well [128]. Thus, one pump photon excites two Tm^{3+} ions to the $^3F_4$ state. The $^3H_4 \rightarrow {}^3F_4$ and $^3H_6 \rightarrow {}^3F_4$ transitions are almost resonant and, therefore, proceed almost without the release of Stokes heat. The quantum efficiency of the process is close to 2 at characteristic concentrations of Tm^{3+} ions (~10^{21} cm^{-3}).

Smooth broadband tuning at the transitions was demonstrated in the 1.85-2.16 μm range in Tm:YAG and Tm:YSGG crystals [129] using a continuous wave Ti:Al_2O_3 laser to simulate diode pumping. In the Tm:YLF crystal, the tuning range lies further in the IR region (2.2-2.46 μm) [130].

Two other transitions in the Tm^{3+} ion (λ_{em}~1.5 μm and λ_{em}~2.3 μm) are not as efficient. Since the upper laser level of both of them ($^3H_4$) depletes during Tm-Tm cross-relaxation processes at high Tm concentrations, to obtain lasing at 1.5 and 2.3 μm, low Tm ions concentrations are required. In YAG crystals, at low concentrations of Tm^{3+} ions, the lifetime of the $^3H_4$ level is about 1 ms; the lifetime of lower laser level of the 2.3 μm transition $^3H_5$ is about 10 μs. That is why the $^3H_4 \rightarrow {^3H_5}$ transition proceeds according to the classic four-level scheme.

A different situation arises with the 1.5 μm transition. The lifetime of the $^3F_4$ state, the lower laser level, is about 10 ms, i.e. an order of magnitude longer than the lifetime of the upper laser level $^3H_4$. That is why the $^3H_4 \rightarrow {^3F_4}$ transition is self-terminated, which usually limits the laser action to the pulse regime only [131]. Elimination of self-termination is possible by codoping with the Ho^{3+} or Tb^{3+} ions which quench the lower laser level $^3F_4$ of Tm^{3+}.

It should be noted that the energy level structure of the Tm^{3+} ion and the intensities of the luminescence transitions in this ion allow the development of cascade and cross-cascade laser schemes [2].

Ytterbium

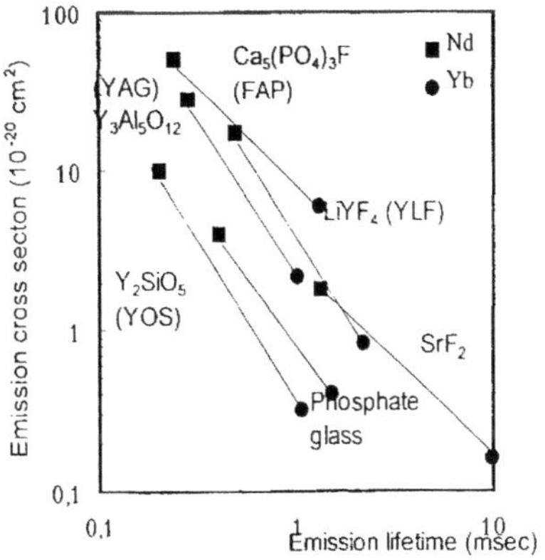

Fig. 9 Energy level diagram of Yb^{3+} ions in YAG

Yb^{3+} ion has only two manifolds: the ground $^2F_{7/2}$ and excited $^2F_{5/2}$ states with an energy difference of about 10 000 cm^{-1}. Stimulated emission of Yb^{3+} ions at room temperature proceeds according to the quasi-three-level scheme. To lower the lasing threshold, the temperature must be decreased. However, even at cryogenic temperatures, the Yb^{3+} ion pumping by broadband flashlamps is inefficient since there is only one absorption band in the Yb^{3+} ion [4]. Even codoping by Nd^{3+} ions as sensitizers does not allow laser action of the Yb^{3+}-laser at room temperature [132]. Thus, Yb^{3+}-materials were of low interest up to the most recent time. However, recent work [133] has demonstrated efficient room temperature operation of the Yb:YAG laser with diode pumping and has regenerated interest in Yb^{3+}-doped materials.

Actually, the emergence of the InGaAs diode emitting in the range of the Yb^{3+} ion absorption band (~950 nm) allows the development of diode-pumped Yb^{3+}-lasers with at least the same efficiency as the AlGaAs-diode-pumped Nd^{3+}-lasers [134,135]. It should be noted that operating in the same spectral region, the Yb^{3+} ion has a number of advantages in comparison to the Nd^{3+} ion. First of all, Yb^{3+} has the least quantum loss among all known ions. For example, Yb:YAG absorbs at 941 nm or 968 nm and emitts at 1.03 μm (see Figure 9), which necessitates the least removal of heat among any one-micron laser dopant [136]. The lifetime of Yb^{3+} upper state in various materials [137] is four times longer in comparison to the Nd^{3+}-doped materials, implying that the Yb-medium can store the same amount of energy using only $^1/_4$ of the pump energy of the diode arrays. On the other hand, the longer lifetime of the Yb^{3+} ion is usually accompanied by a laser emission cross-section of one order of magnitude less than that which the Nd^{3+} ion demonstrates in the same material [135].

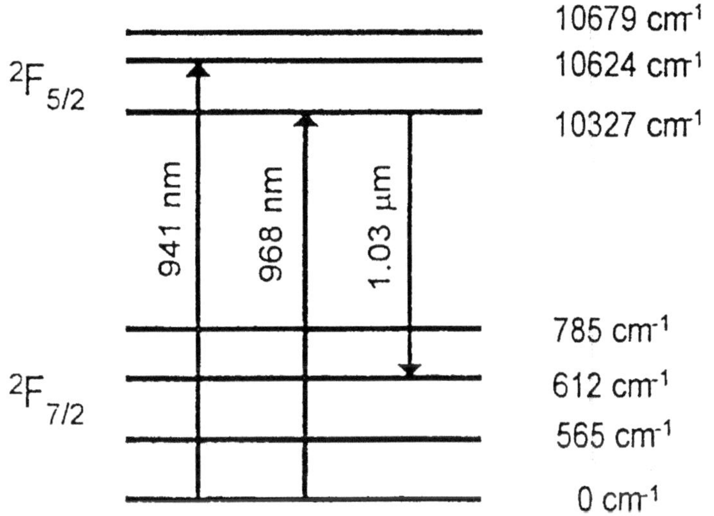

Fig. 10 Emission cross-sections and decay times of Nd^{3+} and Yb^{3+} ions in the different host materials [135]

Figure 10 compares the emission cross-sections σ_{em} and lifetimes τ of Yb^{3+} and Nd^{3+} ions in a number of hosts [135]. It can be easily seen that a larger σ_{em} for Nd^{3+} usually corresponds to a larger emission cross-section for Yb^{3+}. The same tendency can be also seen for lifetime values. This can be explained by the fact that Nd^{3+} and Yb^{3+} usually enter the same crystallographic sites. σ_{em} and τ values for Nd^{3+} are well known for many hosts (see Table 3), so these data can provide the basis for the search of prospective materials doped with Yb^{3+} ions.

It can be seen from Figure 10 that Yb:YAG and Yb:FAP possess the largest values of emission cross-section among Yb^{3+}-systems. The close to centrosymmetrical surrounding in SrF_2 provides the lowest σ_{em} and the largest τ values for both Yb^{3+} and Nd^{3+} . Ytterbium lasers can operate with very high efficiency. It is known [35] that the highest intrinsic efficiency (in the absence of excited state absorption) is determined by the difference in energy between the pump photon and the laser emission photon. In one of the most promising ytterbium media, Yb:FAP ($Yb:Ca_5(PO_4)_3F$), the quantum efficiency is

$$\lambda_{pump}/\lambda_{em}=905\ \text{nm}/1043\ \text{nm}=86.8\ \%.$$

Experimentally obtained in [135], the highest slope efficiency of the Yb:FAP laser was 79.1 %. An important advantage of the ytterbium medium resulting from the specific energy levels of the Yb^{3+} ion is the absence of ESA and up-conversion, i.e. there are no competing processes which could lead to degradation of laser performance.

The possibility of direct pumping of the $^2F_{5/2}$ level of Yb^{3+} by the InGaAs laser diode has invoked interest in the ytterbium ion as an efficient sensitizer of the one-and-a-half-micron luminescence of Er^{3+} at the $^4I_{13/2}\rightarrow{}^4I_{15/2}$ transition [138] and the two-micron luminescence of Ho^{3+} at the $^5I_7\rightarrow{}^5I_8$ transition [139,140]. Use of energy transfer from Yb^{3+} is of great interest to obtain visible emission through up-conversion schemes at the $^1D_2\rightarrow{}^3F_4$, $^1G_4\rightarrow{}^3H_6$, and $^1G_4\rightarrow{}^3F_4$ transitions of Tm^{3+} ions, as well as at the $^5S_2\rightarrow{}^5I_8$ transition of Ho^{3+} (551 nm wavelength) [141].

2.3. ACTINIDES

The actinide group is essentially similar to the group of rare earth elements. The electron configuration of a trivalent actinide is $1s^22s^22p^63s^23p^63d^{10}4s^24p^64d^{10}4f^{14}5s^25p^65d^{10}5f^n6s^26p^6$. Energy levels schemes of trivalent actinides can be found in [3,4].

Unlike lanthanides, the shielding of the $5f$ electrons of actinides is not as efficient, which is why they are more sensitive to external crystal field influences. Due to the larger size of the ions, the crystal field strength for actinides is usually twice as large as that for lanthanides, and the intensities of the absorption are 50-100 times as large for $5f$-$5f$ transitions as for $4f$-$4f$ ones [142].

Actinides are characterized by an increased spin-orbit interaction, the closeness of $5d$- and $5f$-states, and broadening and overlapping of absorption bands. Stronger ion-phonon interaction in actinides increases the probabilities of vibronic transitions. Due to the greater probability of multiphonon non-radiative relaxation, the efficient laser action in actinides is usually limited by transitions between the lower-lying J-states having large energy differences [3]. A specific feature of actinides—radioactivity—prevents the wide use of actinides for laser applications.

U^{3+} is the only actinide ion having demonstrated laser action. U^{3+} has the same number of f-electrons as Nd^{3+}, and it possesses a similar structure of energy levels. Laser action of $U^{3+}:CaF_2$ was obtained at the $^4I_{11/2} \rightarrow {}^4I_{9/2}$ transition of U^{3+} (λ_{em}~2.6 μm) in 1960 [143].

After a long period of time, the interest in actinides has been renewed due to the prospect of developing tunable U^{3+}-lasers. A promising crystal for this purpose is $U^{3+}:LiYF_4$ [144,145]. The emission spectrum of U^{3+} in this crystal at the $^4I_{11/2} \rightarrow {}^4I_{9/2}$ transition is very wide and covers the 1.91-2.97 μm range [142]. A high emission cross-section ($3 \cdot 10^{-19}$ cm^2) and a very strong splitting of the ground state $^4I_{9/2}$ (1115 cm^{-1}, twice as much as for $U^{3+}:CaF_2$) make $U^{3+}:LiYF_4$ a promising material for a four-level scheme of lasing at 2.8 μm with InGaAs diode pumping (λ_{em} ~ 980 nm). In [142,146], continuous wave laser action in U^{3+}:YLF at 2.827 μm was obtained at room temperature pumping with a $Al_2O_3:Ti^{3+}$ laser tuned in the 810-890 nm range.

Divalent uranium ions in alkaline earth fluorides CaF_2, SrF_2, and BaF_2 were used as saturable absorbers for the Q-switching of 1.5 μm laser emission [85,86]. The best results were obtained for the $U^{2+}:CaF_2$ crystal, providing the shortest pulse duration 20 ns without crystal damage. The substantial disadvantage of U^{2+}-fluorides is the considerably higer relaxation times in comparison with other 1.5 μm passive Q-switches, for example, garnets doped with Co^{2+} ions [86].

3. Non-radiative energy transfer in solids

3.1. PAIR INTERACTION

Non-radiative transfer of electron excitation energy between two particles involves an interaction between a donor and an acceptor resulting in decrease of the energy of the donor and the increase of that of the acceptor. Non-radiative energy transfer and radiative energy transfer (absorption by the acceptor of energy radiated by the donor) differ in that in the former a photon is not emitted, and, moreover, a donor has the additional possibility to deactivate and, in so doing, to decrease the lifetime of its excited state. In contrast to radiative energy transfer, non-radiative transfer occurs usually only on distances much shorter than the emission wavelength corresponding to a given transition [148,149].

To determine the probability of resonant dipole-dipole non-radiative energy transfer (i.e. taking place, if the emission spectrum of the donor overlaps the absorption spectrum of the acceptor) in a pair of particles (W_{DA}), Förster proposed to use the quantum-mechanical expression obtained by first order perturbation theory [150,151]:

$$\delta W_{DA} = \frac{2\pi}{\hbar} \cdot \left|V_{DA}\right|^2 \delta\rho(E), \text{ where } V_{DA} = \left\langle \Psi'_D \Psi_A \middle| V_{DA} \middle| \Psi_D \Psi'_A \right\rangle \quad (1)$$

Here $\Psi'_D \Psi_A$ and $\Psi_D \Psi'_A$ are, respectively, "excited donor-unexcited acceptor"and "unexcited donor- excited acceptor" states, V_{DA} - the operator of the donor-acceptor interaction, $\rho(E)$- the distribution of the density of system states corresponding to resonant energy exchange.

We can obtain the total probability W_{DA} by integrating equation (1) over the distribution of system states. Förster had shown that distributions are determined by the emission spectrum of the donor and the absorption spectrum of the acceptor. The probability of resonant dipole-dipole non-radiative energy transfer in pairs of particles can be easily expressed using empirically derived experimental values, namely [148-151]:

$$W_{DA} = \frac{9\chi^2 c^4}{8\pi \tau_D R^6} \int F_D(w) \sigma_A(w) \frac{dw}{w^4} \qquad (2)$$

where σ_A is the absorption cross-section of the transition in the acceptor, $F_D(w)$ is the normalized spectrum of emission of donors ($\int F_D(w)dw=1$), τ_D is the emission lifetime of donor, R is the distance between particles, c is the velocity of light, χ is the orientational factor, and w is the transition frequency.

Furthermore, Dexter [152] extended this theory to the cases of the interaction of higher multipole and exchange interaction.

According to [150-152], if the multipole interaction takes place, the respective probabilities are equal to:

$$W_{DD} = C_{DD} f_{DD}(R); \; W_{DA} = C_{DA} f_{DA}(R) \qquad (3)$$

Here C_{DD} and C_{DA} are microparameters of donor-donor and donor-acceptor interactions; they are numerically equal to the probabilities of interaction of two particles being a unit distance from each other. R is the distance between two particles, $f_{DD}(R) = R^{-S}$, $f_{DA}(R) = R^{-m}$, and S and m are the order of the interactions. In particular, they are equal to 6 for the dipole-dipole interaction, 8 for the dipole-quadrupole interaction, and 10 for the quadrupole- quadrupole interaction.

In the case of the exchange interaction, the probability of non-radiative energy transfer is equal to:

$$W = W(R_0)\exp\left[-\frac{2(R-R_0)}{L}\right] \tag{4}$$

Here R_0 is the minimum possible distance between interacting particles, and L is the characteristic scale of interaction [152-154].

It should be noted that general statements of the theory of resonant non-radiative energy transfer theory in a pair of particles can be formulated by the example of the classic model of two linked harmonic oscillators considered as an analog of donor-acceptor pair. This point of view was elaborated by M.D. Galanin [155].

Besides resonant energy transfer, there is non-resonant transfer taking place if the difference between the position of the luminescence band of the donor and the absorption band of the acceptor is large, such that energy transfer is possible only with the creation or destruction of lattice phonons [156,157].

A great number of papers are devoted to the theoretical examination of the elementary act of the transfer of electron excitation energy, including investigations by the density matrix method (see, for instance [158-176]). Describing the processes of energy transfer by a matrix density permits us phenomenologically to take into account the relaxational components related to the interaction of particles with their environment. In this case, two relaxation times are introduced to characterize di-level system: *longitudinal* - which characterizes the relaxation of diagonal elements of density matrix and *transverse* - which determines the rate of violation of phase correlations. Usually, ignoring fine details, three typical cases of non-radiative interaction depending on the ratio of task parameters are considered:

1. Weak non-coherent interaction (Förster-Dexter-Galanin model) [148-152].

$$\Gamma >> \frac{1}{\tau_A},\ \frac{1}{\tau_D},\ W_{DA};\ \ W_{DA} << \frac{1}{\tau_A} \qquad (5)$$

Here τ_D and τ_A are the lifetimes of donor and acceptor excited states, and Γ is the cross phase relaxation rate determining the spectral width of the emission and absorption bands. Γ is much higher than all rates of the other relaxational processes; this provides a non-coherent type of interaction and the possibility to apply a probabilistic approach to describing the energy transfer processes. The last relation provides a quick relaxation of the excitation in the acceptor, guaranteeing non-reversibility of the donor-acceptor transfer.

2. Strong non-coherent interaction

$$\Gamma >> W_{DA} >> \frac{1}{\tau_A} \qquad (6)$$

In an analogous way quick phase relaxation allows us to apply the probabilistic approach. However, in this case rather fast (in comparison to τ_A^{-1}) back energy transfer from acceptor to donor occurs. Excitation of the donor is followed by the mixing of the excitation between donor and acceptor with a time W_{DA}^{-1}, and then by the decay of *collective* excited states proceeding with a joint lifetime [168,169]. It should be noted that in [161,165,176] both cases were described in the same way by means of the so called "generalized probabilities".

3. Strong coherent interaction (Davidov's model) [159,160]:

$$|W_{DA}| >> \Gamma, \frac{1}{\tau_D}, \frac{1}{\tau_A} \quad (7)$$

In this case coherent oscillations of the population of interacting particles occur.

In solids doped with transition ions, a large phase width of interacting transitions is usually observed, proving the applicability of both the probabilistic approach and the method of adding transfer probabilities in the description of the processes of non-radiative energy transfer [176].

3.2. Energy Transfer in a Continuum of Particles

For both theoretical and applied studies, especially when a new active medium for solid state laser is being analyzed, a very important problem is encountered. How can we average elementary interactions over a disordered continuum of donors and acceptors? It is obvious that the non-radiative relaxation of excitations in a group of particles will be determined by the superposition of a great number of energy transfer elementary acts. If energy transfer takes place among like particles, the process is referred to as a migration of energy in the donor subsystem. In this case, the observed excitation lifetime does not change. If energy transfers from one sort of particles (donors) to another (acceptors), the process is referred to as donor-acceptor non-radiative energy transfer in the group of particles. Here, in the absence of back energy transfer, the observed lifetime of excited state of donors will, in general, decrease since the excited states provide an additional possibility for deactivation [148,149].

Let us see how the problem of averaging of pair interactions over a disordered group of donors and acceptors can be solved in the most commonly occurring case of the weak non-coherent interaction [148-152]. Imaging a volume where several donors and acceptors are randomly located. Let us assume that at time $t=0$, some fraction of the donors are excited (we assume that the probability to be excited is the same for all donors, i.e. both excited and unexcited donors occur randomly in the volume). Further energy transfer from the excited donors to the acceptors may proceed by two following ways (or by a combination of them):

1. By the direct donor-acceptor interaction, when the excited donor transfers its energy directly to the acceptor. In this case, the process of energy transfer is said to have a static nature (excitations disappear at the same location where they appear).

2. By the migration (through donors) of energy to the site of its destruction. In this case, the excitation changes its place and, correspondingly, acceptor environment before it meets an acceptor. Migration accelerates the decay of the excited donor group, delivering the excitation closer to acceptors than in the static case.

3.2.1. Kinetics of the Luminescence of Donors

Static transfer

First, we consider static energy transfer, taking place usually at small donor concentration, when the migrationally stimulated quenching of luminescence is weak. For multipole irreversable energy transfer, if we ignore the finite volume of particles and there is full orientational averaging, the number of donor excitations at moment t is expressed as [148-152,177,178]:

$$N(t) = N(0) \exp\left[-\frac{4}{3}\pi\Gamma\left(1-\frac{3}{m}\right) n_A \, C_{DA}{}^{3/m} \, t^{3/m}\right] \qquad (8)$$

Here n_A is the acceptor concentration (the term $\exp\left(-\frac{t}{\tau_D}\right)$ is dropped for simplicity).

It can be seen that such static disordered decay is non-exponential, its rate decreasing with time. Obviously, this fact is connected to the deexcitation of donors which have a favorable acceptor environment regarding energy transfer, and the slow deactivation of donors having a less favorable acceptor environment.

One of the conditions for the above equations for any particular crystal is a small concentration of acceptors. In [179,180] the exact formula describing static energy transfer and taking into account the particular crystal structure, which is valid for any concentration of acceptors, is obtained:

$$N(t) = N(0)\exp\left[\sum_i \ln\left(1 - x_A + x_A\, e^{-W_{DA(i)}\cdot t}\right)\right] \tag{9}$$

Here x_A is the relative concentration of acceptors, i.e. the ratio of the quantity of acceptors to the quantity of sites of the acceptor sublattice (the probability that the site of the sublattice will be occupied by any acceptor is the same for all sites). The summation is executed over sites of the acceptor sublattice, with the origin located at a donor site. As can be seen, the idea of (9) is to substitute the summation of transfer probabilities over randomly located acceptors for the summation over ordered sites of the crystal lattice in which acceptors can be inserted.

When time t is small: $t << t_1 = W_{DA}^{-1}(R_0)$, (where $W_{DA}(R_0)$ is the probability of interaction between donor and acceptor, when they are at a minimum distance), expression (9) becomes simpler:

$$N(t) = N(0)\ \exp\left[-x_A \sum_i W_{DA(i)}\cdot t\right] \tag{10}$$

i.e., when t is small, the decay of the group of donor excitations is exponential (so called ***static ordered decay***).

It is worth noting that Equation (9) was obtained for a random distribution of donors and acceptors over the sample. In [181-188], the situation when donors and acceptors occupy crystal lattice sites in a not random, but rather correlated, manner was examined. There, the cases when the concentration of acceptors near the donor is more and less than equal to the probability far from the acceptor were examined. By means of the model elaborated in [181-188] a number of "anomalous" experimental facts, for example the particularities of non-radiative energy transfer processes $Nd^{3+} \rightarrow Er^{3+}$ in YAG crystals, were explained. The deviation of activator concentration from equiprobable was also observed in garnet crystals doped with Cr^{3+} and RE^{3+} ions [189-191]. It is obvious that similar situations when interacting particles demonstrate a correlated distribution can also take place for other experimental situations, so, in general, the applicability of Equation (9) for static energy transfer should be validated in every particular case.

Diffusion model

It was mentioned above that energy migration accelerates the donors' excitation decay stimulating the delivery of excitation to sites with the optimum conditions for energy transfer. One of the first attempts to describe the process of the migrationally stimulated quenching of luminescence was by use of the diffusion equation [192-198]. When one considers the diffusion mechanism of transfer, the diffusion and delocalization of excitations in a system of like particles should be distinguished from the diffusionally stimulated energy transfer from the donor subsystem to the acceptor subsystem. Moreover, the possibility to describe the delocalization of excitation in terms of diffusion theory does not yet mean that when quenching impurities are present, diffusionally stimulated quenching is necessarily present [196].

In the general case, to describe the process of energy migration in a system of like particles, the following system of balance equations should be solved [148]:

$$dP_i / dt = -\sum_j \left[W_{ij} P_i(t) - W_{ji} P_j(t)\right] \qquad (11)$$

Here P_i is donor's excitation density at the i-th center, i.e. the probability to observe the excitation at the i-th center; W_{ij} is the probability of energy transfer from the i-th center to the j-th one.

At the initial time (t=0), the probability of excitation at any center is described by the δ-function $:P_i(0)=\delta_{i0}$.

The system of equations (11) allows us to find an exact solution only for an ordered lattice [199]. If t is large, the diffusion solution occurs:

$$P_0(t) = \left(W_{dif} \cdot t\right)^{-3/2} ; \; W_{dif} = 4\pi n^{2/3} D_{od} \qquad (12)$$

Here D_{od} is the diffusion coefficient for the ordered lattice:

$$D_{od} = \frac{1}{6} \sum_i W_{0i} R_{0i}^2 \qquad (13)$$

Here W_{0i} is the probability of interaction of the initial ion and the ith ion, R_{0i} is the distance between the initial ion and the ith ion.

If the lattice is not ordered, a system of equations like (11) can not be solved exactly. To avoid the difficulties of the solution of a system of equations for a random location of particles, the authors of [198] introduced the concept of a *quasi-diffusion* coefficient determining the average time of spreading of a excitations of donors. In [157], the experimental observation of diffusion in a disordered group of like particles was obtained. The long-term asymptote of the spreading excitations for the disordered system was found to be the same as for the ordered system. Experimentally measured values of the diffusion coefficients for the dipole-dipole interaction in the disordered lattice were proportional to $n^{4/3}$, as well as for the ordered lattice.

To observe the diffusion mechanism of migrationally stimulated luminescence, the average distance between donors should be much less then the typical distance for the quenching interaction (the radius of the quenching sphere). In this case, the density of the donor excitation changes little within the average distance between donors, so we can approximate with a continuous distribution of this density [192]. The concurrent influence of direct donor-acceptor interaction and diffusion quenching on the kinetics of donor excitation decay was examined for the first time in [195], where the quenching interactions were included in the initial diffusion equation. For $S=m=6$, the decay of donors excitations at long time proceeds with a constant rate:

$$\overline{W} = 0{,}68 \cdot 4\pi n_A D^{3/4} C_{DA}^{1/4} \tag{14}$$

Here D is the diffusion coefficient: $D = \alpha n_D^{4/3} C_{DD}$ (α is a numerical coefficient).

Inserting D in $\overline{W}$, we obtain:

$$\overline{W} = 0{,}68 \cdot 4\pi \alpha^{3/4} n_D n_A C_{DD}^{3/4} C_{DA}^{1/4} \tag{15}$$

For other S and m values, the expression for $\overline{W}$ will look like [196]:

$$\overline{W} \sim C_{DA}^{\frac{1}{m-2}} n_A C_{DD}^{\frac{m-3}{m-2}} n_D^{\frac{(s-2)(m-3)}{3(m-2)}} \tag{16}$$

The description of migrationally stimulated luminescence quenching via a diffusion processes had been used for a long period of time. However, in [200], it was shown that as a rule, in solids, the conditions required for the correct description of migrationally stimulated luminescence quenching by duffusion terms are not met.

Hopping model

The hopping mechanism of migrational quenching, an alternative to the diffusion one, was suggested in [198,201] and elaborated in a number of works [180,196,202-207]. It assumes that if the characteristic distance of the quenching interaction is considerably smaller than the average distance between donors, the excitation entering the quenching volume leaves during one jump; the interaction with an acceptor can change from weak to strong and conversely. Burstein [196,198,201] obtained an integral equation describing hopping luminescence quenching with the hypothesis that the variation of the excitation quenching probability, when excitation hops among the donors, is a random Markov's process without correlation:

$$N(t) = N_0(t) \cdot R_e(t) - \int_0^t N_0(t-t')\dot{R}_e(t-t')N(t')dt' \quad (17)$$

Here $N_0(t)$ is the function describing the kinetics of static decay, $R_e(t)=\exp(-t/\tau_0)$, τ_0^{-1} is the most probable frequency of excitation jumps along the lattice of donors.

As t gets large, the expression for the rate of "stationary" dipole-dipole migrationally limited quenching is [198,201]:

$$\overline{W} = \pi\left(\frac{2\pi}{3}\right)^{5/2} C_{DA}^{1/2} C_{DD}^{1/2} n_A n_D \quad (18)$$

In the first version of the theory, the process of decay probability fluctuation was assumed to be Markovian and non-correlated. However, this is not true even if the donors form an ordered system. There is always a chance that the excitation will return to the center which it has already occupied *(memory effect)*. One of the way to take the effect into account by correcting the τ_0 -value, which enters in the numerical coefficient in expression (18). In this way in [196], the following expression was obtained:

$$\overline{W}=\frac{4}{3}\pi C_{DA}^{3/m}\, n_A\, (\beta_{DD}C_{DD})^{1-\frac{3}{m}}\cdot n_D^{\frac{S\left(1-\frac{3}{m}\right)}{3}}, \qquad (19)$$

where

$$\beta_{DD}=6\left[\frac{2\pi}{3}\Gamma\left(1-\frac{3}{S}\right)\right]^{S/3} \Big/ S\cdot\Gamma\left(\frac{S}{3}\right) \qquad (20)$$

There were also attempts to take into account the difference between hopping in a disordered system of donors and the equivalent process in an ordered lattice [196]. In this case equation (17) is the same, but the $R_e(t)$ function is replaced by R(t), which takes into account the return of the excitation to the initial center from the closest donor [208]:

$$R(t)=\exp\left\{-\frac{n_D}{2}\int_V\left[1-\exp(-2W_{DD}t)\right]dV\right\} \qquad (21)$$

For $S=m=6$ from (17) we will get:

$$W=\frac{8\pi^3}{9\sqrt{2}}n_A n_D\sqrt{C_{DA}C_{DD}} \qquad (22)$$

which is equivalent to (18) within a numerical constant.

When experimental results are being treated, the actual donor excitation decay is usually divided into three typical regions [209]:

1. Static ordered decay: here, the quenching has a maximum rate:

$$W_{in} = x_A \cdot \sum_i W_{DA_i} \tag{23}$$

   It occurs at $t << t_1$

2. Static disordered decay corrected according to non-stationary migration quenching. It occurs at $t_1 << t << \tau_0$

3. "Stationary" (with a constant rate) migration quenching. It occurs at $t >> \tau_0$

In [210,211], another approach to the description of migrationally stimulated luminescence quenching was elaborated. This method does not require dividing the experimental curves into separate stages, but rather describes the total process in a unified way, taking into account the crystal structure. Although this process is very computer time consuming, it is the most adequate way to determine the microscopic constants of interaction C_{DA} and C_{DD} and the multipolarity of the interaction on the basis of experimental curves of the donor excitation decay.

The growth of the donor concentration causes an increase of concentration quenching, but, at very high donor concentration, luminescence quenching proceeds at the kinetic limit regime [201]; its constant does not change with the concentration of donors. To realize the kinetic limit regime, the characteristic time of migration of excitation should be notably smaller than the reverse rate of quenching at a minimum donor-acceptor distance. In this case, when comparing the two processes, namely, the delivery of excitation to the places with the optimum conditions for transfer and the time for actual transfer directly to the acceptor, the last process is the limiting one *(bottleneck model).*

Since in this case the density of excitations of all donors levels off, the rate of quenching becomes equal to the rate of the initial stage of static decay, i.e. the maximum possible rate of excitation quenching in a group of randomly distributed donors and acceptors (23) [179,180,201,209].

An integral view of (23) for multipole donor-acceptor interaction is [196,201]:

$$W_{in} = \frac{4}{m-3}\pi n_A \frac{C_{DA}}{R_0^{m-3}} \qquad (24)$$

An investigation of non-radiative energy transfer processes in a group of *self-quenched* impurities, when every particle can act as both a donor and acceptor due to the specific energy levels, is of great interest. In this case luminescence quenching takes place by the so called *cross-relaxation* process, when the excitation energy divides approximately in half between two interacting like particles. The cross relaxation process is possible if, in the energy spectrum of a donor, there is a level located approximately half way along excited and ground states. A typical example is the Nd^{3+} ion. It is commonly known that between the metastable level $^4F_{3/2}$ and the ground level $^4I_{9/2}$, the Nd^{3+} ion has several intermediate states, which provide a cross-relaxation exchange of excitation from the $^4F_{3/2}$ level [205,209].

The functional dependence of the rate of the "stationary" jumping migrationally limited self-quenching of luminescence on the microparameters of the interaction and the concentration of activators is usually the same as for hopping quenching. For $s=m=6$ we have: $\overline{W} \sim \sqrt{C_{DA}C_{DD}}\,n^2$ [196,198,204,207,212-216].

Cross-relaxational self-quenching of luminescence can considerably influence the parameters of the active laser medium leading to restrictions on the usable concentrations of the activator. On the other hand [217-219], in some situations, cross relaxation processes can play a positive role increasing laser efficiency, competing with Stokes losses.

3.2.2. *The Kinetics of the Luminescence of Acceptors*

The effect of the sensitization of luminescence when different luminescent dopants serve as energy absorbers and emitters, is of great importance in the development of new efficient laser media [220-223]. To achieve this effect successfully, obviously, a number of conditions should be met. Donors should have intensive absorption bands in the pumping region. The absorbed energy ought to transfer by a non-radiative process with a high quantum yield to acceptors—active particles—which should radiate in a desirable spectral region. The use of the sensitization effect has allowed the study of many efficient laser media. From our point of view, one of the most prominent examples of this effect is the development in the General Physics Institute, Russia of a new class of solid-state lasers on the basis of scandium-gallium garnet crystals doped with Cr^{3+} and RE^{3+} ions [25,223]. The prerequisite to realize the donor potential of Cr^{3+} discussed in [223] is the active participation of the $^4T_2$ level of Cr^{3+} in the energy transfer process, which is provided by the small energy difference between the $^4T_2$ and ^2E excited states of Cr^{3+}. There are a number of crystals, including some garnets, which meet this requirement. This condition provides fast non-radiative energy transfer from Cr^{3+} to RE^{3+} ions with high quantum yield.

To search for efficient laser media based on sensitized materials, one should know how to describe quantitatively the processes of population and decay of the laser levels of the active particles (energy acceptors) when they are interacting with sensitizers (energy donors). The foundations of theory of acceptor luminescence were stated by M. D.

Galanin [148,155,177,220] and developed in [224-229]. The following is a short description of their results.

Let us write a system of kinetic equations for the population of excited states of energy donors and acceptors after the end of an exciting pulse of light [228]:

$$\dot{N}_D = -\frac{N_D}{\tau_D} - \sum_D \sum_A W_{DA} \cdot P_D + \sum_D \sum_A W_{AD} \cdot P_A \qquad (25)$$

$$\dot{N}_A = -\frac{N_A}{\tau_A} + \sum_D \sum_A W_{DA} \cdot P_D - \sum_D \sum_A W_{AD} \cdot P_A \qquad (26)$$

Here N_A and N_D are the number of excited donors or acceptors at any moment, P_A and P_D are the probabilities of any particular donor or acceptor to be excited at any moment.

After combining (25) and (26) and integrating over time, we get the time dependence of the population of the excited states of acceptors:

$$N_A(t) = \exp\left(-\frac{t}{\tau_A}\right)\left\{N_A(0) + \int_0^t \exp\frac{t'}{\tau_A}\left(-\frac{N_D(t')}{\tau_D} - \dot{N}_D(t')\right)dt'\right\}$$

(27)

Knowing the kinetics of the decay of donor excitations, by means of (27), we can get the population kinetics and decay of the upper laser level of the acceptor, including the case when back energy transfer and migration of the excitations over both donor and acceptor subsystems take place.

More general equations, taking into account concentration quenching of luminescence in the acceptor subsystem (which occurs, for example, if

Nd^{3+} ions are being sensitized by other ions, among them Cr^{3+} [25,228]) were obtained in [229,230]. There the system consisted of donors, acceptors-I (A_I), and acceptors-II (A_{II}) serving as quenchers of the luminescence of active acceptors (A_I) was examined. The external source of emission excites acceptors-A_I and donors; the latter can transfer their energy to active particles A_I in a non-radiative manner. For the stationary population of A_I,

$$N_{stat} = N^*_{A_I} \cdot \eta_{A_I} + N^*_D \cdot (1 - \eta_D) \cdot \eta_{A_I} \tag{28}$$

where

$$\eta_{A_I} = \tau_{A_I}^{-1} \cdot \int_0^\infty \exp\left(-\frac{t}{\tau_{A_I}} - \Pi_{A_I}(t) \right) dt \ ,$$

$$\eta_D = \tau_D^{-1} \cdot \int_0^\infty \exp\left(-\frac{t}{\tau_D} - \Pi_D(t) \right) dt$$

Here N^*_{AI} and N^*_D are the numbers of acceptors-I and donors, correspondingly, excited by an external source in a unit of time; $\Pi_D(t)$ and $\Pi_{AI}(t)$ are the functions describing the energy transfer from donors to acceptors-I and the quenching of luminescence of acceptors-I by acceptors-II, respectively.

If the system is excited by a δ-impulse emission, we obtain:

$$N_{A_1}(t) = N_{A_1}(0)\exp\left(-\frac{t}{\tau_{A_1}} - \Pi_{A_1}(t)\right) +$$
$$+ N_D(0)\cdot\int_0^{t} \frac{d\Pi_D(t')}{dt'}\exp\left(-\frac{t'}{\tau_D} - \Pi_D(t') - \frac{t-t'}{\tau_{A_1}} - \Pi_{A_1}(t-t')\right)dt' \tag{29}$$

where

$$N_D(t) = N_D(0)\exp\left(-\frac{t}{\tau_D} - \Pi_D(t)\right),$$
$$N_{A_1}(t) = N_{A_1}(0)\exp\left(-\frac{t}{\tau_{A_1}} - \Pi_{A_1}(t)\right)$$

Knowing the kinetics of decay of the activator excited state after δ-excitation, one can obtain decay kinetics for excitation by an arbitrary excitation function of light $\varphi(t)$:

$$N_{A_1} = \int_0^t \varphi(t')dt'\left\{n_{A_1}\int\frac{f(\nu)}{\nu}\sigma_{A_1}(\nu)d\nu\cdot exp\left[-\frac{t-t'}{\tau_{A_1}} - \Pi_{A_1}(t)\right] +\right.$$
$$+n_D\int\frac{f(\nu)}{\nu}\sigma_D(\nu)d\nu\times\int_{t'}^{t} dt''\left[\frac{d\Pi_D(t''-t')}{d(t''-t')}\times\right.$$
$$\left.\left.\times exp\left[-\frac{t''-t'}{\tau_D} - \Pi_D(t''-t') - \frac{t-t''}{\tau_{A_1}} - \Pi_{A_1}(t-t'')\right]\right]\right\} \tag{30}$$

Here $f(\nu)$ is the function describing the spectral composition of the exciting pulse, and $\sigma(\nu)$ is the absorption cross-section.

The above mentioned equations allow the optimization of acceptor concentration in the doped crystal, if luminescence quenching or self-quenching takes place [230,231].

It should be noted that, in doped crystals, different mechanisms of interionic interactions can concurrently take place. For example, in [230,231], the sensitization of the luminescence of Nd^{3+} ions by Cr^{3+} in scandium garnet crystals was examined, and, after short excitation by light, the initial sections of the decay curves of the excited state ($^4T_2$, ^2E) of Cr^{3+}, a sharp decrease of population (so-called *fast component*) was observed. The suggestion was made that along with the long-distance mechanisms, short-distance ones were also taking place. Obviously, their contribution should increase with an increase in the concentration of acceptors, as was determined experimentally. In [191] the decay of the Cr^{3+} excited state in scandium garnet crystals was examined in detail in the static non-radiative energy transfer approximation. It was shown that fast transfer can be described by the general formula of static quenching. The energy transfer in the examined crystals was characterized by a mixture of different multipolarities.

Since a great amount of experimental work devoted to the investigation of non-radiative energy transfer in condensed media has been published, their detailed review is impossible here. In this chapter, we discussed examples critical to laser development. In 1985, B.M. Antipenko et al. [116] suggested an efficient scheme of a 2-μm Ho laser and implemented it in a YAG crystal. The triple system, consisting of Cr^{3+}, Tm^{3+}, and Ho^{3+} dopants, was investigated. Cr^{3+} ions, which demonstrate in garnet crystals an intense broadband absorption in the visible spectral region, absorb the energy of flashlamp pumping and transfer it non-radiativly with a high quantum yield to the $^3H_4$ level of Tm^{3+}. The cross-relaxation of Tm^{3+} ions causes two Tm^{3+} ions in the $^3F_4$

state instead of one Tm^{3+} in the $^3H_4$ state. The third stage of the process is the migrationally stimulated energy transfer from the Tm^{3+} ions to the upper laser level $^5I_7$ of Ho^{3+}. In [116], the quantum yield of the excitation transfer to the $^5I_7$ level of the Ho^{3+} in YAG:Cr,Tm,Ho crystal was reported to be close to 2. In this scheme, the use of donors and acceptors was realized, and reduced heat release was achieved. Efficient 2-μm Ho lasers based on YAG:Cr,Tm,Ho, YSGG:Cr,Tm,Ho, and GSAG:Cr,Tm,Ho crystals were described [116,232-234].

Let us note that the development of diode lasers emitting in the range of the absorption band of Tm^{3+} ions (~0.8 μm) has resulted in efficient 2-μm holmium lasers pumped by diode lasers directly into the thulium absorption bands [235,236].

3.2.3. *Strong Non-Coherent Interaction*

The above described laser crystals had concentrations of luminescent dopants of $8 \cdot 10^{20}$ cm^{-3} (Tm^{3+}) and $5 \cdot 10^{19}$ cm^{-3} (Ho^{3+}). At these concentrations, the regime of strong non-coherent interaction was established between groups of the ions [168,169,237-240]. The same regime of strong non-coherent interaction was observed for other pairs of ions, e.g., Yb and Er ions [241-244], and in a number of other experimental situations.

As noted above, when such interaction takes place in a pair of particles, the excitation mixes between the donor and acceptor at a rather high rate, and further decay of the "collectivized" excited states proceeds with a joint lifetime. Obviously, to realize this regime in a pair of particles, i.e. the rates of direct and back energy transfer being considerably higher than the rates of intracenter decay of the excited states of donor and acceptor, the distance between particles ought to be rather short, which can be achieved by increasing the concentrations of interacting particles. Evidently, if a strong non-coherent interaction regime is to be realized, it would take place at higher concentrations.

When one wants to design a certain laser scheme where the strong non-coherent interaction regime takes place, he should be able to determine the concentrations of the interacting particles in the excited states (n_1^* and n_2^*), which will define the joint lifetime of the excited states [168,169,237-246]:

$$\tau_{general}^{-1} = \frac{n_1^* A_1 + n_2^* A_2}{n_1^* + n_2^*} \tag{31}$$

Here A_1^{-1} and A_2^{-1} are the intracenter lifetimes of the excited states of the donors and the acceptors, correspondingly.

If, in a group of donors and acceptors, fast migration of energy among like particles exists, one can apply the so called collective kinetic equations [169,242-245] which result in the following expression (if the multipolarities of direct and reverse transfers are the same) :

$$\frac{n_1^*}{n_2^*} = \frac{n_1}{n_2} \cdot \frac{W_{2\to1}}{W_{1\to2}} \tag{32}$$

Here n_1, n_2, $W_{2\to1}$, and $W_{1\to2}$ are, respectively, concentrations and rates of pair energy transfer at a fixed distance between two unlike particles. As was shown in [38,246], the above expression remains true in the general case, when the migration of energy among like particles is not favored over energy transfer between unlike particles, and the collective kinetic equations are not used.

Another method for the description of the excitation distribution in a system of strongly non-coherently interacting donors and acceptors is based on the suggestion that the populations of corresponding Stark sublevels of interacting states follow a Boltzmann distribution in the unified system of energy levels of the particles; this was elaborated on in [247,248].

The state of the art in the theoretical and experimental research of non-radiative energy transfer processes in condense media in a number of cases permits the calculations of particular situations of interionic interactions in doped crystals and glasses, including the processes of quenching and sensitization of luminescence, and assist the search for new active media for solid state lasers with optimum parameters.

4. Up-conversion processes in solids

In principle, the term *up-conversion* can be associated with a number of physical phenomena. Here, we will define up-conversion as the non-coherent transformation of radiation, which excites the optically active particles, into higher values of energy. When the energy of the emitted photon is higher than the energy of exciting photons, this phenomenon is called *anti-Stokes luminescence*. The compounds which demonstrate such anti-Stokes luminescence are referred to as anti-Stokes luminophors. The majority of materials known to be anti-Stokes luminophors are crystals and glasses doped with trivalent rare-earth ions. RE^{3+} ions have a complicated system of energy levels, which allows different mechanisms of the population of upper excited states. Besides, transitions in the system of $4f$-electrons of RE^{3+} ions are forbidden as intraconfigurational, and, although this is partially removed by the crystalline field, there are usually many metastable levels with relatively long lifetimes. This fact results in a high population of excited states of the interacting ions, and, further, serves to enhance the up-conversion process. It should be noted that the great interest in the investigation of up-conversion processes is because they greatly influence the output characteristics of many solid-state lasers.

Let us consider the main up-conversion mechanisms.

4.1. CONSECUTIVE ABSORPTION OF PHOTONS (*GSA* AND *ESA*)

This mechanism, suggested by Blombergen, has provided the basis for the development of IR-emission counters [249]. The basic scheme of this mechanism can be demonstrated by taking a three-level system as an example (1 is the ground state). Emission with frequency ν_{12} is absorbed at the 1→2 transition (*ground state absorption, GSA*) populating level 2, then a second photon (ν_{23} frequency) is absorbed from this level (2→3 transition, excited state absorption, *ESA*) populating level 3. We can write:

$$h\nu_1 + A \rightarrow A^*; A^* + h\nu_2 \rightarrow A^{**} \rightarrow A + h\nu_{A^{**}}$$

(A^* and A^{**} are, respectively, the once- and twice-excited particle A) [250]. In this case, an ion excited by one photon, is additionally excited by another photon. Thus, luminescence from level 3 (3→1 transition) lies in the anti-Stokes region of spectrum with respect to the energy of every absorbed photon. Obviously, if the energy spectrum of the ion contains more intermediate levels, one can also observe luminescence at other transitions, in particular, with frequency smaller than ν_{12} or ν_{23}. This situation took place in one of the first experiments described in the literature on IR photons counting [251]. Pumping light with λ=2.3 μm excited Pr^{3+} ions in the $LaCl_3$ crystal, and populated the $^3H_6$ level of Pr^{3+} ($^3H_4 \rightarrow {}^3H_6$ transition, see Figure 7a). From this level, absorption, with λ=558 nm, took place and the $^3P_2$ level was populated ($^3H_6 \rightarrow {}^3P_2$). Then, intracenter non-radiative relaxation to the $^3P_0$ level occurred, and, from this level, luminescence with λ=618 nm was observed ($^3P_0 \rightarrow {}^3H_6$). Anti-Stokes emission (with respect to both frequencies of the exciting light) from $^3P_0$ was also observed. In [252], when fluoride crystals doped with Pr^{3+} were excited by 4.2 μm emission ($^3H_4 \rightarrow {}^3H_5$) and 540 nm emission ($^3H_5 \rightarrow {}^3P_0$), luminescence with λ=485 nm ($^3P_0 \rightarrow {}^3H_4$) was observed. The same crystals doped with Er^{3+}

demonstrated the red luminescence of Er^{3+} ions (λ=651 nm, $^4F_{9/2} \rightarrow$ $^4I_{15/2}$, see Figure 7a) when excited by 973 nm emission ($^4I_{15/2} \rightarrow ^4I_{11/2}$) and 1.96 μm emission ($^4I_{11/2} \rightarrow ^4F_{9/2}$) [253]. Obviously, if frequencies ν_{12} and ν_{23} are the same or close to each other, the entire process can proceed under the excitation by one-frequency emission. Thus, green luminescence of Er^{3+} ($^4S_{3/2} \rightarrow ^4I_{15/2}$) in some hosts appears with excitation by 970 μm emission due to the resonanse of the transitions $^4I_{15/2} \rightarrow ^4I_{11/2}$ and $^4I_{11/2} \rightarrow ^4F_{7/2}$.

The basis for the Blombergen mechanism of up-conversion is the absorption of pumping light from excited states (*excited state absorption, ESA*), which has no essential differences with absorption from the ground state. This mechanism in combination with other mechanisms of energy transformation is now widely used for the development of up-conversion lasers.

It should be noted that the mechanism of consecutive absorption going on with the participation of real intermediate states differs from the superficially similar mechanism of excited state population by multiphoton absorption of light, when two or more photons are absorbed simultaneously by one ion, without the participation of intermediate states [254]. The latter process is less efficient than the consecutive absorption mechanism, all other factors being equal. To proceed efficiently, the process of multiphoton absorption usually requires a higher density of exciting emission. Actually, the process that uses more real intermediate states has a higher probability, because, according to Dexter [255], "photons spread quickly, and their energy does not remain in the media for a long time in contrast to the energy of real intermediate states".

4.2. STEPWISE SENSITIZATION OF LUMINESCENCE

This mechanism was suggested by Auzel in 1966 [256,257]. Generally, this mechanism is governed by the "conventional" (for example, multipole) non-radiative energy transfer, that was considered before.

Here excited particles are operating as energy acceptors. In 1966, having investigated the green emission of two luminophors, $Na_{0.5}Yb_{0.5}WO_4$:Er(2%) and $Na_{0.5}La_{0.5}WO_4$:Er(2%), under IR excitation, Auzel noted that emission intensity of the first sample was considerably higher than the emission of the second one, and the emission intensity was quadratic with the intensity of the exciting light. Auzel suggested a mechanism for the increase of population of the $^4S_{3/2}$ level of Er^{3+} from which green luminescence ($^4S_{3/2} \rightarrow {}^4I_{15/2}$) was observed. The Yb^{3+} ion absorbs an IR photon and changes to the $^2F_{5/2}$ excited state. Due to migrationally stimulated non-radiative energy transfer, the excitation from the $^2F_{5/2}$ Yb^{3+} state can transfer to the $^4I_{11/2}$ Er^{3+} state. If there is another excited Yb^{3+} near the excited Er^{3+}, a second energy transfer from this Yb^{3+} ion to the previously excited Er^{3+} ion ($^4I_{11/2} \rightarrow {}^4F_{7/2}$) takes place populating $^4F_{7/2}$ Er^{3+}. After non-radiative intracenter relaxation, the $^4S_{3/2}$ level is populated, and from it, green luminescence is observed. This mechanism allows us to easily explain the quadratic dependence of the green luminescence with the exciting IR emission. Actually, the intensity of up-conversion luminescence is proportional to the number of excitation events summed per unit time, i.e., it is the product of the concentrations of excited particles, which are, in their turn, proportional to the intensity of exciting emission.

The up-conversion mechanism suggested by Auzel can be written, according to [250], as:

$$S + S + A + 2h\nu_S \rightarrow S^* + S^* + A \rightarrow S + S^* + A^* \rightarrow S + S + A^{**} \rightarrow S + S + A + h\nu_{A^-}$$

where S is the sensitizer, A is the acceptor.

Auzel has studied other pairs of rare-earth ions, in particular, Yb^{3+}-Tm^{3+} [257]. When tungstate crystals doped with Tm^{3+} and Yb^{3+} were excited by 0.96 μm light, they demonstrated luminescence at λ=0.47 μm. Due to stepwise triple energy transfer from Yb^{3+} to $^3H_6$, $^3F_4$, and $^3H_4$ Tm^{3+} levels, the $^1G_4$ level of Tm^{3+} was populated, and from it blue luminescence was observed. The intensity of the luminescence was proportional to the third power of the intensity of the exciting IR radiation, which confirmed a three-stage process.

Several works by Auzel have provided significant improvement of the intensity of luminophors; additional cases of two-, three- and more-stage processes of stepwise sensitization were discovered and studied. B.M. Antipenko et al. [258-260] observed three-, four- and five-stage sensitization of luminescence of Er^{3+}, Ho^{3+}, and Tm^{3+} ions by Yb^{3+} ions in $BaYb_2F_8$ crystal, when it was excited into the absorption bands of Yb^{3+}. In [261], Er^{3+} ions in $BaYF_5$ were excited by 1.5 μm emission ($^4I_{15/2}$ - $^4I_{13/2}$ transition); this caused the appearance of anti-Stokes luminescence lines at wavelengths up to 410 nm ($^2H_{9/2} \rightarrow {}^4I_{15/2}$ transition) due to stepwise sensitization. A number of other examples of such processes can be found in reviews (see, for instance, [250,262,263]).

It should be noted that the efficiency of stepwise sensitization processes, when concentrations of dopants are properly choosen, can be quite high, which makes them widely usable in up-conversion lasers.

4.3. Cooperative Sensitization of Luminescence

The cooperative sensitization of luminescence consists of the one-stage process of energy transfer from two or more excited donors to one acceptor. This phenomenon was first observed in 1966 by Ovsyankin and Feofilov [264-268]. When excited by IR emission (λ~1 μm), BaF_2 crystals doped with Yb (10%) and Tm (0.5%) displayed visual luminescence of the Tm^{3+} ions, whereas crystals containing no Yb^{3+} did not display this emission. Analysis showed that it was a new phenomenon, *cooperative sensitization of luminescence*. When two Yb^{3+} ions excited by IR pumping transfer concurrently their energy to one Tm^{3+} ion, anti-Stokes visual luminescence at 470 nm ($^1G_4 \rightarrow {}^3H_6$) and 670 nm

($^1G_4 \rightarrow {}^3F_4$) can be observed. The phenomenon is the reverse process for the one Dexter had studied, namely, the phenomenon of the simultaneous transfer of the energy of one donor to two acceptors [269,270]. Authors of [264-268] also observed green anti-Stokes luminescence of Er^{3+} and Ho^{3+} ions ($^4S_{3/2} \rightarrow {}^4I_{15/2}$ and $^5S_2 \rightarrow {}^5I_8$ transitions, respectively) in crystals co-doped with Yb^{3+}. However, in this case, the fact that both Er and Ho ions have intermediate energy levels allowing stepwise sensitization prevents an exact determination of the up-conversion mechanism.

It is interesting to note that the cooperative interaction of excited ions can occur even without third particle-acceptor, with the participation of virtual intermediate states (so called *cooperative luminescence*) [255]. In this case two particles in excited states pass to lower states by means of spontaneous emission of a photon of combined energy. This was described, for example, in [271], where $YbPO_4$ crystal under excitation by 1-μm emission demonstrated blue-green luminescence (497 nm). Since Yb^{3+} ion does not have any real states from which the luminescence could go up, authors of [271] suggested that in that case the cooperative luminescence of Yb^{3+} ions took place.

In the case of the cooperative sensitization of luminescence, the scheme of energy transfer [250] looks like:

$$S + S + A + 2h\nu_S \rightarrow S^* + S^* + A \rightarrow S + S + A^{**} \rightarrow S + S + A + h\nu_{A^{**}}$$

The ion pairs Yb-Tb and Yb-Gd are convenient samples for examining the cooperative sensitization of luminescence due to the specific energy levels preventing consecutive energy transfer from Yb^{3+} to Tb^{3+} and Gd^{3+} to the $^5D_4$ and $^6P_{7/2}$, levels, respectively [255,272]. In [273,274] anti-Stokes luminescence from these levels resulting from the cooperative sensitization of luminescence was reported in fluoride crystals. Blue luminescence from Tb^{3+} ions ($^5D_3 \rightarrow {}^7F_5$ transition, λ=470 nm) was also noticed. In the beginning, cooperative population of the $^5D_4$ level of Tb^{3+} ion was carried out; then, this ion was additionally excited by the interaction of two excited ions, Tb^{3+} ($^5D_4$) and Yb^{3+} ($^2F_{5/2}$), and passed to the $^5D_3$ level of Tb^{3+} [274]. It should be noted, however, that the above mentioned two-photon absorption could make a contribution to anti-Stokes luminescence from the $^5D_4$ level. Another cooperative process is possible, namely, *combinational absorption* [275,276], when the excited Yb^{3+} ion interacts with the pump photons and takes the Tb^{3+} ion to the $^5D_4$ excited state while it drops to the $^2F_{7/2}$ ground state [277]. The theory of this mechanism of up-conversion is given in [255].

According to Dexter [255], to achieve the cooperative sensitization of luminescence, unlike the consecutive sensitization mechanism, strong interacting partners having essentially overlapping wave functions are needed [255,268-270]. This fact is responsible for the considerably higher concentrational dependence of cooperative sensitization of luminescence in comparison to the consecutive one [276]. The latter, all other factors being equal, should, according to Dexter's rules, be more efficient, because real intermediate states are used.

It shoud be noted that in the majority of cases, up-conversion processes can proceed both in the forward and reverse directions. This circumstance, obviously, decreases the quantum yield of anti-Stokes luminescence. The efficiency of anti-Stokes luminescence increases if the up-conversion process is irreversable, due to fast intra- or intercenter relaxation of excitations to lower levels, or due to Stokes shift between the sum of the energies of the initial states and the energy of the state to which the anti-Stokes transfer of excitation takes place [278-281].

Thus, the irreversibility of anti-Stokes excitation processes should appear more strongly in oxide crystals compared to fluoride ones, because the former have a more extensive phonon spectrum [281-288].

When one wants to identify an up-conversion mechanism, information about the relaxational characteristics of the process, including the lifetime of the anti-Stokes luminescence, is very important. As an example, let us consider the kinetics of red and green luminescence of Er^{3+} for different excitation methods [289]. Due to the specific scheme of energy levels of the Er^{3+} ion up-conversion population of $^4S_{3/2}$ and $^4F_{9/2}$ luminescence levels can proceed concurrently by both stepwise sensitization and cooperative sensitization mechanisms. Let us consider that (as is typical for Er ions) the lifetimes of the $^4I_{13/2}$ and $^4I_{11/2}$ interacting levels (τ_i) is notably higher than the lifetimes of the $^4S_{3/2}$ and $^4F_{9/2}$ levels, and τ_i values change little due to up-conversion interaction. Assuming that the populations of upper states $^4S_{3/2}$ and $^4F_{9/2}$, increasing by up-conversion processes, are proportional to the product of the populations of the lower interacting levels, the observed decay time of anti-Stokes luminescence is equal to:

$$\tau^{-1} = \sum \tau_i^{-1}$$

The observed lifetime of green luminescence from the $^4S_{3/2}$ level is equal to:

$$\tau^{ob}_{^4S_{3/2}} = \frac{\tau_{^4I_{11/2}}}{2} \text{ and } \frac{\tau_{^4I_{11/2}}}{3}$$

for two- and three-stage processes, correspondingly, and for red luminescence:

$$\tau_{ob} = \frac{\tau_{^4I_{11/2}} \cdot \tau_{^4I_{13/2}}}{\tau_{^4I_{11/2}} + \tau_{^4I_{13/2}}}$$

for the two-photon process, which was actually proved experimentally. The intensity of the anti-Stokes luminescence is quadratic in the intensity of the excitation for the two-photon processes, and is proportional to the third power of the intensity for a three-photon process; it conforms to the number of excitations, the accumulation of which leads to the population of the upper state [289].

It should be noted that, strictly speaking, in a disordered continuum of particles, the rate of the up-conversional population of upper states does not always have to be proportional to the product of the concentrations of the initial excited particles. For instance, consider the case when the so called *superfast migration of excitation* over donors (such as in the hopping mechanism of donor-acceptor interaction) occurs [201]. The authors of [290] obtained general equations describing a non-linear interaction in the hopping approximation. It provided the basis for obtaining in [291] certain analytical expressions for the static and migrationally stumulated up-conversion quenching of luminescence. Analysis of this has allowed us to determine the region of validity of commonly used empirical expressions. The theoretical description of the interaction of the excited particles in a disordered group can be also found in [292-296]. Paper [297], where the interaction of the excited ions of Nd in scandium borate and gadolinium vanadate crystals was studied, can be considered as an example of the identification and experimental examination of up-conversion processes proceeding by static and migrationally stimulated mechanisms.

4.4. Influence of Up-conversion on Lasing

Now let us consider the influence of up-conversion processes on laser operation. The influence can be twofold. Up-conversion processes influence negatively laser efficiency if they decrease the population of the upper laser level, or result in losses at the lasing wavelength, or decrease the efficiency of the excitation of the upper laser level. On the other hand, if up-conversion processes are favorable for the populating of the upper laser level or the depopulating of the lower laser level their influence on laser efficiency is positive. As applied to the active media of solid state lasers, the range of influence of up-conversion processes on lasing efficiency lies from the total stoppage of lasing (for instance, due to intensive *ESA* from the upper laser level of the laser frequency) to the dominant mechanism for causing an inverse population in the lasing transition (up-conversion lasers).

First, let us consider the most typical examples of the negative influence of up-conversion processes. One of these cases is *non-linear quenching of luminescence* from the upper laser level, i.e. the process of depopulating of this level due to the interaction of excited ions. In this case, the rate of quenching depends on the intensity of excitation. This fact is dependent on the concentration of acceptors along with the intensity of excitation.

One of the first studies of the non-linear quenching of luminescence was carried out by the authors of [298,299], who discovered the dependence of the quenching rate of luminescence from the ^2E level Cr^{3+} in ruby and uranile salts on the pumping intensity. When the intensity of excitation increased, the quenching rate increased, and the quantum yield of Cr^{3+} luminescence dropped. The authors explained this experimental fact by the interaction of excited Cr^{3+} ions. One of the Cr^{3+} ions ("donor") is in the ground state $^4A_2$, and another ("acceptor") additionally excited to the $^4T_1$ state, decayed to the ^2E level. This process resulted in the deactivating of the excited Cr^{3+} ions, which lead to a reduction of the lifetimes and luminescence quantum yield. We note Ref. [300], where the authors have discovered and carried out the quantitative examination of the interaction of excited Cr^{3+} ions in a variety of laser crystals.

Laser media doped with Er^{3+}, Tm^{3+}, Ho^{3+} and Yb^{3+} demonstrate typical examples of the non-linear quenching of luminescence. As was mentioned before, the lower excited states of $^5I_7$ Ho^{3+}, $^5I_6$ Ho^{3+}, $^3F_4$ Tm^{3+}, $^4I_{13/2}$ Er^{3+}, $^4I_{11/2}$ Er^{3+}, and $^2F_{5/2}$ Yb^{3+} demonstrate high metastability, which is why the population of these levels in excited laser crystals can be high. If these levels act as upper laser ones, their interaction with other excited states could cause a reduction of their population and a decrease of laser efficiency. In [301-303] it was shown that the interaction of excited Yb^{3+} ions in the $^2F_{5/2}$ state and Ho^{3+} ions in the $^5I_6$ state in $BaYb_2F_8:Ho^{3+}$ and $LiYbF_4:Ho^{3+}$ crystals reduces considerably the efficiency of 3-μm lasing at the $^5I_6 \rightarrow {}^5I_7$ transition of Ho^{3+} due to the depopulating of this $^5I_6$ upper laser level and the populating of the $^5I_7$ lower laser level. Thus, the interaction of the excited Ho^{3+} ($^5I_6$) and the Yb^{3+} ($^2F_{5/2}$) ions takes Yb^{3+} to the ground state, while Ho^{3+} rises to the $^5S_2$ state; further cross-relaxation from $^5S_2$ results in the populating of the lower laser $^5I_7$ level of the 3-μm Ho laser.

The interaction of excited Er^{3+} ($^4I_{13/2}$), Tm^{3+} ($^3F_4$), and Ho^{3+} ($^5I_7$) ions in $BaEr_2F_8$:Tm,Ho crystal and Tm^{3+} ($^3F_4$) and Ho^{3+} ($^5I_7$) ions in a YAG:Cr,Tm,Ho crystal leads to the degradation of the lasing characteristics of the 2-μm Ho lasers operating at the $^5I_7 \rightarrow {}^5I_8$ transition [304-307]. The interaction of excited Er^{3+} ions ($^4I_{13/2}$) and Yb^{3+} ions ($^2F_{5/2}$) degrades the lasing parameters of 1.5-μm Er lasers operating at the $^4I_{13/2} \rightarrow {}^4I_{15/2}$ transition of Er^{3+} in laser glasses doped with Cr^{3+}, Yb^{3+}, and Er^{3+} [308,309].

Non-linear quenching of luminescence from the upper laser level $^4I_{11/2}$ of the 3-μm Er laser in scandium garnets was discovered and quantitatively examined in [310-312]. In this case, the summation of the excitations of the $^4I_{11/2}$ Er^{3+} level and the further cross-relaxation of the $^4S_{3/2}$ level leads to the populating of the $^4I_{13/2}$ lower laser level in much the same way as in case of the 3-μm lasing of Ho^{3+} ions. The same crystals doped with Cr^{3+}, Er^{3+}, Ho^{3+}, and Tm^{3+} proved to demonstrate non-linear quenching of luminescence associated with both an interaction between RE^{3+} ions and an interaction of RE^{3+} and excited Cr^{3+} ions [313,314].

Up-conversion quenching of luminescence from the upper laser level $^4F_{3/2}$ of Nd^{3+} in scandium borate and vanadate crystals was examined in [297].

Up-conversion processes proceeding by the Blombergen mechanism, i.e. due to *ESA*, can also play very negative role in laser operation.

One of the first ions which has been examined in this respect was Cr^{3+}. The study of *ESA* in different Cr^{3+}-doped materials, including ruby, began more that 30 years ago [315-319]. The study began from the measurements and theoretical interpretations of absorption of Cr^{3+} from the ^2E level in a number of crystals. Development of tunable lasers operating at a $^4T_2 \rightarrow {}^4A_2$ vibronic transition of Cr^{3+} caused increasing interest in *ESA*, since *ESA* from the $^4T_2$ level Cr^{3+} restricts considerably the tuning region of laser emission. *ESA* in GSGG and LLGG laser crystals doped with Cr^{3+} was examined in [13,320]. The shrinkage of the tuning band in the crystals is associated with *ESA*. The short-wave edge of the tuning band (approximately at 780 nm) is determined by both the *ESA* -transitions $^2E \rightarrow {}^2T_2(b)$, $^2T_1(b)$, and $^4T_2 \rightarrow {}^4T_1(b)$, and the absorption from the ground state. The long-wave edge of this band (approximately 850 nm) is determined by the *ESA*-transition $^4T_2 \rightarrow {}^4T_1(a)$. This fact was associated with Jahn-Teller splitting of the T-excited state of Cr^{3+} ions in high-symmetry C_{3i} sites, which leads to the production of additional *ESA*-bands overlapping the gain contour. Thus, the tuning range in these materials is about 100 nm. Crystal hosts containg Cr ions in non-centrosymmetrical sites demonstrate considerably wider tuning range. In $ZnWO_4$:Cr [321] and calcium-gallium-germanate crystals [322], the tuning range of Cr^{3+} in near-IR is about 400 nm.

In [323], the spectroscopic properties of a number of garnets and perovskites doped with Ni^{2+} were studied. Despite the high quantum efficiency of luminescence and large emission cross-section, all attempts to get efficient lasing were unsuccsessful due to *ESA*.

In [324], *ESA*-absorption of Ti^{3+} in Al_2O_3 and in $YAlO_3$ was measured. In the former crystal *ESA* was not observed in a wide spectral region, while in the latter crystal, strong ESA at $\lambda<550$ nm associated with the charge transfer band was discovered; thus $YAlO_3:Ti^{3+}$ in contrast to $Al_2O_3:Ti^3$, is not a good candidate for tunable solid state lasers.

It should be noted that in a number of hosts, very strong *ESA* -bands with charge transfer were discovered for Cr^{3+} [318,320,325-330]. The absorption of short-wave pumping emission causes undesired coloration of the crystal and changes of the thermal regime of the active element [328].

In [331,332], the *ESA* of the $Mn^{2+}:MgAl_2O_4$ crystal was examined. It was established that the cross-section of excited state absorption at the lasing wavelength was two orders of magnitude higher than the cross-section of the laser transition. Thus, despite the intense luminescence of Mn^{2+} in spinel at room temperature, it is impossible to get laser action in the system. In a similar manner, the influence of ESA is the reason for the low efficiency of lasers based on V^{2+} ions in MgF_2 and $KMgF_3$ crystals [333].

In [334], ESA from the upper laser level $^4F_{3/2}$ Nd^{3+} in $YLiF_4$ was studied. It was established that, at typical values of pumping energy, the lifetime of $^4F_{3/2}$ dropped from 570 to 410 μs. When the laser operates in Q-switch mode, about 20 % of stored energy is lost due to ESA. The ESA spectra of Nd^{3+} in various laser materials were studied also in [335].

On the other hand, in numerous cases, up-conversion processes result in the populating of upper laser level and/or the depopulating of lower laser level, which leads to the emergence of new laser situations. In particular, these processes allow the creation of *up-conversion lasers*, where up-conversion processes are the main mechanisms for the achievement of inverse population. Below some of the most typical examples will be described.

The first up-conversion laser was reported in [336] with BaY_2F_8:Yb,Er and BaY_2F_8:Yb,Ho crystals. When the crystals were pumped into the ytterbium absorption bands (about 1 μm) at 77°K, stimulated emission was obtained at the $^4F_{9/2} \rightarrow {}^4I_{15/2}$ transitions of Er^{3+} (670 nm) and the $^5S_2 \rightarrow {}^5I_8$ transaction of Ho^{3+} (551.5 nm). The main mechanism for the creation of inverse population was the stepwise sensitization of Er^{3+} and Ho^{3+} luminescence by Yb^{3+} ions.

The possibility of efficient excitation of lasing from Ho^{3+} in the $BaYb_2F_8$ crystal by stepwise sensitization at room temperature was demonstrated in [258-260,301-307,337], where the development of a number of up-conversion lasers based on Er^{3+}, Tm^{3+}, and Ho^{3+} ions using Yb^{3+} ions as sensitizers and Nd-laser emission as an excitation source was reported.

In [338-342], the phenomenon of the non-linear quenching of luminescence from $^4I_{13/2}$ Er^{3+} in the YAG crystal was discovered and used for creation of the inverse population at the $^4I_{11/2} \rightarrow {}^4I_{13/2}$ transition. It was demonstrated theoretically and experimentally that, in Er:YAG, a laser operating at the self-terminated transition $^4I_{11/2} \rightarrow {}^4I_{13/2}$ (the lifetime of lower laser level exceeds the lifetime of upper laser one) could be achieved. The interaction of two Er^{3+} ions in the $^4I_{13/2}$ state leads to the population of the $^4I_{9/2}$ level, and, after fast intracenter non-radiative relaxation of this level, the upper laser level of the 3-μm laser $^4I_{11/2}$ level is populated. These processes populate the upper laser level and depopulate the lower one increasing considerably the efficiency of the transformation of absorbed energy into the laser emission as well as making possible continuous wave lasing on the self-terminated transition $^4I_{11/2} \rightarrow {}^4I_{13/2}$. In [342-346,] the lasing from the Er^{3+} ions pumped into the $^4I_{13/2}$ lower laser level was reported. In this case, the inverse population was due to nothing but the up-conversion transformation of the absorbed energy. In particular, the authors of [343] reported a CW up-conversion erbium laser operating at 855 nm ($^4S_{3/2} \rightarrow {}^4I_{13/2}$) on the basis of a $CaF_2:Er^{3+}$ crystal pumped into

$^4I_{15/2} \rightarrow ^4I_{13/2}$ transition. The interaction of three Er^{3+} ions in the $^4I_{13/2}$ state leads to the appearance of the Er^{3+} ion in the $^4S_{3/2}$ state. As in the case described before, continuous wave lasing was obtained at the self-terminated transition.

The study of the influence of the non-linear depopulating of the lower laser level $^4I_{13/2}$ on the operation of the 3-μm erbium laser based on various crystals was carried out in [347-350]. Up-conversion lasers based on Er^{3+} ions were reported in [351-355].

In [356], the successive sensitization of the luminescence of Tm^{3+} ions in YF_3:Yb:Tm crystal resulting in anti-Stokes emission with λ=0.81 and λ=0.475 μm was examined.

As in the case of stepwise sensitization, the process of consecutive absorption forms the basis for a number of up-conversion lasers. In [357], laser emission with λ=491, 520, 605, and 635 nm from the $^3P_0$ level of Pr^{3+} was reported in a fluoride fibre doped with Pr^{3+} excited concurrently by emission with λ=1.01 and λ=0.835 μm. A similar addition of optical frequencies by up-conversion processes in $BaYb_2F_8:Er^{3+}$ crystal was observed in [358], where the transformation of the emission of erbium (1.54 μm) and neodymium (1.054 μm) lasers into emission with λ=0.67 μm ($^4F_{9/2} \rightarrow ^4I_{15/2}$) was obtained. In [359], a silicate fibre doped with Tm^{3+} ions was excited by λ=660 nm emission. The authors observed luminescence at $^1D_2 \rightarrow ^3F_4$ (λ_{max}=460 nm), $^1D_2 \rightarrow ^3F_6$ (λ_{max}=370 nm), and $^1G_4 \rightarrow ^3H_6$ (λ_{max}=480 nm) transitions. The first absorption of pump light resulted in populating the $^3H_4$ level, and then the absorption from the $^3H_4$ state populated the $^1D_2$ level. When Tm^{3+} ions were excited by λ=1.06 μm emission, up-conversion luminescence at $^1G_4 \rightarrow ^3H_6$ (467 nm), $^1G_4 \rightarrow ^3F_4$ (650 nm), $^3H_4 \rightarrow ^3H_6$ (780 nm), and $^1D_2 \rightarrow ^3H_6$ (λ_{max}=370 nm) transitions was observed due to three-step absorption from $^3H_6$, $^3F_4$, and $^3H_4$ levels. When the Tm-Yb system was excited by λ=840 nm emission, up-conversion luminescence at $^1G_4 \rightarrow ^3H_6$ (475 nm), $^1G_4 \rightarrow ^3F_4$ (650 nm), $^1G_4 \rightarrow ^3H_5$, and $^3H_4 \rightarrow ^3H_6$ (790 nm) transitions was detected. Here the population of the $^1G_4$ level occured by the three-stage stepwise energy transfer from excited Yb^{3+} ions to the $^3H_6$, $^3F_4$, and $^3H_4$ levels of Tm^{3+}. And, finally, when the Tm-Yb system was excited by λ=1.064 μm emission, up-conversion luminescence at $^1D_2 \rightarrow {}^3H_6$ (370 nm), $^1D_2 \rightarrow {}^3F_4$ (460 nm), and $^1G_4 \rightarrow {}^3H_6$ (475 nm) was discovered.

In [360], up-conversion lasers based on the BaY_2F_8:Yb:Tm crystal operating at 456, 482, 512, 649, and 799 nm pumped by the 960 nm emission into ytterbium absorption bands were reported. Diode pumping has been demonstrated for the 456, 512, and 649 nm transitions.

A green up-conversion laser operating at the $^4S_{3/2} \rightarrow {}^4I_{15/2}$ transition of Er^{3+} in $YLiF_4$ crystal, pumped by 810 and 970 nm emission (absorption from $^4I_{15/2}$ and $^4I_{11/2}$ levels) was reported in [361,362]. CW laser operating at 650, 800, 1500 nm with pumping into the absorption band of Yb (960 nm) was developed on the basis of the Yb,Tm:$YLiF_4$ crystal [363].

In conclusion of this section, we will consider an interesting and efficient combined scheme of the population of laser levels called an *avalanche* process [364-371]. The physical base of the scheme is the absorption of pump energy from excited states and further cross-relaxational "multiplying" of the excitations.

The basic scheme of the development of inverse population due to ESA and cross-relaxational multiplying of excitations will be discussed taking the three-level system (the first level is the ground state) as an example. Absorption of the pump energy from the first excited state (2) populates the second excited state (3). Then, the cross-relaxational interaction of the particle in state 3 and the particle in ground state (1) takes place: 3→2, 1→2. Here, if the quantum efficiency of the cross-relaxational "exchange" is close to 2 (so that this process is principal among all the processes of the depopulation of level 3), instead of *one* excitation leaving level 2, *two* excitations arise at this level due to cross-relaxation, and the total population of level 2 increases. As can be seen, in this mechanism, depopulation of the ground state proceeds due to the non-radiative cross-relaxational "exchange" of energy between like particles. This results in the doubling of the amount of excitations in the absence of the direct interaction of the electromagnetic field with the ground state of the particle.

If the intensity of pumping emission was high enough, practically all particles would had been in state 3. Actually, the less the population of ground state, the lower the rate of cross-relaxational populating of level 2. At the same time, absorption of the pumping energy from level 2 populates level 3. So, the scheme allows in principle an inverse population at all three transitions: 3→1, 2→1, 3→2.

The avalanche phenomenon was discovered in [364] when infrared quantum counters of $LaCl_3$ and $LaBr_3$ crystals doped with Pr^{3+} ions were studied. Pumping with λ=4.5 μm caused the population of the first excited state $^3H_5$ of Pr^{3+} ($^3H_4 \rightarrow ^3H_5$ transition). Then, due to ESA from the $^3H_5$ level ($^3H_5 \rightarrow ^3P_1$) levels, $^3P_1$ and $^3P_0$ were populated, and, finally, red luminescence from these levels was observed ($^3P_1 \rightarrow ^3F_2$, $^3P_0 \rightarrow ^3F_2$, $^3P_0 \rightarrow ^3H_6$). Due to the cross-relaxation processes $^3H_6 \rightarrow ^3H_5$ and $^3H_4 \rightarrow ^3H_5$, efficient population of the $^3H_5$ level took place, which caused, in its turn, a strong increase in the intensity of red luminescence as the intensity of excitation at the $^3H_5 \rightarrow ^3P_1$ transition increased noticeably.

In [365], a number of CW tunable lasers based on Yb^{3+},Pr^{3+}-doped fibers lasing at 491-493 nm ($^3P_0 \rightarrow ^3H_4$), 517-540 nm ($^3P_1$, $^3P_0 \rightarrow ^3H_5$), 605-622 nm ($^3P_0 \rightarrow ^3H_6$), and 635-637 nm ($^3P_0 \rightarrow ^3F_2$) under pumping by 780-885 nm emission were reported. The population of the upper laser levels was carried out using the avalance effect. With diode-laser pumping (852 nm), CW laser operation at 491 nm, 520 nm, 615 nm, and 635 nm was achieved. During the first stage of the process, the $^1G_4$ level was populated due to "usual" energy transfer from Yb to Pr. Then the absorption of pump energy from this level resulted in the population of the upper laser levels $^3P_0$. The next process involves the avalanche mechanism: cross-relaxation $^3P_0 \rightarrow ^1G_4$ Pr^{3+}, $^2F_{7/2} \rightarrow ^2F_{5/2}$ Yb^{3+}, and energy transfer $Yb^{3+} \rightarrow Pr^{3+}$ (the first stage). Two Pr^{3+} ions in the $^1G_4$ state result. The same situation took place in [366], where the up-conversion laser with λ=635 nm (0.8 μm pumping) with a fibre doped with Pr and Yb was reported.

The avalanche scheme was realized in a number of crystals of garnet structure doped with Tm^{3+} and Er^{3+} [368,369]. Absorption of pumping emission from the excited states $^3F_4$ of Tm^{3+} and $^4I_{13/2}$ of Er^{3+} caused the population of the $^3H_4$ and $^4S_{3/2}$ levels, respectively, and cross-relaxation processes from these levels doubled the populations of the initial excited states. In [369] the 2-μm YSGG:Tm^{3+} laser pumped by 1.064 μm emission (absorption from the $^3F_4$ excited state of Tm^{3+}) was demonstrated.

A summary table of rare-earth-doped crystals, in which photon-avalance phenomenon has been observed, is given in review [370].

In this section, the authors have considered only a part of the numerous publications devoted to up-conversion resulting in the populating of excited states of activators. However, the given examples demonstrate the great potential of these methods for the creation of efficient anti-Stokes luminophors, including the active media of up-conversion lasers. Generally, the best results have been obtained by the cooperative using of all the mechanisms for interaction of activator ions with one another, as well as with the electromagnetic field of the pump.

5. Conclusion

In this chapter we have discussed the main spectroscopic and laser properties of luminescent dopants in solids. It can be seen that transition metal ions are of considerable interest for laser applications, and their potentialities are still not exhausted.

A relevant present-day problem is the transformation of the IR-emission of diode lasers into visible emission by using up-conversion processes in crystals and glasses. In addition, a great deal of attention is paid to the development of visible and UV lasers at the $4f$-$5d$ transitions in rare-earth ions. The great potential of divalent rare earth ions as luminescent dopants has not been adequately realized. The use of iron group ions having unusual oxidation states as luminescence dopants is also very promising. The interest in actinide ions has been renewed due to the recent successful experiments on the lasing of U^{3+} ions at room temperature.

The development and study of various schemes of sensitization, including the mechanisms of direct energy transfer onto the upper laser level, are very important for creating high efficiency lasers, including lasers emitting in the eye-safe spectral region.

The examination of the processes of non-radiative energy transfer in multidoped systems described in this chapter are very important, particularly, and have not been adequately described by existed models and theories. Further development of the theory will enhance the understanding of the problems discussed in this chapter and will give an impetus to a large number of new investigations.

6. Acknowledgement

The authors wish to express their great appreciation to Natalia N. Klyatskina for her help in the preparation of this chapter, and for the additional support that she provided in various aspects of this work.

References

[1] Sviridov, D.T., Sviridova, R.K., and Smirnov, Yu.F. 1976. *Optical Spectra of Transition Metal Ions in Crystals*, Nauka, Moscow, (in Russian).

[2] Kaminskii, A.A., 1990: *Laser Crystals. 2nd edition*, Springer, Berlin.

[3] Weber, M.J., 1980: Lanthanide And Actinide Lasers, ACS Symposium Sseries No. 131, Lanthanide And Actinide Chemistry And Spectroscopy, ed. by N.M. Edelstein, publ. by Amer. Chem. Soc., Washington, DC, pp. 275-311.

[4] Moulton, P.F., 1982: Paramagnetic Ion Lasers, *CRC Handbook of Laser Science and Technology*, **1**, Lasers and Masers. ed. by M.J. Weber, CRC Press, Boca Raton, FL, pp. 21-146.

[5] Moulton, P.F., 1985: Tunable Paramagnetic-Ion Lasers, *Laser Handbook*, **5**, ed. by M. Bass and M.L. Stitch, Elsevier Science Publishers, Amsterdam, pp. 203-288.

[6] Henderson, B., Imbusch, G.F., 1988: Optical Processes in Tunable Transition-Metal-Ion Lasers, *Contemp. Phys.*, **29**, pp. 235-272.

[7] Caird, J.A., Payne, S.A., 1989: Crystalline Paramagnetic Ion Lasers, *CRC Handbook of Laser Science and Technology. (Suppl.1)*, ed. by M.J. Weber, CRC Press, Boca Raton, FL, pp. 3-100.

[8] Tanabe, Y., Sugano, S., 1954: On the Absorption Spectra of Complex Ions, *J. Phys. Soc. Japan.*, **9**, pp. 753-780.

[9] Sugano, S., Tanabe, Y., Kamimura, H., 1970: *Multiplets of transition-metal ions in crystals*, Academic Press, New York.

[10] Zharikov, E.V., 1990: Rare-Earth Scandiun Garnets: Problems of Materials Science. *Proc. of General Phys. Inst. Optical dense active media*, **26**, Nauka, Moscow, pp. 50-78.

[11] Caird, J.A., 1986: Chromium Activated Crystals as Tunable Laser Media - What Makes Them Special?, *Tunable Lasers II*, ed. by A.B.Budgor, L.Esterowitz, and L.G.DeShazer, Springer, Berlin, pp. 20-34.

[12] Moulton, P.F., 1985: Recent Advances in Transition-Metal-Doped Lasers. *Tunable Solid State Lasers*, ed. by P.Hammerling, A.B.Budgor, and A.A.Pinto, Springer, Berlin, pp. 4-10.

[13] Petermann, K., 1990: :The Role of Excited-State Absorption in Tunable Solid State Lasers, *Opt. and Quant. Electr.*, **22**, pp.199-218.

[14] Moulton, P.F., 1982: Ti-Doped Sapphire: Tunable Solid-State Laser, *Optics News*, **8**, No 6, p. 9.

[15] Albers, P., Jenssen, H.P., Huber, G., and Kokta, M., 1986: Continuous Wave Tunable Laser Operation of Ti^{3+}-Doped Sapphire at 300 K. *Tunable Solid State Lasers II,* ed. by A.B.Budgor, L.Esterowitz, and L.G.DeShazer, Springer, Berlin, pp. 208-211.

[16] Harrison, J., Finch, A., Rines, D.M., Rines, G.A., and Moulton, P.F., 1991: Low-Threshold, CW, All-Solid-State $Ti:Al_2O_3$ Laser, *Optics Lett.,* **16**, pp. 581-583.

[17] Alimpiev, A.I., Bukin, G.V., Matrosov, V.N., Pestryakov, E.V., Solntsev, V.P., Trunov, V.I., Tsvetkov, E.G., and Chebotaev, V.P., 1986: Tunable $BeAl_2O_4:Ti^{3+}$ Laser, *Kvant.Elektr.,* **13**, pp. 885-886, (in Russian).

[18] Kvapil, J., Koselja, Kvapil, J., M., Perner, B., Skoda, V., Kubelka, J., Hamal, K., and Kubecek, V., 1988: Growth and Stimulated Emission of YAP:Ti, *Czech. J. Phys., B,* **38**, p. 237.

[19] Duerr, U., Brauch, U., Knierim, W., and Schiller, C., 1985: Optical and Laser Properties of Cr^{3+}, Co^{2+}, Ni^{2+} and V^{2+} Doped Perovskites, *Tunable Solid State Lasers,* P.Hammerling, A.B.Budgor, and A.A.Pinto eds., Springer, Berlin, pp. 20-27.

[20] Johnson, L.F., Guggenheim, H.J., and Thomas, R.A., 1967: Phonon Terminated Coherent Emission from V^{2+} Ions in MgF_2, *J. Appl. Phys.,* **38**, pp. 4837-4839.

[21] Page, R.H., DeLoach, L.D., Wilke, G.D., Payne, S.A., and Krupke, W.F., 1995: A New Class of Tunable Mid-IR Lasers Based on Cr^{2+}-Doped II-VI Compounds, *Conf. on Lasers and Electro-Optics (CLEO),* Baltimore, MD, paper CWH5.

[22] Renz, R., and Schulz, H.-J., 1983: The Decay of Infrared Luminescence in II-VI Compound Semiconductors Doped by 3d transition Elements, *J. Phys. C.,* **16**, pp. 4917-4932.

[23] Deloach, L.D., Page, R.H., Wilke, G.D., Payne, S.A., and Krupke, W.F., 1995: Properties of Transition Metal-Doped Zinc Chalcogenide Crystals for Tunable IR Radiation, *OSA Proc. Advanced Solid State Lasers,* **24**, B.Chai and S.A.Payne, eds., pp. 127-131.

[24] Page, R.H., DeLoach, L.D., Schaffers, K.I., Patel, F.D., Payne, S.A., Krupke, W.F., and Burger, A., 1996: Recent Developmnts in Cr^{2+}-

Doped II-VI Compound Lasers, *Advanced Solid State Lasers Meeting*, San-Fransisco, CA, USA, Techn. Digest, pp. 246-248.

[25] Zharikov, E.V., Osiko, V.V., Prokhorov, A.M., and Shcherbakov, I.A., 1984: Crystals of Rare-Earth Gallium Garnets Doped with Chromium as Active Media for Solid-State Lasers, *Bull. Acad. Sci. USSR, Phys., Ser.*, **48**, pp. 81-94.

[26] Zharikov, E.V., 1986: Chromium-Doped Garnet Hosts: Crystal Chemistry Development and Properties. *Tunable Lasers II*, ed. by A.B. Budgor, L. Esterowitz, and L.G. De Shazer, Springer, Berlin, pp. 64-75.

[27] Walling, J.C., Peterson, O.G., Jenssen, H.P., Morris, R.C., and O'Dell, E.W., 1980: Tunable Alexandrite Lasers, *IEEE J. Quant. Electr.*, **QE-16**, pp. 1302-1315.

[28] Struve, B., and Huber, G., 1985: The Effect of Crystal Field Strength on the Optical Spectra of Cr^{3+} in Gallium Garnet Crystals, *Appl. Phys. B*, **36**, pp. 195-202.

[29] Zharikov, E.V., Lavrischev, S.V., Laptev, V.V., Ostroumov, V.G., Saidov, Z.S., Smirnov, V.A., and Scherbakov, I.A., 1984: New Possibilities of Cr^{3+} Ions as Activators of the Active Media of Solid State Lasers, *Sov. J. Quant. Electr.*, **14**, pp. 332-336.

[30] Henderson, B., Yamaga, M., O'Donnell, K.P., 1990: Optical Characterization of Tunable Solid-State Laser Gain Media, *Opt. Quant. Electr.*, **22**, pp.167-198.

[31] Huber, G., and Petermann, K., 1985: Laser Action in Cr-Doped Garnets and Tungstates. *Tunable Solid-State Lasers*, ed. by P. Hammerling, A.B. Budgor, and A.P into, Springer, Berlin, pp. 11-19.

[32] Payne, S.A., Chase, L.L., Newkirk, H.W., Smith, L.K., and Krupke, W.F., 1988: $LiCaAlF_6$:Cr^{3+}: a Promising New Solid-State Laser Material, *IEEE J. Quant. Electr.*, **24**, pp. 2243-2252.

[33] Payne, S.A., Chase, L.L., and Wilke, G.D., 1989: Optical Spectroscopy of the New Laser Materials $LiSrAlF_6$:Cr^{3+} and $LiCaAlF_6$:Cr^{3+}, *J. Lumin.*, **44**, pp. 167-176.

[34] Stadler, M., Chai, B.H.-T., and Bass, M., 1991: Flashlamp Pumped Cr:$LiSrAlF_6$ Laser, *Appl. Phys. Lett.*, **58**, pp. 216-218.

[35] Caird, J.A., Payne, S.A., Staver, P.R., Ramponi, A.J., Chase, L.L., and Krupke, W.F., 1988: Quanrum Electronic Properties of the $Na_3Ga_2Li_3F_{12}$:Cr^{3+} Laser, *IEEE J. Quant. Electr.*, **24**, pp. 1077-1099.

[36] Andrews, L.J., Hitelman, S.M., 1987: Excited State Absorption of Cr^{3+} in Low Ligand Field Hosts. *Spectroscopy of solid-state laser-type materials*, ed. by B. Di Bartolo and G. Armagan. Plenum Press, New York, pp. 515-528.

[37] Zharikov, E.V., Kalitin, S.P., Laptev, V.V., Maier, A.A., Osiko, V.V., 1984: Search and Development of Laser Active Media Based on Gallium Garnets. *Report of General Phys. Inst.* No 135, Moscow, pp. 1-38, (in Russian).

[38] Smirnov, V.A., and Shcherbakov, I.A., 1988: Rare-Earth Scandium Chromium Garnets as Active Media for Solid-State Lasers, *IEEE J. Quant. Electr.*, **24**, pp. 949-959.

[39] Scheps, R., 1992: Cr-Doped Solid State Lasers Pumped By Visible Laser Diodes, *Opt. Mater.*, **1**, pp. 1-9.

[40] Shinn, M.D., Buell, T.A., and Sumida, D.S., 1995: Reanalysis of Cr-Nd Energy Transfer in Scandium Garnets, and Its Impact on Visible Diode-Pumping, *OSA Advanced Solid State Lasers*, **24**, B. Chai and S.A. Payne eds., pp. 538-541.

[41] Caird, J.A., Shinn, M.D., Kirchoff, T.A., 1986: Measurements of Losses and Lasing Efficiency in GSGG:Nd,Cr and YAG:Nd Laser Rods, *Appl. Opt.*, **25**, pp. 4294-4304.

[42] Angert, N.B., Borodin, N.I., Garmash, V.M., Zhitnyuk, V.A., Okhrimchuk, A.G., Siyuchenko, O.G., and Shestakov, A.V., 1988: The Laser Action in Impurity Color Centers in Yttrium-Aluminium Garnet Crystals in the Wavelength Range of 1.35-1.45 μm, *Kvant. Elektr.*, **15**, pp. 113-115, (in Russian).

[43] Zverev, G.M., and Shestakov, A.V., 1989: Tunable Near-Infrared Oxide Crystals Lasers, *OSA Proc. Tunable Solid State Lasers Meeting*, **5**, ed. by M.L. Shand and H.P. Jenssen, pp. 66-70.

[44] Petricevic, V., Gayen, S.K., Alfano, R.R., Yamagishi, K., Anzai, H., and Yamaguchi, Y., 1988: Laser Action in Chromium-Doped Forsterite, *Appl. Phys. Lett.*, **52**, pp. 1040-1042.

[45] Petricevic, V., Gayen, S.K., and Alfano, R.R., 1988: Laser Action in Chromium-Activated Forsterite for Near-Infrared Excitation: Is Cr^{4+} the Lasing Ion?, *Appl. Phys. Lett.*, **53**, pp. 2590-2592.

[46] Moncorge, R., Manaa, H., Boulon, G., 1994: Cr^{4+} and Mn^{5+} Active Centers for New Solid State Laser Materials, *Opt. Mater.*, **4**, pp. 139-151.

[47] Avanesov, A.G., Gostischeva, E.A., Zhorin, V.V., Kuznetsov, Ju.A., Pisarenko, V.F., Zhitnyuk, V.A., Shestakov, A.V., 1991: Spectral and Laser Properties of Chromium Doped Yttrium Silicate - a New Medium for Tunable Lasers, *Proc. VIII All-Union Meeting on Spectroscopy of laser materials* (Krasnodar, Russia), pp. 20 (in Russian).

[48] Deka, C., Chai, B.H.-T., Shimony, Y., Zhang, X.X., Munin, E., and Bass, M., 1992: Laser Performance of Cr:Y_2SiO_5., *Appl. Phys. Lett.*, **61**, pp. 2141.

[49] Bakhurova, N.N., Denisov, A.L., Zharikov, E.V., Smirnov, V.A., and Shcherbakov, I.A., 1994: New Cr^{4+}-Doped Compounds with Mellite Structure, *OSA Proc. on Advansed Solid State Lasers*, **20**, ed. by T.Y. Fan and B.H.-T. Chai, pp. 210-214.

[50] K,ck, S., Hartung, S., Petermann, K., and Huber, G., 1995: Cr^{4+}-Doped $LiAlO_2$ and $LiGaO_2$: A New Potential Laser Materials for the Infrared Spectral Range, *OSA Proc. Advanced Solid State Lasers*, 24, B.H.T. Chai and S.A. Payne eds., pp. 486-490.

[51] Chai, B.H.T., and Zhang, X.X., 1995: Modified Wurtzite Structure Compounds for Broad Band Tunable Laser Sources Near 1.3 μm and 1.55 μm Regions, *Advanced Solid State Lasers Meeting, Memphis, Tennessee, USA,* Techn. Digest, paper WE2, pp. 289-291.

[52] Yamaguchi, Y., Yamagishi, K., and Nobe, Y., 1993: The Behavior of Chromium Ions in Forsterite, *J. Cryst. Growth*, **128**, pp. 996-1000.

[53] Munin, E., Villaverde, A.B., Zhang, X.X., and Bass, M., 1993: Broad Band, Intensity Dependent Absorption in Tetravalent Chromium-Doped Crystals, *Appl. Phys. Lett.*, **63**, p. 2590.

[54] Jia, W., Lu, L., Tissue, B.M., and Yen, W.M., 1991: Valence and Site Occupation of Chromium Ions in Single-Crystal Forsterite Fibers, *J. Cryst. Growth,* **109**, pp. 329-333.

[55] Shimony, Y., Kalisky, Y., and Chai, B.H.T., 1995: Quantitative Studies of Cr^{4+}:YAG as a Saturable Absorber for Nd:YAG Laser, *Opt. Mater.,* **4**, pp. 547-551.

[56] Petricevic, V., Seas, A., and Alfano, R.R., 1990: Effective Gain Measurements in Chromium-Doped Forsterite, *OSA Proc. on Advanced Solid State Lasers,* **6**, H.P.Jenssen and G.Dube eds., pp. 73-78.

[57] Kück, S., Petermann, K., Huber, G., 1991: Spectroscopic Investigation of the Cr^{4+}-Center in YAG, *OSA Proc. on Advanced Solid State Lasers,* **10**, G. Dube and L. Chase, eds., pp. 92-94.

[58] Manaa, H., Moncorge, R., Butashin, A.V., Mill, B., and Kaminskii, A.A., 1993: Luminescence Properties of Cr-Doped $LiNbGeO_5$ Laser Crystal, *OSA Proc. on Advanced Solid State Lasers,* **15**, A.A. Pinto and T.Y. Fan eds., pp. 343-345.

[59] Baryshevski, V.G., Korzhik, M.V., Livshitz, M.G., Tarasov, A.A., Kimaev, A.E., Mishkel, I.I., Meilman, M.L., Minkov, B.I., and Shkadarevich, A.P., 1991: Properties of Forsterite and the Performance of Forsterite Lasers with Lasers and Flashlamp Pumping, *OSA Proc. on Advanced Solid State Lasers,* **10**, G. Dube and L. Chase eds., pp. 26-34.

[60] Kück, S., Koetke, J., Petermann, K., Pohlmann, U., and Huber, G., 1993: Spectroscopic and Laser Studies of Cr^{4+}:YAG and Cr^{4+}:Y_2SiO_5, *OSA Proc. on Advanced Solid State Lasers,* 1993, **15**, A.A. Pinto and T.Y. Fan eds., pp. 334-338.

[61] Zharikov, E.V., Kalitin, S.P., Papin, Yu.M., Prokhorov, A.M., Smirnov, V.A., and Shcherbakov, I.A., 1993: Cr^{4+} Ions as New Efficient Sensitizer for 1.5-3 Micron Laser Materials, *Ibid,* pp. 359-362.

[62] Sigachev, V.B., Doroshenko, M.E., Timoshechkin, M.I., Loutts, G.B., Chai, B.H.-T., 1994: Enegry Transfer from Ce^{3+} and Cr^{4+} to Er^{3+} Ions in Garnet and Yttrium Silicate Crystals, *Conf. on Lasers and Electro-Optics (CLEO),* Anaheim, CA, May 1994, Technical Digest, pp. 124.

[63] Borodin, N.I., Okhrimchuk, A.G., Smolin, D.V., and Shestakov, A.V., 1994: Sensitization of Tm^{3+} Luminescence by Cr^{4+} Ions

at Continuous Wave Two-Micron Lasing in YAG:Tm^{3+}, Cr^{4+}, Ca^{2+} Crystals, *Kvant. Elektr.*, **21**, pp. 1038-1040, (in Russian).

[64] Zharikov, E.V., Zabaznov, A.M., Osiko, V.V., Prokhorov, A.M., Saidov, Z.S., Smirnov, V.A., Shkadarevich, A.P., and Shcherbakov, I.A., 1985: Photochromic Properties of Gadolinium-Scandium-Gallium Garnet Crystal, *PREPRINT of General Physics Institute*, No 238, Moscow, (in Russian).

[65] Andrauskas, D.M., Kennedy, C., 1991: Tetravalent Chromium Solid State Passive Q-Switch for Nd:YAG Laser Systems, *OSA Proc. on Advanced Solid State Lasers*, **10**, G. Dube and L. Chase eds., pp. 393-396.

[66] Demchuk, M.I., Mikhailov, V.P., Zhavoronkov, N.I., Kuleshov, V.V., Prokoshin, P.V., Yumashev, K.V., Livshitz, M.G., and Minkov, B.I., 1992: Chromium-Doped Forsterite as a Solid-State Saturable Absorber:, *Opt. Lett.*, **17**, pp. 929-930.

[67] Mikhailov, V.P., Kuleshov, N.V., Zhavoronkov, N.I., Yumashev, K.V., Prokoshin, P.V., and Shcherbitsky, V.G., 1995: Solid-State Passive Q-Switches and Mode-Lockers Based on d^2-Ion Doped Crystals for Near Infrared and Visible Lasers, *OSA Proc. on Advanced Solid State Lasers*, **24**, B.H.T. Chai and S.A. Payne eds., pp. 449-453.

[68] Demchuk, M.I., Zharikov, E.V., Zabaznov, A.M., Manichev, I.A., Mikhailov, V.P., Prokhorov, A.M., Shkadarevich, A.P., Chernjavskii, A.F., Shcherbakov, I.A., and Yumashev, K.V., 1987: Mode Locking in a Neodimuim Laser with a Gadolinium-Scandium-Gallium Garnet Switch, *Sov. J. Quant. Electr.*, **17**, pp. 266-267, (in Russian).

[69] Kruglik, G.S., Skripko, G.A., and Shkadarevich, A.P., 1984: *Tunable lasers utilizing activated crystals*, Byelorussian Polytechnic Institute, Minsk, Byelorussia, (in Russian).

[70] Zharikov, E.V., Zabaznov, A.M., Prokhorov, A.M., Shkadarevich, A.P., and Shcherbakov, I.A., 1986: Use of GSGG:Cr,Nd Crystals with Photochromic Centers as Active Elements in Solid State Lasers, *Sov. J. Quant. Electr.*, **16**, pp. 1552-1554.

[71] Evstigneev, V.L., Zharikov, E.V., Zavartsev, Yu.D., Nerobov, O.V., Studenikin, P.A., Shcherbakov, I.A., and Yumashev, K.V., 1990:

Influence of Cr^{4+} Saturable Absorber on Energy Extraction in Cr,Nd:YSGG, *OSA Proc. on Advanced Solid State Lasers*, **6**, H.P. Jenssen and G. Dube eds., pp. 187-189.

[72] Moncorge, R., Manaa, H., Deghoul, F., Guyot, Y., Kalisky, Y., Pollack, S.A., Zharikov, E.V., and Kokta, M., 1996: Q-Switch and Excited State Absorption Experiments with Cr^{4+}:LuAG Single Crystals, *Advanced Solid State Lasers Meeting,* (San Francisco, CA, USA), *Technical Digest,* pp. 409-411.

[73] Marfunin, A.S., 1975: *Spectroscopy, Luminescence and Radiation Centers in Minerals,* Nedra, Moscow, (in Russian).

[74] Randall, J.T., 1939: The Fluorescence of Compounds Containing Manganese, *Proc. Roy. Soc.*, **170**, pp. 272-292.

[75] Wortman, D.E., and Morrison, C.A., 1995: Analysis of the Energy Levels of Mn^{2+} in Haloapatite Structures, *Opt. Mater.*, **4**, pp. 487-505.

[76] Clausen, R., and Petermann, K., 1988: Mn^{2+} As a Potential Solid-State Laser Ion, *IEEE J. Quant. Electr.*, **24**, pp. 1114-1117.

[77] Merkle, L.D., Pinto, A., Verdun, H.R., and McIntosh, B., 1992: Laser Action from Mn^{5+} in $Ba_3(VO_4)_2$, *Appl. Phys. Lett.*, **61**, pp. 2386-2388.

[78] Whitmore, M.N., Verdun, H.R., and Singel, D.J., 1993: Electron-Paramagnetic-Resonance Spectroscopy of Manganese-Doped $Ba_3(VO_4)_2$: Identification of Tetrahedral Mn^{5+} and Mn^{4+} Centers, *Phys. Rev. B*, **47**, pp. 11479-11482.

[79] Capobianco, J.A., Cormier, G., Manaa, H., Moncorge, R., Bettinelli, M., Galarneau, P., and Labranche B., 1991: Near-Infrared Fluorescence and Gain Measurements of the Mn^{5+} ($3d^2$) Ion in $Sr_5(PO_4)_3Cl$ and Ca_2PO_4Cl, *OSA Proc. on Advanced Solid State Lasers*, **10**, G. Dube and L. Chase eds., pp. 379-385.

[80] Merkle, L.D., Verdun, H.R., and McIntosh, B., 1993: Spectroscopy and Laser Operation of Mn^{5+}-Doped Vanadates, *OSA Proc. on Advanced Solid State Lasers*, **15**, A.A. Pinto and T.Y. Fan eds., pp. 310-314.

[81] Johnson, L.F., Dietz, R.E., and Guggenheim, H.J., 1964: Spontaneous and Stimulated emission from Co^{2+} Ions in MgF_2 and ZnF_2, *Appl. Phys. Lett.*, **5**, pp. 21-22.

[82] Moulton, P.F., 1985: An Investigation of the Co:MgF_2 Laser system, *IEEE J. Quant. Electr.*, **QE-21**, pp. 1582-1595.

[83] Welford, D., and Moulton, P.F., 1988: Room-Temperature operation of a Co:MgF_2 Laser, *Opt. Lett.*, **13**, pp. 975-977.

[84] Imbulsch, G.F., Donegan, J.,F., and Bergin, F.J., 1987: Transition Metal Ion-Doped Materials of Laser Interest, *Spectroscopy of Solid-State Laser-Type Materials*, B. DiBartolo and G. Armagan eds., Plenum Press, N.Y., pp. 165-178.

[85] Stultz, R.D., Camargo, M.B., Montgomery, S.T., Birnbaum, M., and Spariosu, K., 1994: U^{4+}:SrF_2 Efficient Saturable Absorber Q-Switch for 1.54 μm Erbium:Glass Laser', *Appl. Phys. Lett.*, **64**, pp. 948-950.

[86] Stultz, R.D., Camargo, M.B., and Birnbaum, M., 1995: Divalent Uranium and Cobalt Saturable Absorber Q-Switches at 1.5 μm, *OSA Proc. on Advanced Solid State Lasers*, **24**, B.H.T. Chai and S.A. Payne eds., pp. 460-464.

[87] Johnson, L.F., Dietz, R.E., and Guggenheim, H.J., 1963: Optical Maser Oscillation from Ni^{2+} in MgF_2 Involving Simultaneous Emission of Photons, Phys. Rev. Lett., 11, pp. 318-320.

[88] Skripko, G.A., and Shkadarevich, A.P., 1986: Stimulated Emission from Al_2O_3 Crystals Doped by Titanium and Vanadium Ions, *Physics and Spectroscopy of Laser Crystals*, ed. by A.A. Kaminskii, Nauka, Moscow, pp. 257-268, (in Russian).

[89] Avanesov, A.G., Dvornikova, V.G., Zhorin, V.V., Kuznetsov, Y.A., Kuznetsova, T.G., and Pisarenko, V.F., 1993: Spectroscopy of Forsterite Single Crystals Doped by Nickel And Vanadium Ions, *J. Prikl. Spektr.*, **59**, pp. 152-154, (in Russian).

[90] Mikhailov, V.P., Kuleshov, N.N., Zhavoronkov, N.I., Prokoshin, P.V., Yumashev, K.V., and Sandulenko, V.A., 1993: Optical Absorption and Nonlinear Transmission of Tetrahedral V^{3+} (d^2) in Yttrium Aluminium Garnet, *Opt. Mater.*, **2**, pp. 267-272.

[91] Brunold, T.C., Hazenkamp, M.F., and Guedel, H.U., 1996: CrO_4^{3-} and MnO_4^{2-}: Broadband Emitters in the NIR and potential Laser Systems, *Advanced Solid State Lasers Meeting*, San Fransicco, CA, USA, Techn. Digest, pp. 286-288.

[92] Lupei, V., Boulon, G., Lupei, A., Elejalde, M.J., Brehier, A., and Pedrini, C., 1994: Spectrally Selective Energy Transfer from Fe^{3+} to Tm^{3+} in garnet lattices, *J.Lumen.*, **60&61**, pp. 237-240.

[93] Dixon, G.J., and Johnson, L.F., 1992: Low-Threshold 2-μm Holmium Laser Excited by Nonradiative Energy Transfer from Fe^{3+} in YGG, *Opt. Lett.*, **17**, pp. 1782-1784.

[94] Bagdasarov, Kh.S., Kevorkov, A.M., Melkonyan, T.A., Baryshevskii, V.G., Korzhik, M.V., Kuzmin, V.V., Lishits, M.G., Meilman, M.L., 1989: Iron Impurity as a New Sensitizer of Rare Earth Luminescence, *Doklady AN SSSR*, **305**, pp. 583-586, (in Russian).

[95] Johnson, L.F., Dillon, J.F., and Remeika, L.P., 1970: Optical Properties of Ho^{3+} Ions in Yttrium Gallium Garnet and Yttrium Iron Garnet, *Phys. Rev. B*, **1**, pp. 1935-1346.

[96] Goerlich, P., Karras, H., Koetitz, G., and Lehmann, R., 1964, Spectroscopic Properties of Activated Laser Crystals, *Phys. Stat. Sol.*, **5**, p. 3.

[97] Nepomnyashchikh, A.I., Tokarev, A.G., Chernyago, B.P., and Figura, P.V., 1991: Passive Laser Q-Switches on the Basis of BaF_2:Nd^{2+} Crystals, *Pisma ZhTF*, **17**, pp. 1-3, (in Russian).

[98] Basiev, T.T., Mirov, S.B., Sychev, S.A., 1991: Passive Laser Q-Switches Based on MeF_2:Nd^{2+} (Me-Ca, Sr, Ba) Crystals, *Proc. SPIE, Solid State Lasers And New Laser Materials*, V.V.Osiko, and A.A.Mak eds., **1839**, pp. 182-197.

[99] Schipper, W.J., Meijerink, A., and Blasse, G., 1994: The Luminescence of Tm^{2+} in Strontium Tetraborate, *J. Lumin.*, **62**, pp. 55-59.

[100] Dieke, G.H., Crosswhite, H.M., 1963, The Spectra of Doubly and Triply Ionized Rare Earths, *Appl. Opt.*, **2**, pp. 675-686.

[101] Lizzo, S., Meijerink, A., and Blasse, G., 1994: Luminescence of Divalent Ytterbium in Alkaline Earth Sulphates, *J. Lumin.*, **59**, p. 185.

[102] Waynant, R.W., 1982: Vaccum Ultraviolet Laser Emission from $Nd^{3+}:LaF_3$, *Appl. Phys. B*, **28**, p. 205.

[103] Waynant, R.W., and Klein, P.H., 1985: Vaccum Ultraviolet Laser Emission from $Nd^{3+}:LaF_3$, *Appl. Phys. Lett.*, **46**, p. 14.

[104] Bowman, S.R., Shaw, L.B., Feldman, B.J., Ganem, J., 1995: A Seven Micron Solid State Lasers, *OSA Proc. on Advanced Solid State Lasers*, **24**, B.H-T. Chai and S.A. Payne eds., pp. 116-118.

[105] Dubinskii, M.A., Semashko, V.V., Naumov, A.K., Abdulsabirov, R.Yu., and Korableva, S.L., 1993: Ce^{3+}-Doped Colquirite a New Concept of All-Solid-State Tunable Ultraviolet Laser, *J. Mod. Opt.*, **40**, p. 1.

[106] Pinto J.F., Rosenblatt, G.H., Esterowitz, L., Castillo, V., and Quarles, G.J., 1994: Tunable Solid State Laser Action in $Ce^{3+}:LiSrAlF_6$, *Electron. Lett.*, **30**, p. 240.

[107] Ehrlich, D.J., Moulton, P.F., and Osgood, R.M., Jr, 1979: Ultraviolet Solid-State Ce:YLF Laser at 325 nm, *Opt. Lett.*, **4**, pp. 184-186.

[108] Sarukura, N., Lin, Z., Sewaga, Y., Semasko, V.V., Naumov, A.K., Korableva, S.L., Abdulsabirov, R.Y., and Dubinskii, M.A., 1995: Direct and Passive Subnanosecond Pulse-Train Generation from a Self-Injection-Seeded Ultraviolet Solid-State Laser, *Opt. Lett.*, **20**, pp. 599-601.

[109] Sigachev, V.B., Doroshenko, M.E., Timoshechkin, M.I., Loutts, G.B., and Chai, B.H.-T., 1994: Energy Transfer from Ce^{3+} and Cr^{3+} to Er^{3+} ions in Garnet and Yttrium Silicate Crystals, *Conf. Lasers and Electro-Optics (CLEO)*, Anaheim, CA, May, 1994, Technical Digest, p. 124.

[110] Sandrok, T., Danger, T., Heumann, E., Huber, G., and Chai, B.H.T., 1994: Efficient Continuous Wave-Laser Emission of Pr^{3+}-Doped Fluorides at Room Temperature, *Appl. Phys. B.*, **58**, p. 149.

[111] Lawson, J.K., Payne, S.A., 1993: Excited-State Absorption of Pr^{3+}-Doped Fluoride Crystals, *Opt. Mater.*, **2**, pp. 225-232.

[112] Wong, S.K., Mathieu, P., Pace, P., 1990: Eye-Safe Nd:YAG Laser, *Appl. Phys.*, **57**, p. 650.

[113] Hodgson, N., Golding, D.J., and Eisel, D., 1994: Efficient Nd:YAG Laser Operating at a Wavelength of 1.444 μm with 100 W, 5 Joules per Pulse Output, *OSA Proc. on Advanced Solid State Lasers*, **20**, T.Y. Fan and B.H.-T. Chai eds., pp. 6-10.

[114] Payne, S.A., Chase, L.L., Smith, L.K., Kway, W.L., and Krupke, W.F., 1992: Infrared Cross-Section Measurements for Crystals Doped with Er^{3+}, Tm^{3+}, and Ho^{3+}, *IEEE J. Quant. Electr.*, **28**, pp. 2619-2630.

[115] Johnson, L.F., Geusic, J.E., and Van Uitert, L.G., 1966: Efficient, High Power Coherent Emission from Ho^{3+} Ions in Yttrium Aluminium Garnet, Assisted by Energy Transfer, *Appl. Phys. Lett.*, **8**, pp. 200-202.

[116] Antipenko, B.M., Glebov, A.S., Kiseleva, T.I., and Pismennyi, V.A., 1985: 2.12-μm Ho:YAG Laser, *Pisma. Zh. Tech.Phys.*, **11**, pp. 682-685, (in Russian).

[117] Bowman, S.R., Rabinovich, W.S., Feldman, B.J., Winings, M.J., 1990: Tuning the 3-μm Ho:$YAlO_3$ Laser, *OSA Proc. on Advanced Solid State Lasers*, **6**, ed. by H.P. Jenssen and J. Dube, pp. 254-256.

[118] Kaminskii, A.A., and Antipenko, B.M., 1989: *Multilevel Operating Schemes of Crystalline Lasers*, Nauka, Moscow, (in Russian).

[119] Kuo, Y.K., Birnbaum, M., Chen, W., Yue, K., and Huang, M.F., 1995: Passive Q-Switching of the Tm,Cr:YAG 2 μm Laser with a Ho:YLF Solid-State Saturable Absorber, *OSA Proc. Advanced Solid State Lasers*, **24**, B.H.T. Chai and S.A. PAyne eds., pp. 445-448.

[120] Zharikov, E.V., Zhekov, V.I., Kulevskii, L.A., Murina, T.M., Osiko, V.V., Prokhorov, A.M., Savel'ev, A.D., Smirnov, V.V., Starikov, B.P., and Timoshechkin, M.I., 1975: Stimulated Emission from Er^{3+} Ions in Yttrium Aluminium Garnet Crystals at λ=2.94 μm, *Sov. J. Quant. Electr.*, **4**, pp. 1039-1040.

[121] Laser on yttrium-erbium-aluminium garnet crystals (λ=2.94 μm). Proc. of General Physics Institute, 1989, **19**, ed. by T.M. Murina, Nauka, Moscow, pp. 1-152, (in Russian).

[122] Zharikov, E.V., Il'ichev, N.N., Kalitin, S.P., Laptev, V.V., Malyutin, A.A., Osiko, V.V., Pashinin, P.P., Prokhorov, A.M., Saidov, Z.S., Smirnov, V.A., Umyskov, A.F., and Shcherbakov, I.A., 1986: Spectral, Luminescence, and Lasing Properties of Yttrium Scandium

Gallium Garnet Crystal Activated with Chromium and Erbium, *Sov. J. Quant. Electr.*, **16**, pp. 635-640.

[123] Herbert, T., Risk, W.P., Macfarlane, R.M., and Lenth, W., 1990: Diode-Laser-Pumped 551-nm Upconversion Laser in $YLiF_4:Er^{3+}$, *OSA Proc. on Advanced Solid State Lasers*, **6**, H.P. Jenssen and G. Dube eds., pp. 379-383.

[124] White, K.O., Schleusener, S.A., 1972: Coincidence of Er:YAG Laser Emission with Methane Absorption at 1645.1 nm, *Appl.Phys.Lett.*, **21**, pp. 419-420.

[125] Stange, H., Petermann, K., Huber, G., and Duczynski, E.W., 1989: Continuous Wave 1.6 μm Laser Action in Er Doped Garnets at Room Temperature, *Appl. Phys. B*, **49**, pp. 269-273.

[126] Spariosu, K., and Birnbaum, M., 1992: Room-Temperature 1.644 Micron Er:YAG Lasers, *OSA Proc. on Advanced Solid State Lasers*, **13**, L.L. Chase and A.A. Pinto eds., pp. 127-130.

[127] Spariosu, K., Stultz, R.D., and Birnbaum, M., 1993: $Er:Ca_5(PO_4)_3F$ Saturable Absorber Q Switch for the Er:Glass Laser at 1.53 μm, *Appl. Phys. Lett.*, **62**, pp. 2763-2765.

[128] Antipenko, B.M., 1984: Cross-Relaxation Pumping Schemes of Laser Transitions, *Zh.Tekhn.Fiz.*, **54**, pp. 385-388, (in Russian).

[129] Stoneman, R.C., and Esterowitz, L., 1989: Room Temperature CW Tm^{3+}:YAG and Tm^{3+}:YSGG Lasers Continuously Tunable Over the Range 1.85-2.16 μm, *OSA Proc. Tunable Solid State Lasers*, **5**, ed.by M.L. Shand and H.P. Jenssen, pp. 157-160.

[130] Pinto, J.F., Esterowitz, L., and Rosenblatt, G.H., 1994: Tm^{3+}:YLF Laser Continuously Tunable between 2.20 and 2.40 μm, *Opt. Lett.*, **19**, p. 883.

[131] Stoneman, R.C., and Esterowitz, L., 1990: Laser-Pumped Room-Temperature 1.50 μm $Tm^{3+}:YLiF_4$ Continuous-Wave Cascade Laser, *OSA Proc. on Advanced Solid State Lasers*, **6**, ed. by H.P. Jenssen and G. Dube, pp. 267-270.

[132] Robinson, M., and Asawa, C.K., 1967: Stimulated Emission from Nd^{3+} and Yb^{3+} in Noncubic Sites of Neodymium- and Ytterbium- Doped CaF_2, *J. Appl. Phys.*, **38**, pp. 4495-4501.

[133] Lacovara, P., Wang, C.A., Choi, H.K., Aggarwal, R.L., and Fan, T.Y., 1991: Room-Temperature InGaAs Diode-Pumped Yb:YAG Laser, *OSA Proc. on Advanced Solid State Lasers*, **10**, ed. by G. Dube and L. Chase, pp. 283-285.

[134] Fan, T.Y., 1992: Optimizing the Efficiency and Stored Energy in Quasi-Three-Level Lasers, *IEEE J. Quant. Electr.*, **28**, pp. 2692-2697.

[135] Payne, S.A., Smith, L.K., DeLoach, L.D., Kway, W.L., Tassano, J.B., and Krupke, W.F., 1994: Laser, Optical, and Thermomechanical Properties of Yb-Doped Fluorapatite. *IEEE J. Quant. Electr.*, **30**, pp. 170-179.

[136] Fan, T.Y., 1993: Heat Generation in Nd:YAG and Yb:YAG, *IEEE J. Quant. Electr.*, **29**, pp. 1457-1459.

[137] DeLoach, L.D., Payne, S.A., Chase, L.L., Smith, L.K., Kway, W.L., and Krupke, W.F., 1993: Evaluation of Absorption and Emission Properties of Yb^{3+} Doped Crystals for Laser Applications, *IEEE J. Quant. Electr.*, **29**, pp. 1179-1191.

[138] Schweizer, T., Heumann, E., Heine, F., and Huber, G., 1994: Room Temperature Continuous Wave Er,Yb:$Y_3Al_5O_{12}$ Laser at 1.6 μm with Yb-Er Pump Energy Transfer, *Conf. on Lasers and Electro-Optics (CLEO/Europe'94)*, Amsterdam, Aug. 1994, Techn.Digest, paper CFC 3, p. 389.

[139] Antipenko, B.M., Nikitichev, A.A., 1991: Diode-Pumped Operating Scheme for Ho:YAG Laser, *Proc. SPIE, Solid State Lasers and New Laser Materials*, V.V. Osiko and A.A. Mak eds., **1839**, pp. 173-174.

[140] Nikitichev, A.A., Pismennyi, V.A., 1995: 1 W CW 2.12 μm Lamp Pumped Room Temperature YAG:Yb,Ho Laser, *8-th Laser Optics Conf.*, St.Petersburg, June 1995, Techn. Digest, vol. II, pp. 20-21.

[141] Jarman, R.H., Wallenberg, A.J., and Thrash, R.J., 1993: Growth and Fabrication of Single Crystal Yb,Tm:BaY_2F_8 Fibres for Up-Conversion Visible Laser Operation, *Proc. SPIE, Growth, Characterization and Application of Laser Host and Nonlinear Crystals II*, **1863**, ed. B.H.T. Chai, pp. 106-113.

[142] Louis, M., Simoni, E., Hubert, S., Auzel, F., Meichenin, D., and Gesland, J.Y., 1995: A New CW Laser at 2.82 μm $U^{3+}/LiYF_4$: Laser

Parameters, *OSA Proc. on Advanced Solid State Lasers*, **24**, B.H.-T. Chai and S.A. Payne eds., pp. 141-145.

[143] Sorokin, P.P., Stevenson, M.J., 1960: Stimulated Infrared Emission from Trivalent Uranium, *Phys. Rev. Lett.*, **5**, pp. 557-559.

[144] Quarles, G.J., Esterowitz, L., Rosenblatt, G.H., Uhrin, R., and Belt, R.F., 1992: Crystal Growth and Spectroscopic Properties of U^{3+}:$LiYF_4$, *OSA Proc. on Advanced Solid State Lasers*, **13**, ed. by L.L. Chase and A.A. Pinto, pp. 306-309.

[145] Jenssen, H.P., Noginov, M.A., Cassanho, A., 1993: U:YLF, a Prospective 2.8 μm Laser Crystal, *OSA Proc. on Advanced Solid State Lasers*, **15**, ed. by A.A. Pinto and T.Y. Fan, pp. 463-467.

[146] Meichenin, D., Auzel, F., Hubert, S., Simoni, E., Louis, M., and Gesland, J.Y., 1994: A New CW Room Temperature Laser at 2.82 μm: U^{3+}/$LiYF_4$, *Electron. Lett.*, **30**, p. 1257.

[147] Akhmetov, N.S., 1988, *General and inorganic chemistry*, Vysshaja shkola, Moscow, (in Russian).

[148] Agranovich, V.M., and Galanin, M.D., 1978: *Electron Excitation Energy Transfer in Condenced Media*, Nauka, Moscow, (in Russian) .

[149] Ermolaev, V.L., Bodunov, E.N., Sveshnikova, E.B., and Shahverdov, T.A., 1977: *Non-Radiative Transfer of Electron Excitation Energy*, Nauka, Leningrad, (in Russian).

[150] Förster, Th., 1948: Zwischenmolekulare Energiewanderund und Fluoreszenz, *Ann. Phys.*, **2**, pp.53-75.

[151] Förster, Th., 1949: Experimentelle und Theoretische Untersuchung des Zwischenmolekularen Übergangs von Elektronenanregungsenergie, *Ztschr. Naturforsch.A.*, **4**, pp.321-327.

[152] Dexter, D.L., 1953: A Theory of Sensitized Luminescence in Solids, *J. Chem. Phys.*, **21**, pp. 836-850.

[153] Ynokuti, M., and Hirayama, F., 1965: Influence of Energy Transfer by the Exchange Mechanism on Donor Luminescence, *J. Chem. Phys.*, **43**, pp. 1978-1989.

[154] Ermolaev, V.L., 1963: Energy Transfer with the Participation of Triplet State in Organical Systems, *Uspekhi Fiz. Nauk*, **80**, pp. 3-40, (in Russian) .

[155] Galanin, M.D., 1960: Resonance Energy Transfer in Luminescent Solutions, *Proc. Lebedev Phys. Inst. USSR Acad.Sci.*, **12**, pp. 3-53, (in Russian) .

[156] Vasiliev, I.V., Zverev, G.M., Kolodnyi, G.Ya., and Onishchenko, A.M., 1969: On Non-Resonance Energy Transfer Among RE^{3+} Dopants, *J. Exp. Teor. Fiz*, **56**, pp. 122-133, (in Russian).

[157] Alimov, O.K., Ashurov, M.X., Basiev, T.T., Kirpichenkova, E.O., and Murav'ev, V.V., 1987: Electron Excitation Energy Transfer Among Dopant Ions in Disordered Media, *Proc. General Phys. Inst. USSR Acad.Sci.*, Nauka, Moscow, **9**, pp. 50-147, (in Russian).

[158] Dexter, D.L., Förster, Th., and Knox, R.S., 1969: The Radiationless Transfer of Energy of Electron Excitation Between Impuriry Molecules in Crystals, *Phys. Stat. Solidi*, **34**, pp. K159-K162.

[159] Davidov, A.S., 1963: *Wave mechanics*, Physmathgis, Moscow, (in Russian).

[160] Davidov, A.S., 1968: The Radiationless Transfer of Energy of Electron Excitation Between Impurity Molecules in Crystals, *Phys. Stat. Solidi*, **30**, pp. 357-366.

[161] Burshtein, A.I., 1965: Energy Transfer Kinetics in Three-Level System, *J. Exp. Teor. Khim.*, **1**, pp. 563-573, (in Russian) .

[162] Burshtein, A.I., 1984: Balance Description of Quenching and Energy Transfer, *Khim. Fiz.*, **3**, pp. 1059-1068, (in Russian).

[163] Stepanov, B.E., 1982: On Calculation of Probability of Energy Transfer Between Molecules in Condense Media, *J. Khim. Fiz.*, **1**, pp. 1620-1624, (in Russian).

[164] Gamurar, V.Ya., Perlin, Yu.E., and Cukerblat, B.S., 1969: Non-Radiative Energy Transfer in Doped Crystals, *Fiz. Tverd. Tela*, **11**, pp. 1193-1199, (in Russian).

[165] Trifonov, E.D., and Shehtman, V.L., 1969: On Theory of Non-Radiative Transitions, *Fiz. Tverd. Tela*, **11**, pp. 2984-2991, (in Russian).

[166] Agabekyan, A.S., and Malikyan, A.O., 1969: Investigations of Non-Radiative Energy Transfer by Method of Density Matrix, *Proceedings of I All-Union Seminar On Non-Radiative Energy Transfer In Condense Media*, Lori, Erevan, pp. 5-16, (in Russian).

[167] Agabekyan, A.S., and Melikyan, A.O., 1972: On Criteria of Applicability of Theory of Resonance Energy Transfer, *Opt. Spektr.*, **32**, pp. 288-295, (in Russian) .

[168] Agabekyan, A.S., 1975: Capture of Excitation in a System of Linked Atoms, *Opt. Spektr.*, **38**, N1, pp. 180-181, (in Russian).

[169] Agabekyan, A.S., 1981: Energy Transfer in Conditions of Strong Noncoherent Interaction, *J. Prikl. Spektr.*, **34**, pp. 707-711, (in Russian).

[170] Agabekyan, A.S., 1986: Physical Processes of Elemental Act of Resonance Energy Transfer, *J. Khim. Fiz.*, **5**, pp. 583-588, (in Russian).

[171] Kustov, E.F., 1972: Energy Transfer Among Dopants in Crystals, *Opt. Spektr.*, **32**, pp. 323-330, (in Russian) .

[172] Shehtman, V.L., and Trifonov, E.D., 1970: On the Theory of Cooperative Luminescence, *Phys. Stat. Solidi*, **B41**, pp. 855-860.

[173] Bozko, A.N., and Pukhov, K.K., 1973: Energy Transfer in Condense Media, *Fiz. Tverd.Tela*, **15**, pp. 980-988, (in Russian).

[174] Trifonov, E.D., and Shehtman, V.L., 1973: On the Theory of Energy Transfer in Crystals, *Spectroscopy of Crystals*, Nauka, Leningrad, pp. 139-142, (in Russian).

[175] Davidov, A.S., and Serikov, A.A., 1973: The Theory of Migration of Excitation Energy in Crystal, *Izv. AN SSSR, Ser. Fiz.*, **37**, pp. 474-478, (in Russian).

[176] Burshtein, A.I., and Pusep, A.Yu., 1974: Collective Quenching of Luminescence, *Fiz. Tverd.Tela*, **16**, pp. 2318-2326, (in Russian).

[177] Galanin, M.D., 1955: On the Influence of Concentration on Luminescence of Solutions, *J. Exp.Teor. Fiz.*, **28**, pp. 485-495, (in Russian).

[178] Rozman, I.M., 1958: On the Theory of Fluorescence Quenching in Solutions, *Opt. Spektr.*, **4**, pp. 536-538, (in Russian).

[179] Golubov, S.N., and Konobeev, Yu.V., 1971: On the Procedure of Averaging in the Theory of Resonanse Energy Transfer,. *Fiz. Tverd. Tela*, **13**, pp. 3185-3190, (in Russian).

[180] Sakun, V.P., 1972: Kinetics of Energy Transfer in Crystals, *Fiz. Tverd.Tela*, **14**, pp. 2199-2210, (in Russian).

[181] Rotman, S.R., and Hartman, F.X., 1988: Non-Radiative Energy Transfer in Non-Uniform Codoped Laser Crystals, *Chem. Phys. Lett.*, **152**, N 4-5, pp. 311-317.

[182] Rotman, S.R., 1989: Non-Radiative Energy Transfer in Nd-YAG - Evidence for the Correlated Placement of Ions, *Appl. Phys. Lett.*, **54**, pp. 2053-2055.

[183] Hartmann, F.X., and Rotman, S.R., 1989: Energy Transfer by the Exchange Interaction in Non-Uniform Codoped Solid-State Crystals, *Chem. Phys. Lett.*, **163**, N4, pp. 437-442.

[184] Rotman, S.R., 1989: Ambiguities in the Analysis of Non-Radiative Energy Transfer Data in Solid State Laser Materials, *Appl. Phys. B.*, **49**, pp. 59-64.

[185] Rotman, S.R., and Airenberg, G., 1990: Energy Transfer to Acceptors in Solids with Dynamically Changing Effective Concentrations, *IEEE J. Quant. Electr.*, **26**, N2, pp. 210-212.

[186] Rotman, S.R., 1990: Modeling Non-Radiative Energy Transfer with Multiple Mechanisms in Solid-State Materials, *Chem. Phys. Lett.*, **173**, N4, pp. 349-353.

[187] Rotman, S.R., 1990: Analysis of Neodimium-To-Erbium Energy Transfer in Yttrium Aluminium Garnet with Nonuniform-Distribution Model, *Opt. Lett.*, **15**, N4, pp. 230-232.

[188] Hartmann, F.X., and Rotman, S.R., 1991: Symmetry in Non-Radiative Energy Transfer: Fast Transfer in Co-Doped Solid-State Fluorescent Materials, *Chem. Phys. Lett.*, **184**, N5, pp. 537-543.

[189] Kalisky, Y., Rotman, S.R., Boulon, G., Pedrini, C., and Bremer, A., 1994: Spectroscopy, Modelling, and Laser Operation of Holmium Doped Laser Crystals, *Opt. Mater.*, **4**, pp. 25-30.

[190] Rotman, S.R., Eyal, A., Kalisky, Y., Bremer, A., Pedrini, C., and Boulon, G., 1994: Modelling Unusually Fast Non-Radiative Energy Transfer in Crystals, *Opt. Mater.*, **4**, pp. 31-33.

[191] Basiev, T.T., Orlovskii, Yu.V., Ostroumov, B.G., Privis, Yu.S., and Shcherbakov, I.A., 1995: The Nature of Electron Excitation Energy Transfer from Cr^{3+} to Rare-Earth Ions in Garnet Crystals, *J. Kvant. Elektr.*, **22**, N8, pp. 759-764, (in Russian) .

[192] Heber, J., 1971: Excitation-Diffusion-Limited Energy Transfer in Solids, *Phys. Stat. Solidi B*, **48**, pp. 319-346.

[193] Shehtman, V.L., 1972: Influence of Diffusion of Exitons on the Energy Transfer in Crystals, *Opt. Spektr.*, **33**, pp. 284-291, (in Russian).

[194] Agranovich, V.M., 1968: *Theory of Exitons*, Nauka, Moscow, (in Russian).

[195] Yokota, M., and Tanimoto, O., 1967: Effect of Diffusion on Energy Transfer by Resonance, *J. Phys. Soc. Jap.*, **22**, pp. 779-785.

[196] Burshtein, A.I., 1984: Concentration Quenching of Non-Coherent Excitations in Solutions, *Uspekhi Fiz. Nauk*, **143**, pp. 553-600, (in Russian).

[197] Doktorov, A.B., Kipriyanov, A.A., and Burshtein, A.I., 1978: Influence of Kinetics of Particles Approaching in Solutions on Energy Transfer Between Them, *J. Exp. Teor. Fiz.*, **74**, pp. 1184-1198, (in Russian) .

[198] Artamonova, M.V., Briskina, Ch.M., Burshtein, A.I., Zusman, L.D., and Skleznev, A.G., 1972: Study of Time Dependence of Nd^{3+} Luminescence and Avaluation of Electron Excitation Migration of These Ions in Glass, *J. Exp. Theor. Fiz.*, **62**, pp. 863-871, (in Russian).

[199] Huber, D.L., Hamilton, D.S., and Barnet, B., 1977: Time-Dependent Effects in Fluorescent Line Narrowing, *Phys. Rev. B-Solid State*, **16**, pp. 4642-4650.

[200] Perepechko, S.N., Privis, Yu.S., Sakun, V.P., Smirnov, V.A., and Shcherbakov, I.A., 1983: Non-Modelling Description of Evolution of Donors' Excited State When They Are Interacting with Acceptors in Cubic Crystal, *Dokl. AN SSSR*, **271**, pp. 615-618, (in Russian).

[201] Burshtein, A.I., 1972: Jumping Mechanism of Energy Transfer, *J. Exp. Teor. Fiz.*, **62**, pp. 1695-1701, (in Russian).

[202] Zusman, L.D., 1974: Quenching of Luminescence at Presence of Migration of Excitations in Solid Solutions, *Opt. Spektr.*, **36**, pp. 497-502, (in Russian).

[203] Zusman, L.D., 1977: Kinetics of Luminescence Quenching for Jumping Mechanism of Quenching, *J. Exp. Teor. Fiz.*, **73**, pp. 662-670, (in Russian).

[204] Sakun, V.P., 1979: Kinetics of Jumping Mechanism of Luminescence Quenching, *Fiz. Tverd.Tela*, **21**, pp. 662-668, (in Russian).

[205] Basiev, T.T., Voron'ko, Yu.K., Mamedov, T.G., Osiko, V.V., and Shcherbakov, I.A., 1975: Prosesses of Relaxation of Excitations of RE^{3+} Metastable Levels in Crystals, *Spectroscopy of Crystals*, Nauka, Moscow, pp. 155-160, (in Russian).

[206] Vugmeister, B.E., 1976: Spin Diffusion in Disordered Paramagnetic Systems, *Fiz. Tverd. Tela*, **18**, pp. 819-825, (in Russian) .

[207] Bodunov, E.N., Malyshev, V.A., 1987: Theory of Concentrational Self-Quenching, *Opt. Spektr.*, **62**, pp. 1280-1286, (in Russian) .

[208] Huber, D.L., 1981: Dynamics of Incoherent Transfer, *Laser Spectroscopy of Solids*, W.M.Yen and P.M.Selzer eds., Springer-Verlag, Berlin, Heidelberg, New-York, pp. 83-111.

[209] Voron'ko, Yu.K., Mamedov, T.G., Osiko, V.V., Sakun, V.P., and Shcherbakov, I.A., 1976: Study of Nonradiative Relaxation of Excitation Energy in Condense Media with High Concentration of Activators, *J. Exp. Teor. Fiz.*, **71**, N2, pp. 478-496, (in Russian).

[210] Bondar', I.A., Krutikov A.V., Mezenceva, L.P., Perepechko, S.N., Smirnov, V.A., and Shcherbakov, I.A., 1983: Study of Energy Transfer Processes in Rare Earth Pentaphosphate Crystals, *Fiz. Tverd. Tela*, **25**, pp. 1983-1988, (in Russian).

[211] Privis, Yu.S., Smirnov, V.A., and Shcherbakov, I.A., 1984: A Consideration of Crystal Structure in Model of Jumping Quenching of Luminescence, *J. Exp. Teor. Fiz.*, **87**, pp. 589-595, (in Russian).

[212] Burshtein, A.I., 1983: Concentrational Self-Quenching, *J. Exp. Teor. Fiz.*, **84**, pp. 2001-2012, (in Russian).

[213] Vugmeister, B.E., 1983: Self-Quenching of Electron Excitation in Solid Solutions, *Fiz. Tverd. Tela*, **25**, pp. 2796-2798, (in Russian).

[214] Burstein, A.I., and Sakun, V.P., 1983: Migrational Quenching in a System of Centers with Cross-Relaxation, *Chem. Phys. Lett.*, **103**, pp. 205-208.

[215] Tkachuk, A.M., and Klokhishner, S.I., 1986: Self-Quenching of Rare-Earth Ions Luminescence in Crystals, *Opt. Spektr.*, **61**, pp. 84-90, (in Russian).

[216] Sakun, V.P., 1987: On Kinetic Similarity of Processes of Concentrational Quenching and Self-Quenching of Luminescence in Disordered Doped Systems, *J. Khim. Fiz.*, **6**, pp. 1032-1037, (in Russian).

[217] Antipenko, B.M., 1980: Cooperative Mechanism of Excitation of Active Ions Ho^{3+} in $BaYb_2F_8$, *Pisma J. Tekhn. Fiz.*, **6**, pp. 968-972, (in Russian).

[218] Zhekov, V.I., Zubov, B.V., Lobachev, V.A., Murina, T.M., Prokhorov, A.M., Shevel, A.F., 1980: 'The Mechanism of Production of Inverse Population between $^4I_{11/2}$ and $^4I_{13/2}$ levels of Er^{3+} Ion in $Y_3Al_5O_{12}$, *J. Kvant. Elektr.*, **7**, N4, pp. 749-753, (in Russian) .

[219] Antipenko, B.M., 1984: Spectroscopic Schemes of Excitation of Laser Transition Having High Quantum Efficiency of Pumping Bands, *Izv. AN SSSR, Ser. Fiz.*, **48**, N7, pp. 1373-1378, (in Russian) .

[220] Galanin, M.D., 1951: Quenching by Absorbing Substances and Sensitized Fluorescence in Solutions, *Izv. AN SSSR, Ser. Fiz.*, **15**, pp. 543-549, (in Russian).

[221] Johnson, L.F., van Uitert, L.G., Rubin, J.J., and Thomas, R.A., 1964: Energy Transfer from Er^{3+} to Tm^{3+} and Ho^{3+} Ions in Crystals, *Phys. Rev.*, **133A**, N1, pp. 494-498.

[222] Tolstoi, M.N., 1970: Non-Radiative Energy Transfer Between Rare-Earth Ions in Crystals and Glasses, *Spectroscopy of crystals*, Nauka, Moscow, pp. 124-135, (in Russian).

[223] Osiko, V.V., Prokhorov, A.M., and Shcherbakov, I.A., 1980: Active Media of Solid State Lasers, *Izv. AN SSSR, Ser. Fiz.*, **44**, pp. 1698-1715, (in Russian).

[224] Golubov, S.I., 1976: A Theory of Sensitized Fluorescence in Molecular Crystals, *Ph.D. thesis*, p. 23, (in Russian).

[225] Mikhelashvili, M.S., Rozman, I.M., and Cukhaya, G.S., 1974: Regarding Kinetics of Luminescence of Rigid Solution in Presence of Energy Transfer, *Opt. Spektr.*, **36**, pp. 352-354, (in Russian).

[226] Alimov, O.K., Ashurov, M.X., Basiev, T.T., and Voron'ko, Yu.K., 1982: Spectroscopy of Selective Excitation of Yb^{3+} and Nd^{3+} in Phosphate Glasses and Interionic Interactions, *Preprint Lebedev Phys. Inst. USSR Acad.Sci.*, N 77, pp. 41, (in Russian).

[227] Gapontsev, V.P., and Sverchkov, Yu.E., 1979: Kinetics of Luminescence of Doped Centers in Disorder Media, *Preprint Radio Electr. Inst. USSR Acad. Sci.*, N 6, pp. 23, (in Russian).

[228] Avanesov, A.G., Denker, B.I., Osiko, V.V., Ostroumov, V.G., Sakun, V.P., Smirnov, V.A., and Shcherbakov, I.A., 1982: Sensitization of Luminescence and Its Application for Increasing of the Efficiency of Active Media for Solid State Lasers, *Kvant. Elektr.*, **9**, pp. 681-688, (in Russian).

[229] Smirnov, V.A., and Shcherbakov, I.A., 1985: Kinetics of Energy Transfer and Potentialities of Cr^{3+} Ions as Active Particles for Solid State Lasers, *Spectroscopy Of Crystals*, Nauka, Leningrad, pp. 21-42, (in Russian).

[230] Zharikov, E.V. Laptev, V.V., Ostroumov, V.G., Privis, Yu.S., Smirnov, V.A., and Shcherbakov, I.A., 1984: Study of New Active Laser Medium - GSGG:Cr,Nd, *Kvant. Elektr.*, **11**, pp. 1565-1575, (in Russian).

[231] Ostroumov, V.G., Privis, Yu.S., Smirnov, V.A., and Shcherbakov, I.A., 1986: Sensitizing of Nd^{3+} Luminescence by Cr^{3+} in Gallium Garnets, *J. Opt. Soc. Amer., B.*, **3**, pp. 81-93.

[232] Alpat'ev, A.N., Zharikov, E.V., Kalitin, S.P., Umyskov, A.F., and Shcherbakov, I.A., 1987: Efficient 2.088 μm Room Temperature Laser on YSGG:Cr,Tm,Ho Crystal, *Kvant. Elektr.*, **14**, N5, pp. 922-923, (in Russian).

[233] Alpat'ev, A.N., Zharikov, E.V., Noginov, M.A., Ostroumov, V.G., Saidov, Z.S., Smirnov, V.A., Umyskov, A.F., and Shcherbakov, I.A., 1988: YSGG Crystals Doped with Er^{3+}, Tm^{3+}, Ho^{3+} Ions as Active Media for IR Lasers with High Efficiency, *Izv. AN SSSR, Ser. Fiz.*, **52**, pp. 342-347, (in Russian).

[234] Alpat'ev, A.N., Zharikov, E.V., Kalitin, S.P., Smirnov, V.A., Umyskov, A.F., and Shcherbakov, I.A., 1989: The Generation of Gigantic Pulse in TEM_{00} mode (λ=2.088 μm) in YSGG:Cr,Tm,Ho Crystal Using Flashlamp Pumping, *Kvant. Elektr.*, **16**, N4, pp. 672-673, (in Russian) .

[235] Allen, R., Esterowitz, L., Goldberg, L., Weller, L., and Storm, M., 1986: Diode-Pumped 2 μm Holmium Laser, *Electron. Lett.*, **22**, pp. 947-949.

[236] Fan, T.Y., Huber, G., Byer, R.L., and Mitzcherlich, P., 1987: Continuous Wave Operation at 2.1 μm of a Diode-Laser Pumped Tm-Sensitized $Ho:Y_3Al_5O_{12}$ Laser at 300 K, *Opt. Lett.*, **12**, pp. 678-680.

[237] Antipenko, B.M., Glebov, A.S., Kiseleva, T.I., and Pis'mennyi, V.A., 1986: New Spectroscopic Scheme of Active Media for 2-μm Region, *Opt. Spektr.*, **60**, pp. 153-157, (in Russian).

[238] Antipenko, B.M., Glebov, A.S., Krutova, L.I., Solncev, V.M., and Sukhareva, L.K., 1986: Active Media for 2-μm Lasers on GSGG Crystals, *Kvant. Elektr.*, **13**, N7, pp. 1521-1523, (in Russian) .

[239] Antipenko, B.M., Krutova, L.I., and Sukhareva, L.K., 1986: Cascade Lasing on GSGG:Cr,Tm,Ho Crystals, *Opt. Spektr.*, **61**, N13, pp. 659-661, (in Russian) .

[240] Sorokina, I.T., Umyskov, A.F., Smirnov, V.A., and Shcherbakov, I.A., 1994: Tm-Tm and Tm-Ho Energy Transfer Studies in YSGG and GSAG Crystals at the Resonant Excitation of Tm and Ho Ions, *OSA Proc. Advanced Solid State Lasers*, **20**, T.Y. Fan and B.H.-T. Chai eds., pp. 136-140.

[241] Basiev, T.T., Voron'ko, Yu.K., Osiko, V.V., Prokhorov, A.M., and Shcherbakov, I.A., 1976: Expiremental Observation of Excitation Capture in a System of Strong Interacting Particles, *J. Exp. Teor. Fiz.*, **70**, pp. 1225-1233, (in Russian).

[242] Basiev, T.T., Voron'ko, Yu.K., Zharikov, E.V., Osiko, V.V., and Timoshechkin, M.I., 1979: Capture of Excitation of Strong Noncoherent Interaction of Er^{3+} and Yb^{3+} Ions in YAG Crystals, *Kratk. Soobshch. Fiz.*, **8**, pp. 9-12, (in Russian).

[243] Glushko, A.A., Osiko, V.V., Timofeev, Yu.P., and Shcherbakov, I.A., 1980: Kinetics of Population and Decay of High Excited States of TR^{3+} Ions in Terms of Strong Noncoherent Interaction, *J. Exp. Teor. Fiz.*, **79**, pp. 194-206, (in Russian).

[244] Noginov M.A., Smirnov, V.A., Talybov, A.I., and Shcherbakov, I.A., 1989: Rare Earth Scandium Garnets Doped with Cr^{3+} Ions as Active Media for 1.5 μm Er Lasers, *J. Prikl. Spektr.*, **51**, N5, pp. 766-772, (in Russian) .

[245] Ashurov, M.X., Voron'ko, Yu.K., Osiko, V.V., and Sobol, A.A., 1982: Deverse Non-Radiative Energy Transfer in a System of Strong Interacting Particles, *J. Exp.Teor. Fiz.*, **83**, N3, pp. 922-932, (in Russian).

[246] Noginov M.A., Ostroumov, V.G., Saidov, Z.S., Smirnov, V.A., and Shcherbakov, I.A., 1986: Distribution of Excitation in Disordered System of Strong Interacting Particles, *Preprint General Phys. Inst. USSR Acad. Sci.*, N 193, (in Russian).

[247] Antipenko, B.M., Buchenkov, V.A., Glebov, A.S.,Kiseleva, T.I., Nikitichev, A.A., and Pismennyi, V.A., 1988: The Spectroscopy of YSGG:Cr,Tm,Ho Laser Crystals, *Opt. Spektr.*, **64**, N6, pp. 1295-1298, (in Russian) .

[248] Zubenko, D.A., 1992: Interaction of Excited Cr^{3+} Ions and Rare Earth Elements in Laser Crystals, *Ph.D. Thesis*, Moscow, (in Russian) .

[249] Blombergen, N., 1959: Solid State Infra-Red Quantum Counters, *Phys. Rev. Lett.*, **2**, N 3, pp. 84-85.

[250] Yu.P.Chukova, 1980: Anti-Stokes Luminescence and New Possibilities of Its Application, Sov. Radio, Moscow, 1980, (in Russian) .

[251] Porter, J.F., 1961: Fluorescence Excitation by Absorbtion of Two-Consecutive Photons, *Phys. Rev. Lett*, **7**, pp. 414-415.

[252] Brown, M.R., and Shand, W.A., 1963: Infra-Red Quantum Counter Action in Pr-Doped Fluoride Lattices, *Phys. Rev. Lett.*, **11**, pp. 366-368.

[253] Brown, M.R., and Shand, W.A., 1964: Infra-Red Quantum Counter Action in Er-Doped Fluoride Lattices, *Phys. Rev. Lett.*, **12**, pp. 367-369.

[254] Singh, S., and Geusic, J.E., 1966: Observation and Saturation of a Multiphoton Process in $NdCl_3$, *Phys. Rev. Lett.*, **17**, pp. 865-868.

[255] Dexter, D.L., 1973: The New Studies of Sensitized Luminescence and Energy Transfer in Solids, *Izv. AN SSSR, Ser. Fiz.*, **37**, N2, pp. 257-261 (in Russian).

[256] Auzel, F., 1966: Compteur Quantique par Transfert D'energie Entre Deux Ions de Terres Rares Dans un Tungstate Mixte et Dans un Verre, *C.R. Acad. Sc. Paris, B*, **262**, pp. 1016-1019.

[257] Auzel, F., 1966: Compteur Quantique par Transfert D'energie de Tm^{3+} Dans un Tungstate Mixte et Dans un Verre Germanate, *C.R. Acad. Sc. Paris, B*, **263**, pp. 819-821.

[258] Antipenko, B.M., Mak, A.A., Raba, O.B., Sinicin, B.V., and Uvarova, T.V., 1982: The New Schemes of Excitation of Laser Transitions, *J. Tekhn. Fiz.*, **52**, pp. 521-522 (in Russian).

[259] Antipenko, B.M., Mak, A.A., Raba, O.B., Sinicin, B.V., and Uvarova, T.V., 1982: The Scheme of Step Sensitization of Laser Transitions of Er^{3+} in $BaYb_2F_8$, *Kvant. Elektr.*, **9**, pp. 1614-1619 (in Russian).

[260] Antipenko, B.M., Mak, A.A., Nikolaev V.B., Raba, O.B., Seyranyan, K.B., and Uvarova, T.V., 1984: Analysis of Laser Situations in $BaYb_2F_8$:Er^{3+}, *Opt. Spektr.*, **56**, pp. 484-489 (in Russian).

[261] Johnson, L.F., Guggenheim, H.T., Rich, T.C., and Ostermayer, F.W., 1972: Infra-Red-to Visible Conversion by Rare-Earth Ions in Crystals, *J. Appl. Phys.*, **43**, N3, pp. 1125-1137.

[262] Auzel, F.,1973; Materials and Divices Using Double-Pumped Phosphors with Energy Transfer, *Proc. IEEE*, **61**, pp. 758-786.

[263] Lenth, W., and Macfarlane, R.M., 1992: Up-Conversion Lasers, *Opt. Photon. News*, **3**, pp. 8-15.

[264] Ovsyankin, V.V., and Feofilov, P.P., 1966: On Mechanism of Summation of Electron Excitations in Doped Crystals, *Pisma J. Exp.Teor. Fiz.*, **3**, pp. 494-497, (in Russian) .

[265] Ovsyankin, V.V., and Feofilov, P.P., 1966: 'Cooperative Sensitization of Luminescence in Crystals Doped with RE^{3+} Ions, *Pisma J. Exp.Teor. Fiz.*, **4**, pp. 471-474, (in Russian) .

[266] Ovsyankin, V.V., and Feofilov, P.P., 1967: Cooperative Luminescence in Condense Media, *J. Prikl. Specrosc.*, **11**, pp. 498-506 (in Russian).

[267] Ovsyankin, V.V., and Feofilov, P.P., 1967: Cooperative Lumunescence in Solids, *J. Prikl. Spektr.*, **6**, pp. 1828-1833, (in Russian).

[268] Ovsyankin, V.V., and Feofilov, P.P., 1973: Cooperative Processes in Luminescence Systems, *Izv.AN SSSR, Ser. Fiz.*, 1973, **37**, pp. 262-272 (in Russian).

[269] Dexter, D.L.,1957: Possibility of Luminescent Quantum Yields Greater Then Unity, *Phys. Rev.*, **108**, pp. 630-633.

[270] Dexter, D.L.,1962: Cooperative Optical Absorption in Solids, *Phys. Rev.*, **126**, pp. 1962-1967.

[271] Nakazawa, E., and Shionoya, Sh., 1970: Cooperative Luminescence in $YbPO_4$, *Phys. Rev. Lett.*, **25**, pp. 1710-1712.

[272] Miyakawa, T., and Dexter, D.L., 1970: Cooperative and Stepwise Excitation of Luminescence Trivalent Rareearths Ions in Yb^{3+}-Sensitized Crystals, *Phys. Rev. B*, **1**, pp. 70-80.

[273] Livanona, L.D., Saitkulov, I.G., and Stolov, A.l., 1968: Cumulative Luminescence in CaF_2:Gd^{3+},Yb^{3+} Systems, *Opt. Spektr.*, **25**, pp. 609-610 (in Russian).

[274] Livanona, L.D., Saitkulov, I.G., and Stolov, A.l., 1968: The Processes of Summation in CaF_2 and SrF_2 Single Crystals Doped with Tb^{3+} and Yb^{3+}, *Fiz.Tverd.Tela*, **11**, pp. 918-923 (in Russian).

[275] Varsanyi, F., and Dieke, G.H., 1961: Ion-Pair Resonance Mechanism of Energy Transfer in Rareearth Crystal Fluorescence. *Phys. Rev.L ett.*, **7**, N12, pp. 442-443.

[276] Kushida, T., 1973: Probabilities of Energy Transfer and Cooperative Processes in RE^{3+} Ions in Solids, *Izv. AN SSSR, Ser. Fiz.*, **37**, N2, pp. 273-283, (in Russian).

[277] Antipenko, B.M., Dmitruk, A.V., Karapetyan, G.O., Zubkova, V.S., Kosyakov, V.I., Mak, A.A., and Mikhailova, N.N., 1973: 'Transformation of Nd-Laser Emission into Luminescence of Tb^{3+} in Glasses Doped with Tb^{3+} and Yb^{3+}, *Opt. Spektr.*, **35**, N3, pp. 540-545, (in Russian).

[278] Feofilov, P.P., 1972: Cooperative Optical Phenomena in Doped Crystals, *Physics of Doped Centers in Crystals*, Tallinn, pp. 539-563, (in Russian).

[279] Kazaryan, A.K., Timofeev, Yu.P., and Fok, M.V., 1986: Anti-Stokes Transformation of Emission in Solids Doped with RE^{3+} Ions, *Proc. Lebedev Phys. Inst. USSR Acad. Sci.*, Nauka, pp. 4-65, (in Russian).

[280] Glushko, A.A., Osiko, V.V., Timofeev, Yu.P., and Shcherbakov, I.A., 1980: Influence of Crystal Lattice on the Processes of Population of High Excited Levels of RE^{3+} Ions at IR-Excitation, *J. Exp. Teor. Phys*, **78**, pp. 53-61.

[281] Timofeev, Yu.P., Daschkevich, E.I., Kazaryan, A.K., Kut'enkov, A.A., Nguen, K.T., and Fridman, S.A., 1988: Summation and Multiplying of Electron Excitations of RE^{3+} Ions in Cooperative Luminophoros, *Izv. AN SSSR, Ser. Fiz.*, **52**, N4, pp. 791-796, (in Russian) .

[282] Miyakava, T., and Dexter, D.L., 1970: Phonon Sidebands, Multiphonon Relaxation of Excited States, and Phonon Assisted Energy Transfer Between Ions in Solids, *Phys. Rev., B*, **1**, pp. 2961-2969.

[283] Mita, Y, 1972: Luminescence Processes in Yb^{3+} Sensitized Rareearth Phosphors, *J.aAppl. Phys.*, **43**, N4, pp. 1772-1778.

[284] Kuroda, H., Shinoga, Sh., and Kushida, T., 1972: Mechanism and Controlled Factors of Infrared-To-Visible Conversion Processes in Er^{3+} and Yb^{3+} - Doped Phosphors, *J. Phys. Soc. Jap.*, **33**, N1, pp. 125-141.

[285] Elcombe, M.M., and Pryor, Q.W., 1970: The Lattice Dynamics of Calcium Fluoride, *J. Phys. Chem.*, **3**, pp. 492-499.

[286] Riseberg, L.A., and Moos, H.W., 1968: Multiphonon Orbit-Lattice Relaxation of Excited States of Rare Earth Ions in Crystals, *Phys. Rev.*, **174**., pp. 429-438.

[287] Yamada, N., Shinoga, Sh., and Kushida, T., 1972: Assisted Energy Transfer Between Trivalent Rare-Earth Ions, *J. Phys. Soc. Jap.*, **12**, pp. 1577-1586.

[288] Hurrel, J.P., and Porto, S.P., 1968: Optical Phonons of Yttrium Aluminium Garnet, *Phys. Rev.*, **173**, pp. 851-856.

[289] Ovsyankin, V.V., and Fefilov, P.P., 1970: Cooperative Luminescence in Crystals Doped with RE-Activators, *Spectroscopy of Crystals*, Nauka, pp. 135-143 (in Russian).

[290] Sverchkov, S.E., and Sverchkov, Yu.E., 1987: Non-Linear Jumping Quenching of Luminescence of Doped Centers in Solids, *Preprint General Phys. Inst. USSR Acad. Sci.*, N 283, p. 18, (in Russian) .

[291] Zubenko, D.A., Noginov, M.A., Smirnov, V.A., and Shcherbakov, I.A., 1993: Jumping Mechanism of Non-Linear Quenching Luminescence, *Preprint General Phys. Inst. USSR Acad. Sci.*, N 13, pp. 13, (in Russian) .

[292] Bobenko, S.D, Benderskii, V.A., Goldanskii, V.I., Lavrushko, A.G., and Tushinskii, V.P., 1971: Singlet Exciton Annigilation in Antracene Crystals, *Phys. Stat. Solidi (B)*, **45**, N1, pp. 91-97.

[293] Bobenko, S.D, Benderskii, V.A., Goldanskii, V.I., Lavrushko, A.G., and Tushinskii, V.P., 1971: Annigilation of Singlet Excited States in Antracene Solutions, *Chem. Phys. Lett.*, **8**, N5, pp. 598-603.

[294] Kilin, S.F., Mikhelashvili, M.S., and Rosman, I.M., 1978: Kinetics of Non-Linear Quenching of Luminescence, *Izv. AN SSSR, Ser. Fiz.*, **42**, N2, pp. 414-419, (in Russian) .

[295] Agranovitch, V.M., Efremov, N.A., and Zakhidov, A.A., 1980: Exciton Luminescence of Molecular Crystals at High Excitation Level, *Izv. AN SSSR, Ser. Fiz.*, **44**, N4, pp. 759-764, (in Russian).

[296] Pollack, S.A., and Chang, D.B., 1990: Up-Conversion-Pumped Population Kinetics for $^4I_{13/2}$ and $^4I_{11/2}$ States of Er^{3+} Ions in Several Host Crystals, *IEEE J. Quant. Electr.*, **22**, pp. 75-94.

[297] Ostroumov, V., Jenssen, T., Meyn, J.-P., Huber, G.,and Noginov, M.A., 1995: Concentration Quenching and Up-Conversion of Neodimium Ions in $LaSc_3(BO_3)_4$ and $GdVO_4$ Crystals, *OSA Proc. Advanced Solid State Lasers*, **24**, B.H.T. Chai and S.A. Payne eds., pp. 509-513.

[298] Tolstoi, N.A., and Abramov, A.P., 1966: On Possibilities of Theoretical Interpretation of Non-Linear Quenching of Luminescence, *Opt. Spektr.* **20**, N3, pp. 496-498, (in Russian).

[299] Tolstoi, N.A., and Abramov, A.P., 1966: Fluorescence of Uranil Salts at High Excitation Level, *Opt. Spektr.*, **21**, N2, pp. 171-177, (in Russian).

[300] Zubenko, D.A., Noginov, M.A., Ostroumov, V.G., Smirnov, V.A., and Shcherbakov, I.A., 1992: Processes of Interaction of Excited Cr^{3+}-Ions in Laser Crystals, *J. Prikl. Spektr.*, **54**, N1, pp. 77-81, (in Russian).

[301] Antipenko, B.M., Sinicin, B.V., and Uvarova, T.V., 1980: Laser Re-Emitter on $BaYb_2F_8$:Ho; Three-Micron Lasing Channel, *Kvant. Elektr.*, **7**, N9, pp. 2019-2022, (in Russian).

[302] Antipenko, B.M., Vorihalov, I.V., Sinicin, B.V., and Uvarova, T.V., 1980: Laser Convertor on $BaYb_2F_8$:Ho, *Kvant. Elektr.*, **7**, N1, pp. 197-199, (in Russian).

[303] Antipenko, B.M., Podkolzina, I.G., and Tomashevich, Yu.V., 1980: $LiYF_4$:Ho as Active Medium for Laser Convertor, *Kvant. Elektr.*, **7**, N3, pp. 647-649, (in Russian).

[304] Antipenko, B.M., Mak, A.A., Sinicin, B.V., and Raba, O.B., Uvarova, T.V., 1983: New Schemes of Excitation of Laser Transitions, *J. Tekhn. Fiz.*, **3**, N3, pp. 521-522, (in Russian).

[305] Antipenko, B.M., Glebov, A.S., and Dumbravyan, R.V., 1988: Physics of Energy Storage in Active Medium $BaEr_2F_8$:Tm,Ho, *Kvant. Elektr.*, **15**, N6, pp. 1277-1281, (in Russian).

[306] Antipenko, B.M., Buchenkov, V.A., Glebov, A.S., Kiseleva, T.I., Nikitichev, A.A., and Pismennyi, V.A., 1988: Spectroscopy of YAG-Cr,Tm,Ho Laser Crystals, *Opt. Spektr.*, **64**, N6, pp. 1295-1298, (in Russian).

[307] Antipenko, B.M., Buchenkov, V.A., Kiseleva, T.I., and Pismennyi, V.A., 1987: Interpretation of Temperature Dependence of Generation Threshold of YAG:Cr,Tm,Ho, *Opt. Spektr.*, **63**, N2, pp. 393-397, (in Russian).

[308] Guseva, E.D., Dmitryuk, A.V., Kosyakov, V.I., and Makushkhin, B.V., 1974: Energy Transfer in Glasses Doped with Yb and Er, *Proc. Simposium Spectroscopic properties of glasses,* Leningrad, pp. 109-111, (in Russian).

[309] Maticin, S.M., and Izineev, A.A., 1981: Non-Linear Quenching of Luminescence in Laser Erbium Glasses, *Proc. All-Union Meeting on Luminescence,* Leningrad, p. 38, (in Russian) .

[310] Noginov, M.A., Smirnov, V.A., Umyskov, A.F., and Shcherbakov, I.A., 1988: Interaction of Excited Er^{3+} Ions in YSGG Crystals, *Preprint General Phys. Inst. USSR Acad.Sci.,* N 34, p. 14, (in Russian).

[311] Noginov, M.A., Smirnov, V.A., and Shcherbakov, I.A., 1988: Measurement of Interaction Constants for Excited Er^{3+} Ions in YSGG Crystals, *Preprint General Phys. Inst. USSR Acad. Sci.,* N 76, pp. 21, (in Russian).

[312] Gross, R., Denisov, A.L., Zharikov, E.V. Zubenko, D.A., Noginov, M.A., Reimann, U., Smirnov, V.A., Huber, G., and Shcherbakov, I.A., 1991: Depopulaton of Lower Laser Level $^4I_{13/2}$ Er^{3+} in YSGG:Cr,Er, *Laser Phys.,* **1**, pp. 52-56.

[313] Vintner, E., Zubenko, D.A., Noginov, M.A., Prokhorov, A.M., Smirnov, V.A., and Shcherbakov, I.A., 1990: Processes of Interaction of Excited Cr^{3+} and RE^{3+} Ions in Scandium Garnet Crystals, *Kratk. Soobshch. Fiz.,* **10**, pp. 24-26, (in Russian) .

[314] Noginov, M.A., Prokhorov, A.M., Smirnov, V.A., Shcherbakov, I.A., Winter, E., and Zubenko, D.A., 1990: Interaction of Excited Chromium and Rareearth Ions in Scandium Garnet Crystals, *OSA Proc. Advanced Solid State Lasers,* **6**, eds. H.P. Jenssen and G. Dube, pp. 183-186.

[315] Maiman, T.H., 1960: Optical and Microwave-Optical Experiments in Ruby:, *Phys. Rev. Lett.,* **4**, p. 564.

[316] Gires, F., Mayer, G., 1962. Measurements of the Excited Absorption Spectrum of Ruby for Optical Maser, *C.R. Acad. Sci,* **254**, pp. 659-661.

[317] Ivanov, A.P. Rubinov, A.N., and Mihnov, S.A., 1964: Methods of Examination of Substance in State of Strongly Disturbed

Thermodynamic Equilibrium, *Opt. Spektr.*, **17**, N3, pp. 597-606, (in Russian).

[318] Sevast'yanov, B.K., 1975: Spectroscopy of Excited Crystals Doped with Cr^{3+} Ions, *Spectroscopy of Crystals*, Nauka, pp. 122-154, (in Russian) .

[319] Valyashko, E.G., and Timoshenkov, V.A., 1969: Absorption Spectra of Ginger Ruby in Excited State, *J. Prikl. Spektr.*, **11**, N1, pp. 73-75, (in Russian).

[320] Petermann, K., and Stange, H., 1990: Limitation of Tunability by Excited-State Absorption in Cr-Doped Garnets, *OSA Proc. Advanced Solid State Lasers*, **6**, eds. H.P. Jenssen and G. Dube, pp. 59-72.

[321] Kolbe, W., Petermann, K., and Huber, G., 1985: Broad-Band Emission and Laser Action of Cr^{3+}-Doped Zinc-Tungstate at 1 µm Wavelength, *IEEE J. Quant. Electr.*, **QE-21**, N10, pp. 1596.

[322] Kaminskii, A.A., Butashin, A.V., Demidovich, A.A., Koptev, V.G., Mill, B.V., and Shkadarevich, A.P., 1989: Broad-Band Tunable Stimulated Emission from Octahedral Cr^{3+} Ions in New Acentric Crystals with Ca-Gallo-Germanate Structure, *Phys. Stat. Sol. (a)*, **112**, pp. 197.

[323] Koetke, J., Huber, G., and Petermann, K., 1991: Spectroscopy of Ni^{2+}-Doped Garnets and Perovskites for Solid State Lasers, *J. Luminesc.*, **48-49**, pp. 564-568.

[324] Wegner, T., and Petermann, K., 1989: Excited State Absorption of Ti^{3+}:$YAlO_3$, *Appl. Phys.*, **B49**, p. 275.

[325] Sevast'yanov, B.K., Remigailo, Yu.L., Orekhova, V.P., Matrosov, V.P., Tsvetkov, V.B., and Bukin, G.V., 1981: Spectral and Lasing Properties of Alexandrite Laser, *Dokl. AN SSSR*, **256**, pp. 373-376, (in Russian).

[326] Ashurov, M.X., Rustamov, I.R., Smirnov, V.A., Umyskov, A.F., and Shcherbakov, I.A., 1988: Mechanismes of Creating of Color Centers in Cr-Doped Crystals of Scandium Garnets, *Fiz. Tverd. Tela*, **30**, pp. 2296-2302, (in Russian).

[327] Alpat'ev, A.N., Zharikov, E.V. Zagumennii, A.I., Kalitin, S.P., Lutts, G.B., Rustamov, I.R., Smirnov, V.A., Umyskov, A.F., and

Shcherbakov, I.A., 1989: Influence of Spectral Properties of Exciting Emission on Specral-Luminescent and Lasing Properties of YSGG:Cr,Tm,Ho and GSAG:Cr,Tm,Ho Crystals, *Preprint General Phys. Inst. USSR Acad. Sci.*, N38, p. 9, (in Russian) .

[328] Naselskii, S.P., Rustamov, I.R., Smirnov, V.A., Umyskov, A.F., and Shcherbakov, I.A., 1989: Influence of Spectral Properties of Exciting Emission on Lasing and Specral-Luminescent Properties of $YAlO_3$:Cr Crystals, *Preprint General Phys. Inst. USSR Acad.Sci.*, N45, pp. 12, (in Russian).

[329] Andrews, L.J., Hitelman, S.M., Kokta, M., and Gable, D., 1986: Excited State Absorption of Cr^{3+} in K_2NaScF_6 and $Gd_3Ga_2(MO_4)_3$, M=Ga,Al, *J. Chem. Phys.*, 84, pp. 5229.

[330] Lee, H.W., Rayne, S.A., and Chase, L.L., 1989: Excited State Absorption of Cr^{3+} in $LiCaAlF_6$, *Phys. Rev.*, **B39**, p. 8907.

[331] Petermann, K., Clausen, R., Heunmann, E., and Ledig, M., 1989: Time Resolved Excited State Absorption of Mn^{2+}, *Opt. Comm.*, **70**, p. 483.

[332] Clausen, R., and Petermann, K., 1988: Mn^{2+} as a Potential Solid State Laser Ion, *IEEE J. Quant. Electr.*, **24**, pp. 1114-1117.

[333] Payne, S.A., Chase, L.L., and Wilke, G.D., 1988: Excited State Absorption Spectra of V^{2+} in $KMgF_3$ and MgF_2, *Phys. Rev.*, **B37**, p. 998.

[334] Seelert, W., Kortz, H.P., and Yen, W.M., 1992: Excited State Absorption and $^4F_{3/2}$ Lifetime Shortening in Diode Pumped Nd:YLF Q-Switch Lasers, *OSA Proc. Advanced Solid State Lasers*, **13**, eds.L.L. Chase and A.A. Pinto, pp. 209-211.

[335] Burlot, R., Moncorge, R., Manaa, H., Guyot, Y., Boulon, G., and Garcia-Solé, 1995: Excited State Absorption Spectroscopy of the Nd^{3+}-Doped Nonlinear $LiNbO_3$ and $LiTaO_3$ Single Crystals, *Advanced Solid State Lasers Meeting*, Technical Digest, paper MF5, p. 17.

[336] Johnson, L.f., and Guggenheim, H.J., 1971: Infrared-Pumped Visible Laser, *Appl. Phys. Lett.*, **19**, pp. 44-47.

[337] Mak, A.A., and Antipenko, B.M., 1982: Rare-Earth Transformations of Nd-Lasers Spectra, *J. Prik I. Spektr.*, **37**, pp. 1029-1045, (in Russian).

[338] Basiev, T.T., Zharikov, E.V. Zhekov, V.I., Murina, T.M., Osiko, V.V., Prokhorov, A.M., Starikov, B.M., Timoshechkin, M.I., and Shcherbakov, I.A., 1976: Radiative and Non-Radiative Transfers of Er^{3+} in Mixed Yttrium-Erbium-Aluminium Garnets, *Kvant.E lektr.*, **3**, pp. 1471-1477, (in Russian) .

[339] Ashurov, M.X., Basiev, T.T., Voron'ko, Yu.K., Zharikov, E.V. Zhekov, V.I., Murina, T.M., Osiko, V.V., Timoshechkin, M.I., and Shcherbakov, I.A., 1978: Non-Radiative Losses at $^4I_{11/2}$ - $^4I_{13/2}$ Er^{3+} Transition in $Y_3Al_5O_{12}$, $Gd_3Sc_2Al_3O_{12}$, $Y_3Ga_5O_{12}$, $Gd_3Ga_5O_{12}$, CaF_2, *Kvant. Elektr.*, **5**, pp. 1028-1033, (in Russian).

[340] Bagdasarov, H.S., Zhekov, V.I., Lobachev, V.A., Murina, T.M., and Prokhorov, A.M., 1983: Stationary Regime of Lasing of $Y_3Al_5O_{12}:Er^{3+}$ (λ=2.94 μm, T=300 K), *Kvant. Elektr.*, **10**, pp. 452-454, (in Russian).

[341] Zhekov, V.I., Lobachev, V.A., Murina, T.M., and Prokhorov, A.M., 1984: Cooperative Processes in Yttrium-Erbium-Aluminium Garnet Crystals, *Kvant.Elektr.*, **11**, pp. 189-191, 1984, (in Russian); Zhekov, V.I., Murina, T.M., Prokhorov, A.M., Studenikin, M.I., Georgesku, S., Lupei, V., and Ursu, I., 1986: Cooperative Processes in $Y_3Al_5O_{12}:Er^{3+}$ Crystals, *Kvant. Elektr.*, **16**, pp. 274-276, (in Russian).

[342] Zhekov, V.I., Lobachev, V.A., Murina, T.M., Popov, A.V., Prokhorov, A.M., and Studenikin, M.I., 1990: Laser Action of $Y_3Al_5O_{12}$:Er (λ=2.94 μm) at Excitation into Lower Laser Level, *Proc. All Union 6-th Laser Optics Conference,* Leningrad, pp. 42, (in Russian).

[343] Ping, Xie, and Stephen C.Rand, 1990: Continuous-Wave Trio Up-Conversion Lasers, *Appl. Phys. Lett.*, **57**, pp. 1182-1184.

[344] Pollack, S.A., and Chang, D.B., 1988: Ion-Pair Up-Conversion Laser Emission in Er^{3+} Ions in YAG, YLF, SrF_2, CaF_2 Crystals, *J. Appl. Phys.*, **64**, pp. 2885-2893.

[345] Pollack, S.A., and Chang, D.B., 1988: Up-Conversion Pumped Laser Emission in YAG, YLF, SrF_2, CaF_2 Crystals, *J. Appl. Phys.*, **64**, pp. 4077-4086.

[346] Pollack, S.A., and Chang, D.B., 1990: Up-Conversion Pumped Population Kinetics for $^4I_{13/2}$ and $^4I_{11/2}$ States of Er^{3+} Ions in Several Host Crystals, *Opt. and Quant. Electr.*, **22**, pp. 75-94.

[347] Noginov, M.A., Smirnov, V.A., and Shcherbakov, I.A., 1990; Non-Linear Population Processes of Er^{3+} Laser Levels in Chromium-Doped Garnet Crystals, *Opt. and Quant. Electr.*, **22**, pp. 61-74.

[348] Noginov, M.A., Smirnov, V.A., Umyskov, A.F., and Shcherbakov, I.A., 1988: Interaction of Excited Er^{3+} Ions in YSGG Crystals, *Preprint General Phys. Inst. USSR Acad.Sci.*, N 34, p. 14, (in Russian) .

[349] Kalitin, S.P., Noginov, M.A., Ostroumov, V.G., Smirnov, V.A., Umyskov, A.F., and Shcherbakov, I.A., 1988: Nonlinear Interactions of Excited Ions in YSGG:Cr,Er Laser Crystal, *Proc. of 3rd Internation. Sympos. Optika-88*, Budapest, Hungary, Sept.13-16, pp. 187-192.

[350] Gross, R., Denisov, A.L., Zharikov, E.V. Zubenko, D.A., Noginov, M.A., Reimman, U., Smirnov, V.A., Huber, G., and Shcherbakov, I.A., 1991: Depopulation of Lower Laser Level $^4I_{13/2}$ Er^{3+} in YSGG:Cr,Er, *Laser Phys.*, **1**, N6, pp. 52-56.

[351] Pollack, S.A., Chang, D., and Moise, N., 1986: Up-Conversion Pumped Infrared Erbium Laser, *J. Appl. Phys.*, **60**, pp. 4077-4086.

[352] Pollack, S.A., Chang, D., and Birnbaum, M., 1989: Threefold Upconversion Pumped Laser at 0.85, 1.23, and 1.73 μm in Er:YLF Pumped with 1.53 μm Er:Glass Laser, *Appl. Phys. Lett.*, **54**, pp. 869-871.

[353] Pollack, S.A., Chang, D., Birnbaum, M., and Kokta, M., 1991: Upconversion-Pumped 2.8-2.9 μm Lasing of Er^{3+} Ion in Garnets, *J. Appl. Phys.*, **70**, pp. 7227-7239.

[354] McFarlane, R.A., Robinson, M., Pollack, S.A., chang, D.B., and Jenssen, H.P., 1989: Visible and Infrared Laser Operation by Upconversion Pumping of Erbium-Doped Fluorides, *OSA Proc. Tunable Solid State Lasers*, **5**, M.L. Shand and H.P. enssen eds., pp. 179-186.

[355] Gapontsev, V.P., Zhabotinskii, M.E., Izyneev, A.A., Kravchenko, V.B., and Rudnitskii, Yu.P., 1973: Effective Conversion of 1.054-1.54 μm Stimulated Emission, *Pisma J. Exp. Teor. iz*, **18**, pp. 251-253, (in Russian).

[356] Ostermayer, F.W., Jr., J.P. van der Ziel, Marcos, H.M., Van Uitert, L.G., and Geusic, J.E., 1971: Frequency Upconversion in YF_3:Yb:Tm, *Phys.Rev. B*, **3**, pp. 2698-2705.

[357] Smart, R.G., Hanna, D.C., Tropper, A.C., Davey, S.T., Carter, S.F., and Szebesta, D., 1991: CW Room Temperature Upconversion Lasing at Blue, Green and Red Wavelengths in Infrared-Pumped Pr^{3+} -Doped Fluoride Fibre, *Electron. Lett.*, **27**, pp. 1307-1308.

[358] Antipenko, B.M., Voronin, S.P., and Privalova, T.A., 1987: Summation of Optical Frequencies by Cooperative Processes, *Opt. Spektr.*, **63**, pp. 1297-1298, (in Russian).

[359] Hanna, D.C., Percival, R.M., Perry, I.R., Smart, R.G., Townsend, J.E., and Tropper, A.C., 1990: Frequency Upconversion in Tm- and Yb,Tm-Doped Silica Fibers, *Opt. Comm.*, **78**, pp. 187-194.

[360] Trash, R.J., and Johnson, L.F., 1994: Upconversion Laser Emission from Yb^{3+} -Sensitized Tm^{3+} in BaY_2F_6, *J. Opt. Soc. Am. B*, **11**, pp. 881-885.

[361] Heine, F., Heumann, E., Danger, T., Schweizer, T., Huber, G., and Chai, B.H.-T., 1994: Green Upconversion Continuous Wave Er^{3+}:$LiYF_4$ Laser at Room Temperature, *Appl. Phys. Lett.*, **65**, N4, pp. 383-384.

[362] Heine, F., Heumann, E., Möbert, P., Huber, G., and Chai, B.H.-T., 1995: Room Temperature CW Green Upconversion Er^{3+}:YLF_4-Laser Pumped Near 970 nm, *OSA Proc. Advanced Solid State Lasers*, **24**, B.H.T. Chai and S.A. Payne eds., pp. 74-76.

[363] Heine, F. Ostroumov, V., Heumann, E., Jenssen, T., Huber, G., and Chai, B.H.-T., 1995: CW Yb,Tm:LiYF4 Upconversion Laser at 650, 800 and 1500 nm, *OSA Proc. Advanced Solid State Lasers*, **24**, B.H.T. Chai and S.A. Payne eds., pp. 77-79.

[364] Chivian, J.S., Case, W.E., and Eden, D.D., 1979: The Photon Avalanche: a New Penomenon in Pr^{3+}-Based Infrared Quantum Counters, *Appl. Phys. Lett.*, **35**, N2, p. 124, 1979.

[365] Ping, Xie, and Gosnell, T.R., 1995: Diode-Pumped, Continuous-Wave, Blue, Green, Orange, and Red Upconversion Fiber Lasers Operating at Room Temperature, *OSA Proc. Advanced Solid State Lasers*, **24**, B.H.T. Chai and S.A. Payne eds., pp. 101-105.

[366] Allain, J.Y., Monerïe, M., and Poignaut, H., 1991: Red Upconversion Yb-Sensitized Pr Fluoride Fiber Laser Pumped in 0.8 μm Region, *Electron. Lett.*, **27**, pp. 1156-1157.

[367] Koch, M.E., Kueny, A.W., and Case, W.E., 1990: Photon Avalanche Upconversion Laser at 644 nm, *Appl. Phys. Lett.*, **56**, pp. 1083-1085.

[368] Noginov, M.A., Prokhorov, A.M., Sarkissyan, G.K., Shcherbakov, I.A., and Smirnov, V.A., 1991: New Pumping Schemes of Solid State Lasers Due to the Transitions Starting from Active Ions Excited States, *Laser Phys.*, **1**, pp. 615-624.

[369] Noginov, M.A., and Shcherbakov, I.A., 1992: ESA-Pumped Room Temperature Laser, *OSA Proc. Advanced Solid State Lasers*, **13**, L.L. Chase and A.A. Pinto eds., pp. 156-160.

[370] Joubert, M.-F., Guy, S., Jacquier, B., and Linares, C., 1994: The Photon-Avalanche Effect:Review, Model and Application, *Opt. Mater.*, **4**, pp. 43-49.

[371] Pollnau, M., Lüthy, W., and Weber, H.P., 1995: The Possibilities and Limits of Avalanche Lasing on the Green Er^{3+}: $LiYF_4$ Transition, *OSA Proc. Advanced Solid State Lasers*, **24**, B.H.T. Chai and S.A. Payne eds., pp. 523-527.

THE EFFECT OF DEFECTS ON INORGANIC LUMINESCENT MATERIALS

S. R. Rotman

Ben-Gurion University of the Negev
Department of Electrical and Computer Engineering
P.O. Box 653, 84105 Beer-Sheva, ISRAEL

Abstract

Defects present in host crystals can have a strong effect on the luminescent properties of dopant ions in these materials. These effects manifest both in the optical and electrical properties of these crystals.

In this chapter, we will present optical and electrical techniques to determine the defect state of a crystal host and its luminescent dopants. First, we will consider how electrical conductivity measurements can determine the defect structure of the host crystal and its relationship to the optical properties of the dopants. Then, we will discuss how different dopants, and sometimes even the same dopant, can interfere with the expected property of a luminescent ion. These interactions can be seen by site selection spectroscopy and by the dynamic analysis of energy transfer between ions. Finally, we will show how cathodoluminescent measurements demonstrate the competition between defects and dopants for energy in luminescent properties. We present the experimental methods in this chapter as examples of the increasingly sophisticated techniques being used to understand the effect of defects on luminescence activity.

THE EFFECT OF DEFECTS ON INORGANIC LUMINESCENT MATERIALS

S. R. Rotman

Contents

The Effect of Defects on Inorganic Luminescent Materials

S. R. Rotman
Ben-Gurion University of the Negev
Department of Electrical and Computer Engineering
P.O. Box 653, 84105 Beer-Sheva, ISRAEL

1. Introduction

The most basic analysis of wide-band crystals doped with luminescent ions assumes that 1) the ions enter in the fixed (desired) valence state, 2) the sites exhibit the expected symmetries and 3) the dopant ions do not interact with each other. However, many interesting effects occur due to the violation of the above assumptions. The most recent Cr^{+4}:YAG infrared laser is based on chromiium ions being forced into the +4 valence state in a +3 ionic site [1]. Energy transfer between neodymium and chromium is a crucial step in the operation of the neodymium-chromium-doped gadolinum scandium gallium garnet laser [2]. Energy transfer between neighboring identical ions is essential to the principle of the upconversion laser [3].

In a recent paper [4], D. Vivien listed four ways that the lattice matrix and its defects can influence the properties of dopants beyond the standard crystal field effect. They are as follows:

1. Control of the Dopant Oxidation State: the non-stochiometry of a host crystal can often have an effect on the valence of the dopants, due to the need for charge compensation.

2. Control of Site Occupancy: the matrix can affect the particular site that a dopant chooses; in particular, a multitude of sites which are slightly different one from the other can be caused by defects in the host lattice.

3. Optimization of the Concentration of Dopants: the matrix and its defects will determine the optimum concentration of dopants for the best luminescence performance.

4. Optimization of the Host Matrix Chemical Composition: for the desired luminescent properties, the chemical composition of the host matrix may be critical. The stochiometric composition may not be the best.

In the above categories, the effects are primarily based on the material properties of the defects of the lattice; many techniques exist to understand the dopants' effects. Our work over the last fifteen years has centered on the use of optical and electrical techniques to determine the interactions governing the material properties of these crystals. This will be the subject of this chapter.

To describe this topic, rather than briefly reviewing countless experimental techniques performed on different materials, we will go into detail into three experimental studies which have been performed by our group. In each case, we will present the technique on very specific materials. We will also highlight relevant work from other groups; however, we cannot review all the research done and remain within reasonable limits. In Section II, we will review defect-dopant interactions by examining nickel-doped yttrium aluminum garnet (Ni:YAG). We have been one of the pioneering groups at relating the electrical behavior of the material with the optical characteristics. In Section III, we present dopant-dopant interactions through the optical measurement of multi-sites, sites which are subtly varied through the presence of neighboring ions and of crystal defects; we present a new model for understanding the energy transfer properties. In Section IV, we present the deleterious effect of defects by examining models for cathodoluminescent energy flow. We again have picked specific crystals: cerium-doped yttrium aluminum garnet (Ce:YAG) and europium- and manganese-doped lanthanum magnesium hexaaluminates (Eu,Mn:LMA). Each of these sections corresponds to one of the first three effects mentioned in the aforementioned Vivien paper. It is our hope that by presenting these techniques in detail on materials with which we have extensive experience that we will be able to both illustrate these interactions and to enable other researchers to apply these techniques to their own experimental projects.

2. Defect-Dopant Interactions

2.1. Introduction

When an ion enters a crystal site, there is a valence associated both with the dopant and with the crystal site. When these valences are not equal, a compensating mechanism must be available to maintain crystal neutrality. This compensation may take the form of vacancies and interstitials in the crystal structure or additional valence changes in other dopants. These defects are not unalterable; the crystal chemistry can subtly (and occasionally not so subtly) change as a function of the growth conditions or subsequent annealing history.

In general, any number of defect types are possible. A general popular review of such defects can be found in Ref. [5]. In this chapter, we will present a specific case in which the valence of a particular ion (Ni) can be altered via high temperature annealing. These changes can be measured qualitatively both through optical measurements and, perhaps surprisingly, electrical measurements. The laser host, Yttrium Aluminum Garnet (YAG), is affected through its defect structure; a qualitative evaluation of its crystal chemistry is necessary to understand the nickel valence changes. While we present a single example in this section, we suggest that it opens exciting possibilities for the careful control of the optical properties of laser and phosphor crystals.

2.2. Crystal Chemistry Related to Optical and Electrical Properties (Example: Undoped YAG)

2.2.1. Basic Relations

As stated above, we will analyze the luminescence properties of nickel-doped YAG as a function of the annealing conditions under which the crystal was prepared. We must start with the basic crystal chemistry of YAG.

$Y_3Al_5O_{12}$ crystallizes in the garnet structure [6]. There are 180 atoms in the unit cubic cell; it can be best pictured as a collection of smaller cation-oxygen units. There are three cation sites in the crystal. The yttrium ion occupies the dodecahedral site with eight oxygens surrounding it. 60% of the aluminum ions occupy tetrahedral sites with four surrounding oxygens and 40% occupy octahedral sites with six surrounding oxygens.

Given the existence of four atom sublattices in the garnet structure, the number of possible defect types is in principle very large and therefore difficult to establish unambiguously. Fortunately, a series of detailed studies by Larsen and Metselaar [7] has been completed on yttrium ion garnet (YIG), a system with near identical structure to YAG (they form a complete series of solid solutions), which strongly support a defect model based on doubly ionized oxygen vacancies. Although the studies of the transport properties as a function of temperature and oxygen partial pressure do not uniquely confirm the oxygen vacancy role in YAG, as we show, they are fully consistent with this defect picture.

Appendix A gives the mass-action and neutrality equations which govern the concentrations of oxygen vacancies ($[V_0^{\cdot\cdot}]$), electrons (n), holes (p) and acceptors ($[A']$) as a function of the partial pressure of oxygen. Table I shows the solutions n p, and $[V_0^{\cdot\cdot}]$ for each of three possible neutrality conditions: when the oxygen vacancies are neutralized by electrons in the conduction band, the oxygen vacancies and electrons are proportional to $P_{O_2}^{-1/6}$ and the holes to $P_{O_2}^{+1/6}$, while if the number of oxygen vacancies is fixed by acceptors, the number of electrons and holes is instead proportional to $P_{O_2}^{-1/4}$ and $P_{O_2}^{+1/4}$, respectively. Finally, when the concentration of holes is fixed by the acceptor concentrations, oxygen vacancies follow a $P_{O_2}^{-1/2}$ dependence. Since the carrier mobilities are generally found to be P_{O_2} independent, the corresponding partial conductivites will follow the same laws indicated in Table I above. The relevant activation energies are E_n, the energy of formation of an oxygen vacancy, and E_g, the bandgap energy of YAG.

Note that while this table considers one inadvertent acceptor, in truth the effective concentration of acceptors equals the numbers of acceptors (and there may be several types) minus the number of donors weighted by their different valences.

2.2.2 Electrical Conductivity Measurements

If the defects were optically active, their concentration as a function of temperature and annealing conditions would be immediately measurable. However, even when they are not, electrial measurements can (perhaps surprisingly) also be used to monitor the defect structure [8].

The measured total conductivity is often dominated by one of the partial conductivities. Under neutrality conditions where electrons balance the number of oxygen vacancies, one finds *n*- and *p*-type conduction proportional to $P_{O_2}^{-1/6}$ and $P_{O_2}^{+1/6}$, respectively. The ionic conductivity is not observed since the number of electrons equals the number of vacancies while the mobility of the electrons should be orders of magnitude greater than the mobility of the ions. Under these circumstances, the activation energy for electron and hole conduction should be $E_1/3$ and E_g - $E_1/3$, respectively, assuming nonactivated mobilities (Table 1).

Under conditions where the number of vacancies is controlled by inadvertent acceptors dopants, three possible regimes could be observed (see Table 1). These include *n*- and *p*- type regimes, proportional to $P_{O_2}^{-1/4}$ and $P_{O_2}^{+1/4}$, respectively, and possibly, ionic conduction near the center of the defect regime. This conductivity would be P_{O_2} independent and its activation energy would be the activation energy of oxygen vacancy mobility alone while the activation energies for electron and hole conduction would be $E_1/2$ and $E_g - E_1/2$, respectively. Finally, in the last regime, only P_{O_2} - independent p-type electronic conductivity with weak temperature dependence should be observed.

The different possibilities for electro-neutrality yield different temperature and P_{O_2} dependencies. The experimental data will allow us to identify the applicable defect regimes. We will use the data from undoped YAG to illustrate this point.

Fig. 1 shows three isotherms of log σ vs log P_{O_2} obtained for YAG. Under the most reducing atmospheres, YAG is n-type with a partial pressure oxygen dependence of approximately $P_{O_2}^{-1/4}$. Conversion to p-type conduction is evident at the higher partial pressures of oxygen ($1 > P_{O_2} > 10^{-4}$). A weakly P_{O_2}-dependent intermediate regime is also observed.

In an attempt to deconvolute the various contributions to the total conductivity, extrapolated values of the n-type curve were subtracted from the total conductivity. The difference between the total conductivity and the n- type conductivity is plotted in Fig. 2. A flat regime at intermediate partial pressures of oxygen, identified as ionic, is clearly present. Only at the highest partial pressures does conduction become p- type. Quantitative analysis of the data produces the following results:

$$E_1 = 7.8\text{eV},$$
$$E_g = 6.1\text{eV},$$
$$E_{\text{mig}} = 2.3\text{eV},$$

where E_{mig} is the ionic mobility activation energy.

The band gap has been reported [6] to be between 6.3 - 6.6 eV, a value in close agreement with our derived value for the thermal band gap E_g, thus lending further credence to our model.

2.2.3. Summary

Our electrical measurement technique has succeeded; the high-temperature equilibrium transport studies of YAG have shown it to be a mixed ionic-electronic conductor [8]. The atmosphere dependent electrical conductivity is consistent with a defect model dominated by doubly-ionized oxygen vacancies

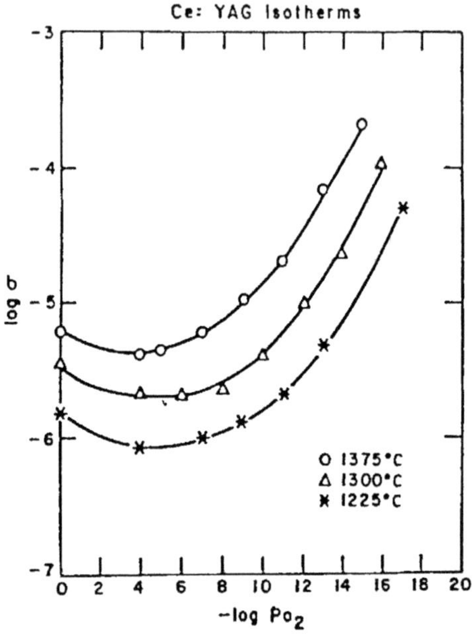

Figure 1. ac conductivity isotherms for YAG, as a function of P_{O_2}. Ce^{+3} ions present in the crystal do not affect the conductivity (after Ref. [8]).

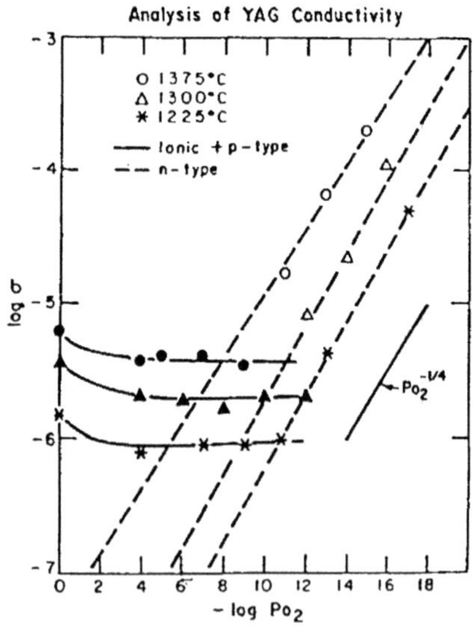

Figure 2. ac conductivity isotherms of YAG separated into n-type and ionic components, as a function of P_{O_2} (after Ref. [8]).

fixed in number by background acceptor impurities. We must now see how this knowledge can be used to understand luminescent behavior when the crystal is optically active.

2.3. Crystal Chemistry Related to Optical and Electrical Properties (Example: Nickel-Doped YAG)

2.3.1 Basic Equations

We are now ready to consider the addition of nickel to YAG to illustrate the effect of aliovalent dopants on the conductivity and optical properties. The presence of a multi-valent ion causes a profound change in the crystal chemistry. Nickel, based on size consideration, should enter tetrahedral or octahedral sites; it has been found to normally reside on octahedral sites in garnets [9]. The neutrality and mass action equations may be found in Appendix B. The solutions are considerably more complex, with the need to solve for p, n, $[V_O^{\cdot\cdot}]$, $[Ni'_{Al}]$ (the Ni^{+2} concentration) and $[Ni^x_{Al}]$ (the Ni^{+3} concentration). When this is done (see Table 2), we can now consider the electrical conductivity of this material.

2.3.2. Electrical Conductivity of Ni:YAG

Fig. 3 shows four isotherms of log σ vs log P_{O_2} obtained for Ni:YAG at temperatures varying from 1130 to 1340°C with P_{O_2} ranging from 1 to 10^{-20} atm. Under the most oxidizing atmospheres, YAG exhibits a decreasing conductivity with decreasing P_{O_2}, characteristic of *p*-type conduction, followed by a region of relatively weak P_{O_2} sensitivity at somewhat lower P_{O_2}. These observations are consistent with predictions of our defect model in region II of Fig. 4 in which $p \propto P_{O_2}^{+1/4}$ and $[V_O^{\cdot\cdot}] \propto P_{O_2}^0$. Further, it is interesting to note that the first plateau in Fig. 5 is followed by an increase in σ with decreasing P_{O_2}, ultimately saturating into a second P_{O_2}-independent plateau of higher conductivity. This follows qualitatively the predictions for $[V_O^{\cdot\cdot}]$ as one moves successively through regions II, III, and IV with decreasing P_{O_2}.

The electrical conductivity of Ni:YAG, measured under high-temperature equilibrium conditions, exhibited results consistent with a defect model in which the concentrations of doubly ionized oxygen vacancies is controlled both by background acceptor impurities and the P_{O_2}-dependent concentration of divalent nickel. The first P_{O_2}-independent regime occurs when $[Ni'_{Al}] << [A']$ and the number of oxygen vacancies is fixed by the background acceptor impurity alone

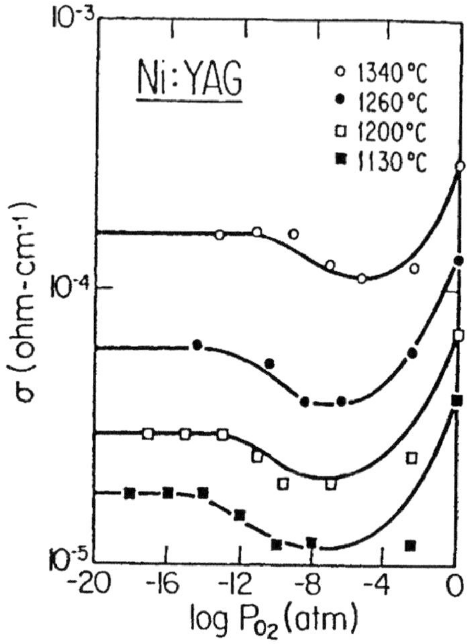

Figure 3. ac conductivity isotherms for Ni:YAG, as a function of P_{O_2} (after Ref. [9]).

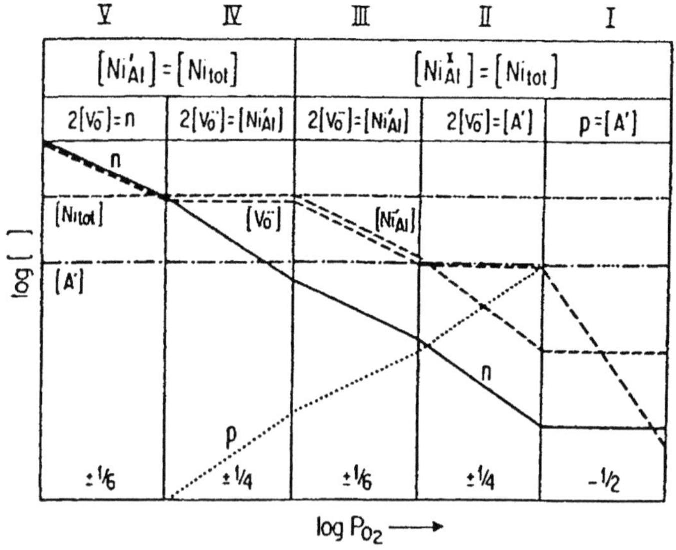

Figure 4. Defect diagram for an oxide doped with nickel exhibits two oxidation states. The top row designates the defect region, the second row indicates the predominant oxidation state of Ni, and the third row designates the applicable neutrality condition (after Ref. [9]).

(region II in Fig. 4). The second P_{O_2}-independent regime represents the case when $[\mathrm{Ni}'_{\mathrm{Al}}]=[\mathrm{Ni}_{\mathrm{tot}}]$ and $2[V_O^{\cdot\cdot}]=[\mathrm{Ni}_{\mathrm{tot}}]+[A']$ (region IV in Fig. 4). Based on the ratio of the magnitude of conductivities within the two P_{O_2}-independent plateaus, we conclude that indeed:

$$[A']/[\mathrm{Ni}_{\mathrm{tot}}]=2,$$

and that Fig. 5 is an accurate reflection of the P_{O_2} dependence of $[\mathrm{Ni}'_{\mathrm{Al}}]$ and $[V_O^{\cdot\cdot}]$. Further analysis of this data [9] yields values of 1.6 and 2.0 - 2.2 eV for E_2 and E_{mig} (nickel ionization and ionic mobility activation energy, respectively).

In summary, high-temperature equilibrium transport studies on nickel-doped YAG show it to be a mixed ionic-electronic conductor [9]. Electrical conductivity data are consistent with a defect model dominated by doubly ionized oxygen vacancies which are fixed in concentration by the number of Ni^{+2} ions and the number of background acceptor impurities. A *p*-type conductivity dominates at oxidizing atmospheres, while the ionic conductivity predominates under more reducing conditions. The nickel changes valence with a change in P_{O_2}; the energy of ionization E_2 is found to be 1.6 eV.

2.3.3. *Optical Measurements of Ni:YAG*

Since Ni^{+3} is optically active, we can expect that the changes seen electrically would be mirrored in optical changes. This is indeed the case [10].

Fig. 6 shows the absorption spectra for Ni:YAG. The primary feature is a band peaking at 420 nm which gives Ni:YAG its characteristic yellow color. Upon reduction, the band disappears; the remaining bands at 340 and 460 nm are due to trace cerium absorption as determined by the location, size, and shape of the bands (Fig. 7). This reaction is reversible; reoxidation causes the 420 nm absorption band to reappear. Optical absorption measurements at higher energies show much stronger absorption extending to the band edge at approximately 190 nm.

In summary, optical studies have confirmed that nickel changes valence in Ni:YAG from Ni^{+3} to Ni^{+2} under reduction. Thus, we have confirmed the changes seen and explained in the electrical conductivity measurements.

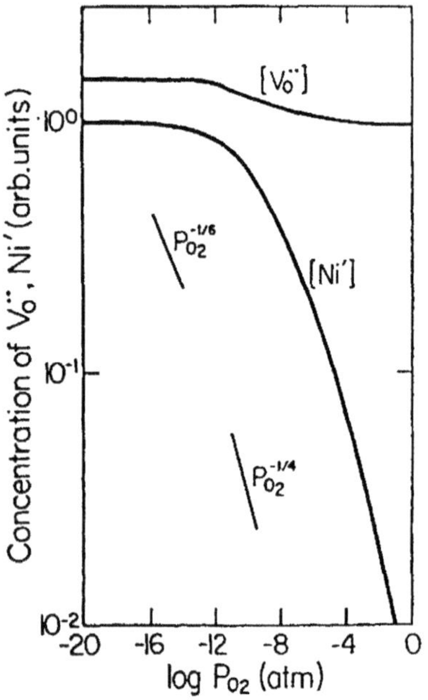

Figure 5. Ratio of $[Ni^{+2}]$ to $[Ni_{tot}]$ and $V_o^{··}$ as a function of P_{O_2} (after Ref. [9]).

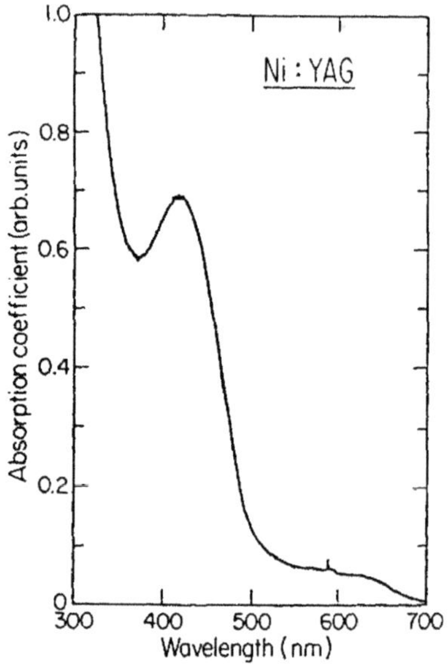

Figure 6. Optical absorption spectrum of oxidized Ni:YAG (after Ref. [10]).

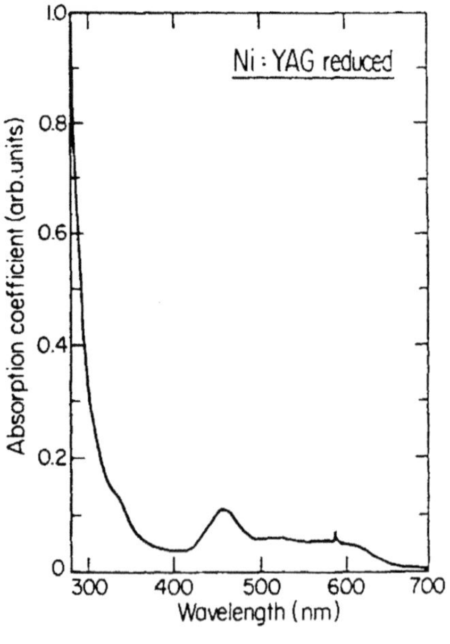

Figure 7. Optical absorption of reduced Ni:YAG (after Ref. [10]).

2.4. Conclusions

We had two goals in this section. First, we wished to show that the valence state of dopant ions in crystals was determined both by the host crystal and by its defects. A systematic method to change these valences (with an eye towards optimization) is possible. Second, electrical conductivity measurements can often be used to measure crystal structure changes, even those unobservable by optical means. While we have detailed this method in one particular crystal, nickel-doped YAG (Ni:YAG), it should be applicable to many cases.

3. Dopant-Dopant Interactions

3.1. Introduction

In our previous section, we considered how the defect structure of the crystal affects the valence of the luminescent dopants. In that analysis, the neutrality of the crystal was maintained by a long distance interaction between the defects and the dopant; we assumed that the dopants still were located in the "ideal" crystal sites with the standard crystal field interactions. Often, a fellow dopant can act as a defect in changing the luminescent property of a second dopant. In this section, we will consider cases in which the dopants are affected by defects and/or other dopants, subtly changing the site symmetry and not so subtly affecting their dynamic spectroscopic behavior. When several "almost similar" sites exist, we will refer to them as multi-sites with a main site and several sub-sites.

3.2. Energy Transfer Model Development

The investigation of multi-sites in crystals has been going on for several years [11]-[18]. However, while the spectra received were being understood, no model existed for analyzing the dynamic time-dependent behavior of the sites, particularly when energy transfer between dopants was possible. We have recently developed such a model; we present it in Appendix C. This model can thus be used to understand multi-site behavior in these crystals. We present this model in this chapter through the examples on actual laser materials.

3.3. Site Selection Spectroscopy

3.3.1. Neodymium-chromium doped YAG (Nd,Cr:YAG)

Since yttrium aluminum garnet (YAG) is perhaps the most famous of the solid-state laser crystals, we will begin by analyzing the spectroscopy of luminescent dopants in this host. In work centered at the Univesity of Lyons, the spectra of chromium and neodymium ions have been studied [11]-[16]. Shown in Fig. 8 is the chromium emission spectrum (the $^2E \rightarrow {}^4A_2$ transition) in Cr:YAG [15].

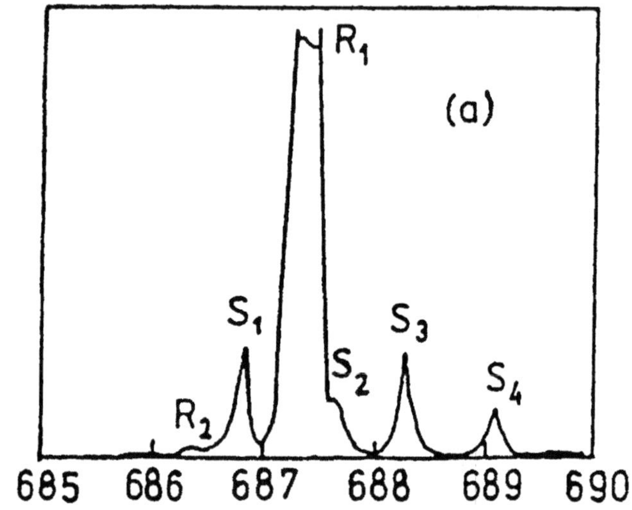

Figure 8. Cr^{+3} emission spectrum of Cr:YAG at 4.2°K under excitation at 532 nm (after Ref. [16]).

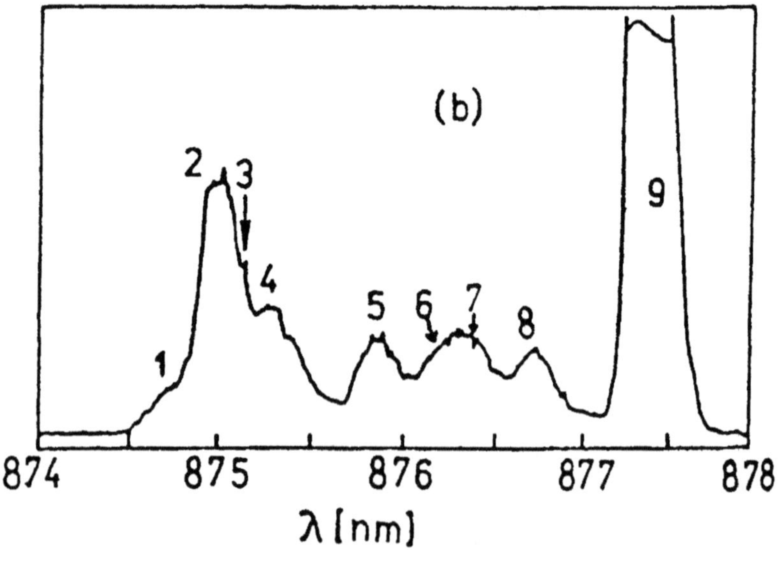

Figure 9. Nd^{+3} emission spectrum of Nd,Cr:YAG at 4.2°K under excitation at 532 nm (after Ref. [16]).

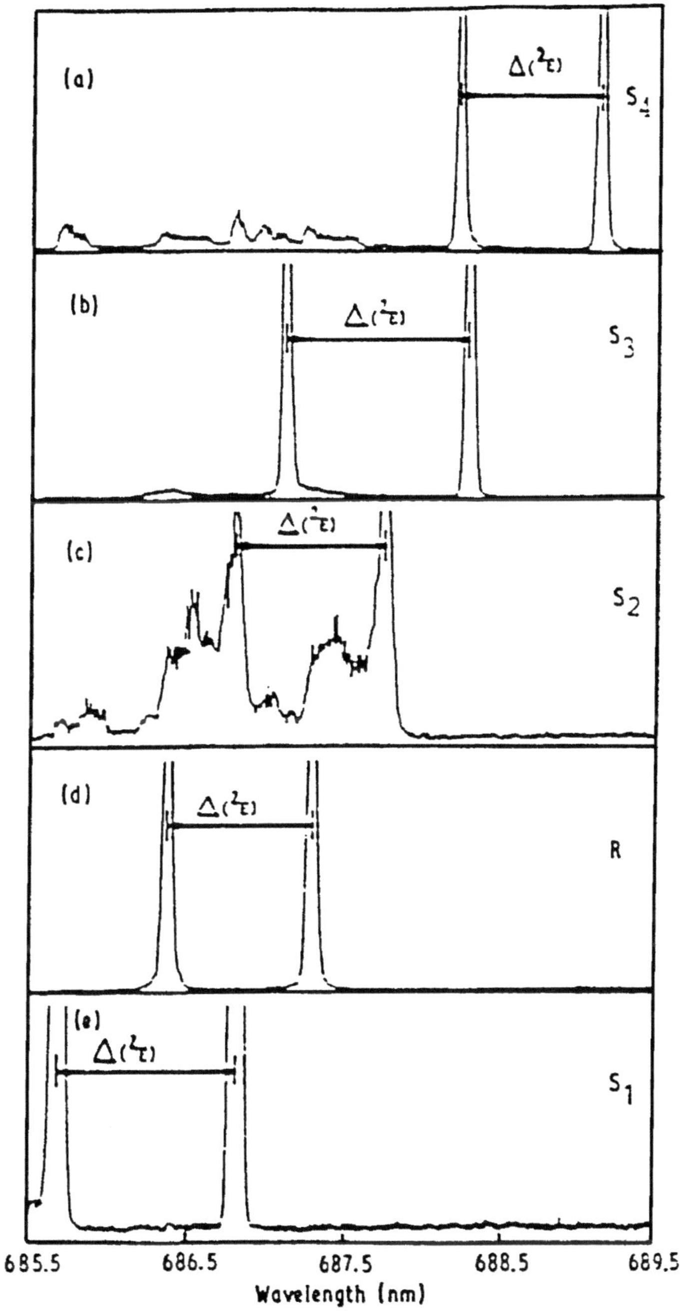

Figure 10. Excitation spectra of Cr:YAG at 4.2°K (the R- and S- lines) (after Ref. [14]).

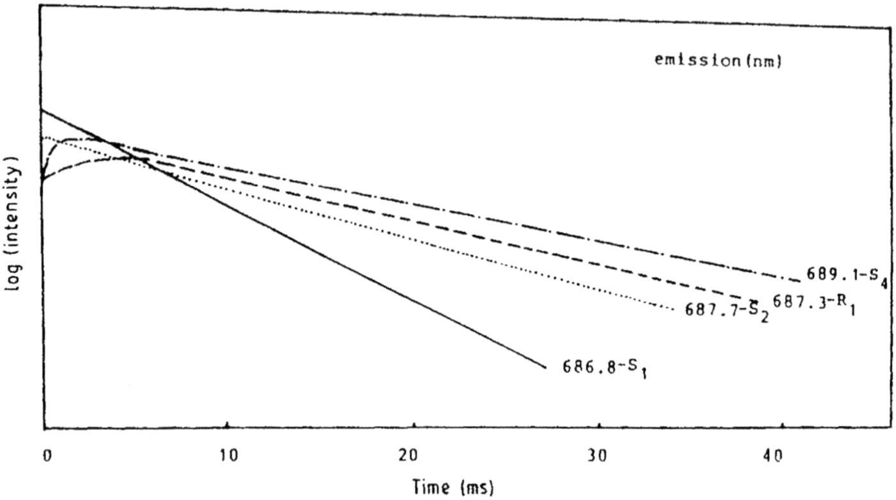

Figure 11. Decay curve of R and S line emissions of Cr:YAG at 4.2°K after excitation at 686.8 nm (the S_1 line) (after Ref. [14]).

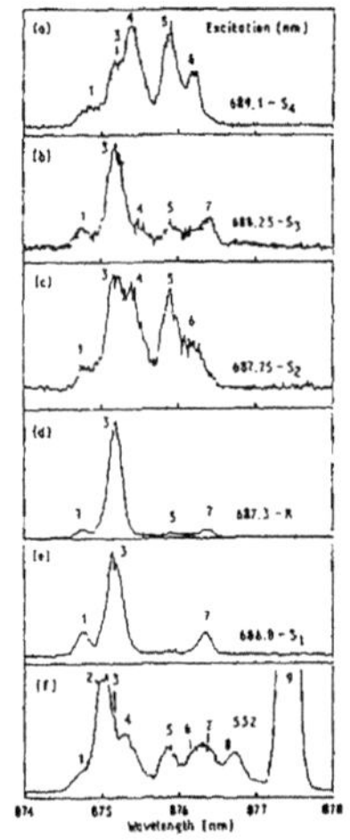

Figure 12. Excitation spectra of Nd^{+3} ions in Cr,Nd:YAG at 4.2°K. after excitation at several Cr^{+3} lines. Cr^{+3} lines excited are indicated on graph (after Ref. [14]).

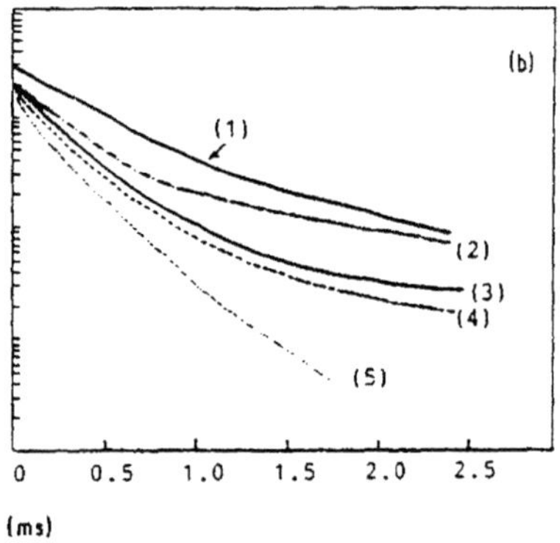

Figure 13. Decay curves of Nd^{+3}ions in CrNd:YAG at 4.2°K under excitation at (1) 686.8, (2) 687.3, (3) 532, (4) 688.25 and 687.75 nm. Emission being observed is at 874.8 nm. (after Ref. [14]).

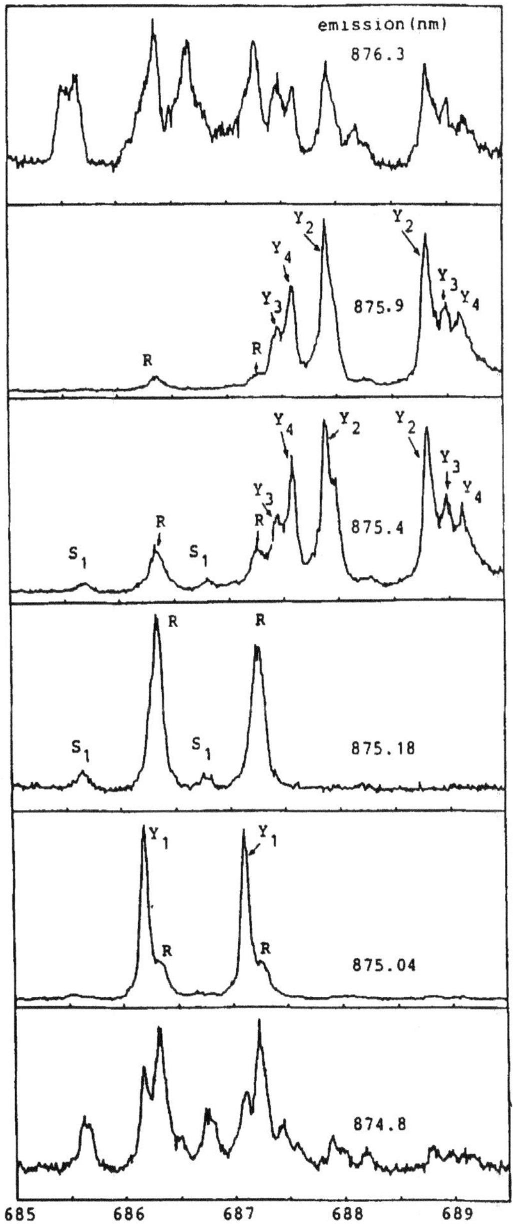

Figure 14. Excitation spectra of Nd^{+3} ions in Cr,Nd:YAG at 4.2°K. Nd^{+3} lines observed are indicated on each graph (after Ref. [14]).

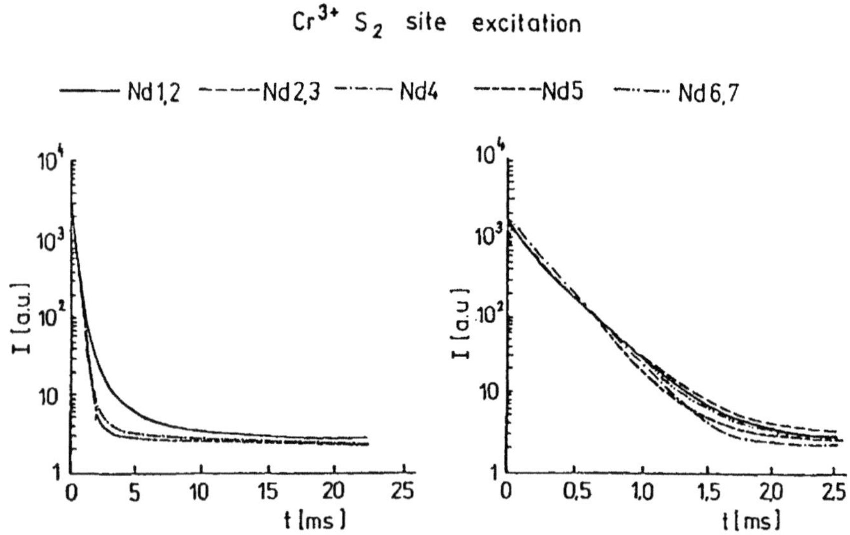

Figure 15. Nd^{+3} decay for some of the Nd^{+3} multisites at 4.2°K after excitation of the Cr^{+3} S_2 site (after Ref. (15)].

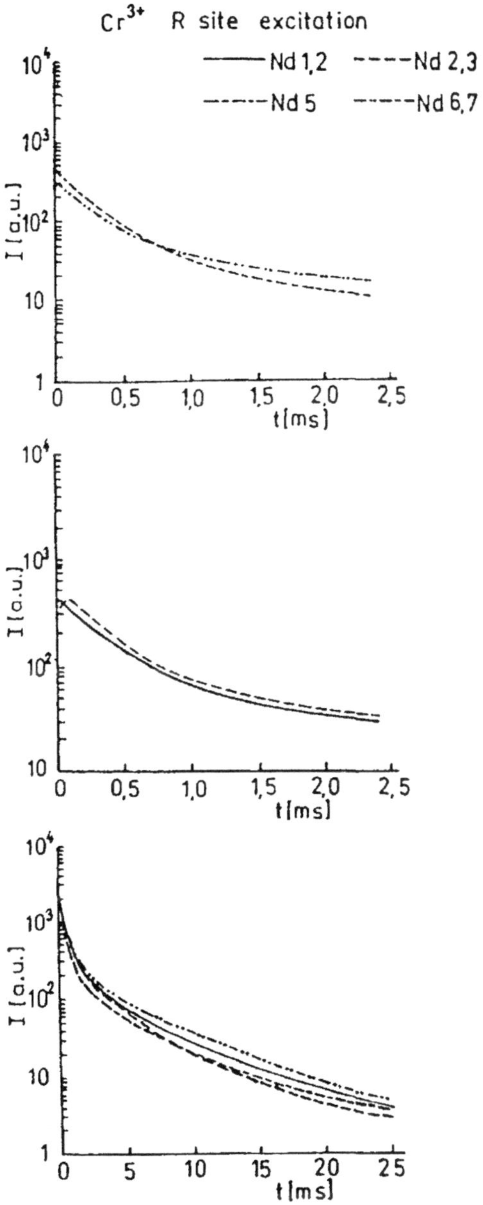

Figure 16. Nd^{+3} decay for some of the Nd^{+3} multisites at 4.2°K after excitation of the Cr^{+3} R site (after Ref. [15]).

The R_1 and R_2 lines are the expected emissions from the main site; the S_1 - S_4 lines represent the emissions from sites which do not have similar symmetries. Nd^{+3} exhibits a similar behavior in neodymium-chromium-doped YAG (Nd,Cr:YAG) with a main site (labeled 9 in Fig. 9) being flanked by eight sub-sites.

The spectroscopic behavior of each site can be quite different. For example, the excitation spectra [14] for each of the chromium sites is shown in Fig. 10; the time decay of these sites is shown in Fig. 11. Several possible reasons have been offered for these chromium multi-sites such as the existence of different chromium isotopes and disorder between the cations of the host lattice; however, no definite cause has been found.

In doubly-doped crystals, the energy transfer properties of the dopants can be critical to the laser or phosphorescent performance. The ions located in multi-sites can show greatly different energy transfer performance. In Fig. 12 are shown the Nd^{+3} emission spectra of the ${}^4F_{3/2} \rightarrow {}^4I_{9/2}$ transition under narrow band excitation of each of the chromium multi-sites [14]. It is clear that radically different Nd^{+3} sites are excited as each different Cr^{+3} site is excited. Fig. 13 shows the different rates of decay of the Nd^{+3} sub-site at 874.8 nm when the various Cr^{+3} sub-sites are pumped.

While in general it is difficult to determine what causes a sub-site to exist, there is a particular case where the origin of the multi-site is possible to determine; this occurs when a sub-site appears upon the doping of the crystal with a second dopant. For example [14], shown in Fig. 14 are the different excitation spectra of several of the Nd^{+3} multi-sites when the Cr^{+3} sites are excited in Nd,Cr:YAG. Of particular interest are the "Y" lines, chromium excitation lines which transfer energy effectively to the neodymium multi-sites. These lines do not appear in Cr:YAG; they are believed to be chromium ions affected by the presence of the neodymium ions. This is a very identifiable population of ions; chromium and neodymium ions located next to one another can be expected to transfer energy rapidly between each other. In Ref. [15], the researchers noted the great differences that existed in the neodymium excited-state temporal response as different chromium sub-sites were narrowly pumped. Fig. 15 shows the temporal response of the different neodymium sites when the S_2 chromium site is pumped. This clearly contrasts to the behavior of the neodymium ion decay when the main R line is pumped (Fig. 16). Additional decays can be found in Ref. [15].

When a standard analysis is performed, none of the curves can be adequately fit. Qualitatively, the fast decay of Fig. 15 is indicative of an almost instantaneous transfer of energy between the neodymium S_2 sub-site and the neodymium ions. Mares at al. [15] postulate that this may indicate an exchange interaction between these chromium ions and the neodymium ions. In our next section, we will consider energy transfer between such pairs. This will represent a new model for the analysis of energy transfer between such multi-sites.

3.3.2. Chromium, Thulium, Holmium-Doped Ytrium Aluminum Garnet (Cr,Tm, Ho:YAG)

3.3.2.1. Introduction:

Studies on Cr,Tm,Ho:YAG have mirrored the studies on Cr,Nd:YAG. In particular, the chromium and thulium ions have been studied and found to exhibit a multi-site structure similar qualitatively to that of chromium and neodymium in Cr,Nd:YAG [17]-[18]. (In particular [17], the chromium multi-site was identical in both Cr,Nd:YAG and Cr,Tm,Ho:YAG.)

In Ref. [17], several Cr^{+3} ion sites were identified. In particular, excitation wavelengths of 687.3 and 687.2 nm are preferentially exciting unpaired and paired chromium ions; the pairing is vis-à-vis a neighboring thulium ion. Similarly, in Ref. [18] the emission of Tm^{+3} at 809 and 806.6 nm are determined from unpaired and paired thulium ions, respectively. These are the wavelengths used for analyzing Cr^{+3} pumping and Tm^{+3} emission.

In Figs. 17 - 18 are shown the decay from Tm^{+3} ions; the experimental procedure is described in Ref. 17. Fig. 17 represents the decay of an unpaired Tm^{+3} ion when the unpaired Cr^{+3} ions are pumped. As we would expect, the standard Förster-Dexter model reasonably well fits the data. Fig. 18 shows the decay of the paired Tm^{+3} ions when the paired Cr^{+3} ions are pumped. In this case, as we would expect, a strong positional correlation (represented by a high value of A in the model of Ref. 18) is measured. The fit to the correlated placement model is shown in Fig. 18 (See Ref. 18 for the actual parameters calculated.)

3.3.3. Europium, Yttium-Doped Ceria

Site selection spectroscopy can also be used to identify the species contributing to ionic conductivity. To explain this method, we will consider as our example yttrium-doped ceria, as analyzed by Yugami et al. [19].

When yttrium ions are inserted into ceria, we would expect the oxygen vacancy concentration, and hence the ionic conductivity, to rise linearly with yttrium concentration. As can be seen in Fig. 19, this is true only till approximately a 3% concentration. After this point, the conduction σ drops rapidly with a rise in the oxygen vacancy mobility activation energy. The obvious question is why.

Site selection spectroscopy can provide an answer. Trace amounts of Eu^{3+} can be inserted into the crystal to be used as a probe. Eu^{3+} ions exhibit sharp lines which are very sensitive to the crystal field. Fig. 20 exhibits this effect. The top curve is the overall emission of Eu^{+3} (the $^5D_0$ - $^7F_1$ transition) between

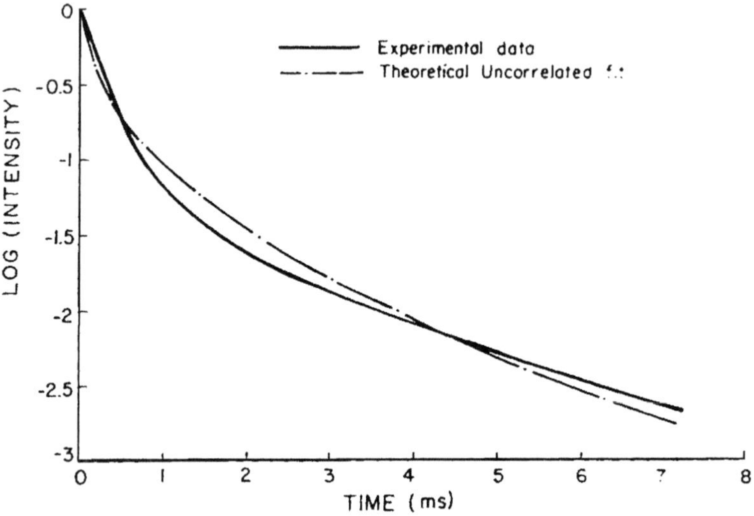

Figure 17. 809 nm (unpaired) Tm^{+3} ionic emission after pumping (unpaired) Cr^{+3} ions at 687.3 nm. (after Ref. [18]).

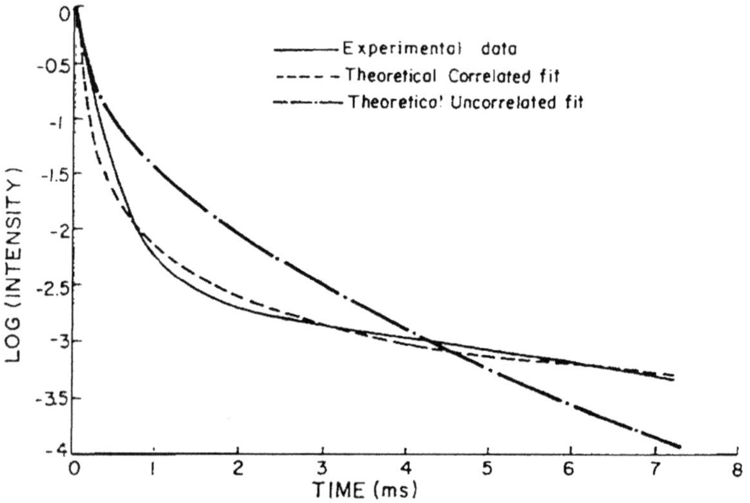

Figure 18. 806.6 nm (paired Tm^{+3} ionic emission after pumping (paired) Cr^{+3} ions, at 687.2 nm (after Ref. ([18]).

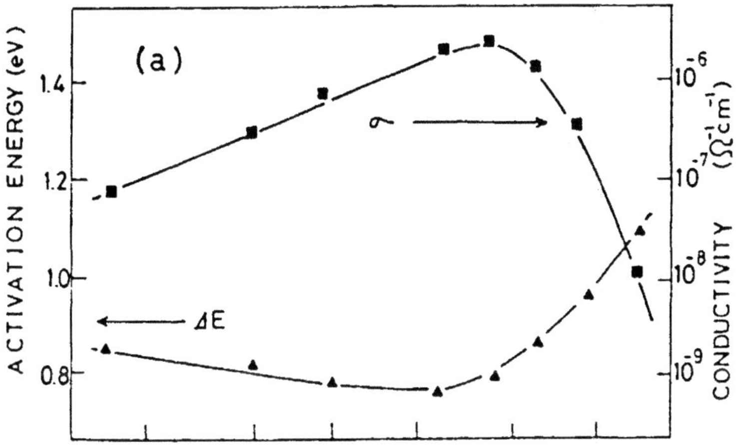

Figure 19. Dependence of the ionic conductivity and activation energy of ceria on Y_2O_3 dopant concentration (after Ref. [19]).

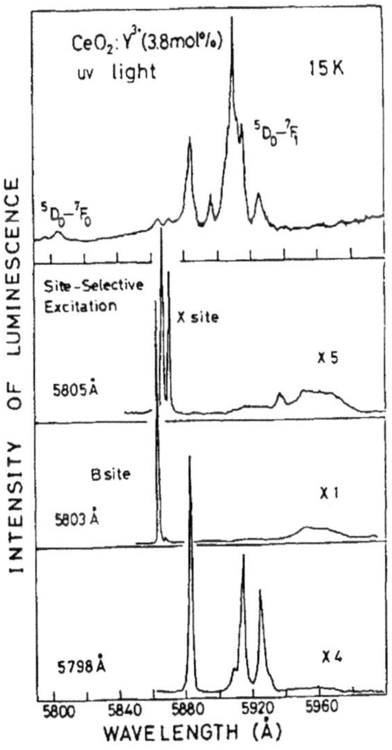

Figure 20. Luminescence of Eu^{+3} ions in 0.4 mol.% yttria-doped ceria sample (after Ref. [19]).

5800 and 5960 Angstroms, when excited with uv light. The lower graphs show the emissions resulting when the crystal is pumped with very narrow lines at 5809 Å, 5806 Å, 5803 Å and 5798 Å. There is clearly a different spectrum for each wavelength. These results are for a 0.4% yttrium doping; qualitatively different results for 3.8% doping can be seen in Fig. 21.

The different emission spectra for the slightly shifted pump indicates that different subpopulation of the europium ions are being excited. In Fig. 22 we see the calculated energy splitting of the various levels.

Yugami now makes the following assignments: for reasons of point-group symmetry, he assigns the C site (and its spectrum) to the perfect Eu^{+3} site, with eight surrounding electrons in Fig. 23. The A_1, A_2 and B sites are Eu^{+3} sites with a single missing nearest neighbor oxygen. The D site is also a Eu^{+3} site with a missing oxygen ion; however, the oxygen vacancy is located in a next-nearest neighbor site. The X site is an oxygen vacancy-europium ion pair which is strongly perturbed by other defects.

We have grouped the A_1, A_2 and B site together as a single oxygen vacancy next to an Eu^{+3} ion. The difference between the sites, according to Yugami, is the closeness of the Y^{+3} dopants to the oxygen vacancy. In order of decreasing distance, we have the A_1, A_2 and B sites.

The importance of this work is as follows: as the concentration of Y^{+3} changes in the crystal, the various types of sites change in relative concentration. In particular, the relative concentrations of the different sites change. This can be seen in Fig. 24. As we would expect, the B site, which has a smaller Y^{+3} to oxygen distance, gains relative to the A sites as the Y^{+3} concentration initially increases.

Now compare these results to the original results for the conductivity shown in Fig. 19. There appears to be a decrease in conductivity activation energy as the B site concentration increases with Y^{+3} concentration. The further decrease of the B site concentration due to the production of X sites, increases dramatically the activation energy. It appears that an optimum distance occurs for the Y^{+3} -oxygen vacancy pair when the movement of the oxygen vacancy is easiest.

The particular material here is not as significant as the method. The site selection spectroscopy has allowed us to pick out dopants not merely with a particular nearest-neighbor configuration, but even with a particular next neighbor dopant. We can relate the formation of such complexes to particular sites. This is indeed a potentially powerful technique for the study of solid-state ionic conductors.

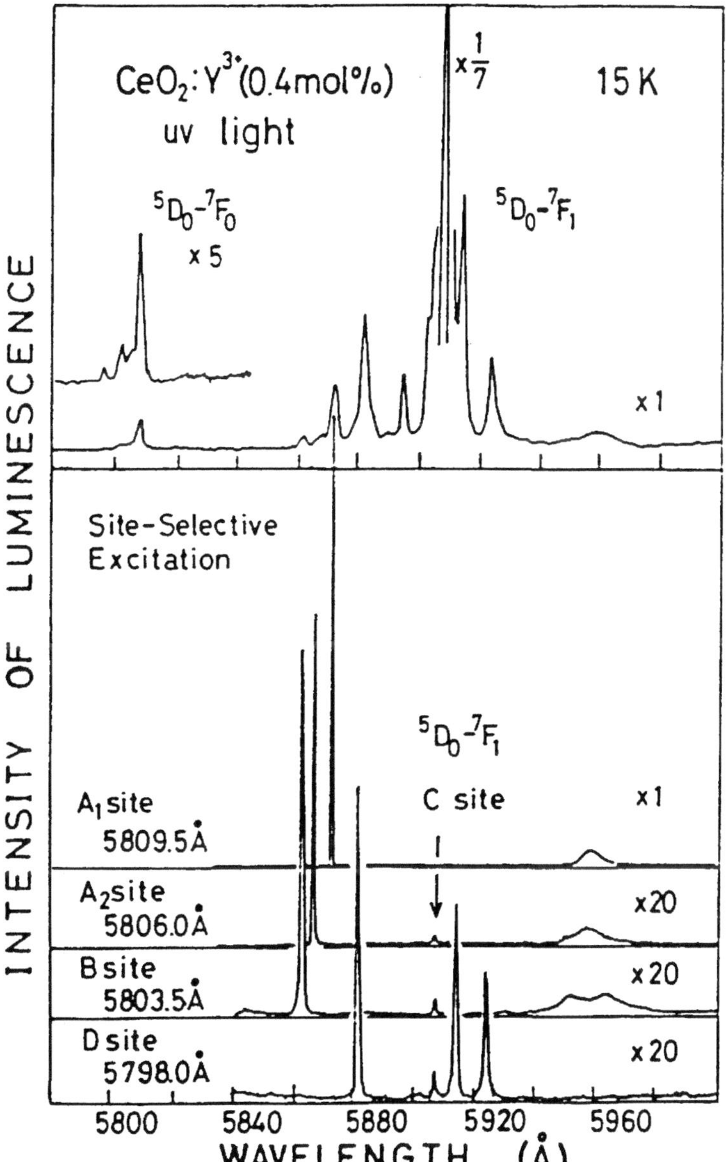

Figure 21. Luminescence of Eu^{+3} ions in five sites in the ceria crystal (after Ref. [19]).

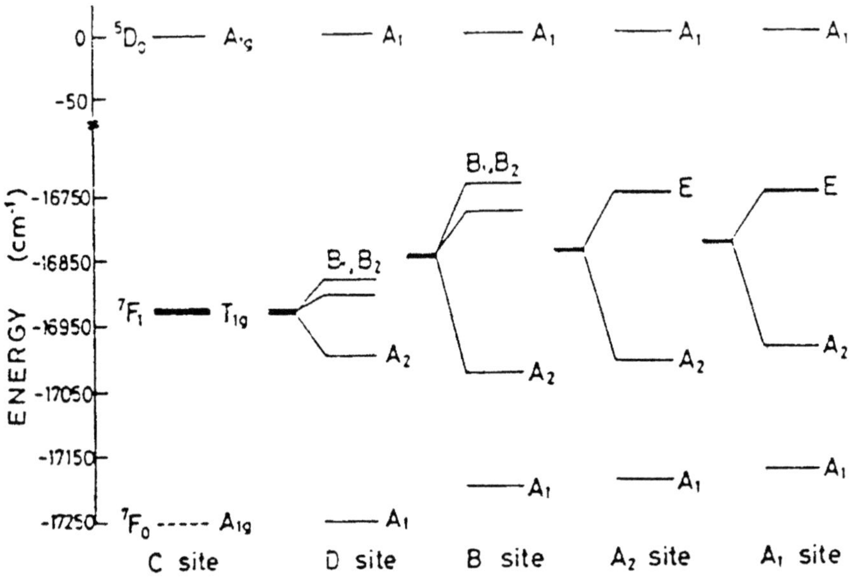

Figure 22. Energy diagrams of Eu^{+3} ions in five sites in the ceria crystal (after Ref. [19]).

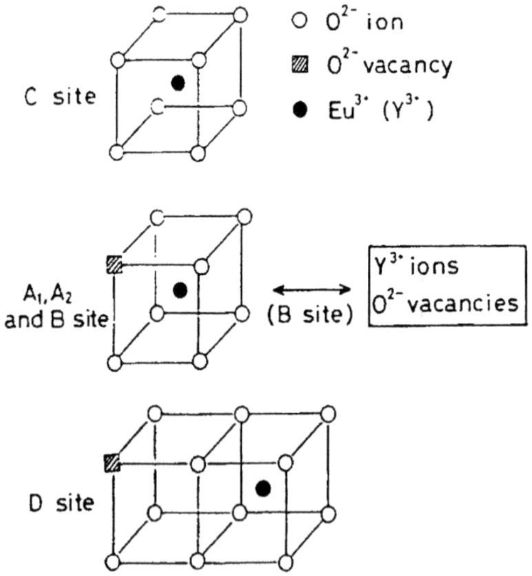

Figure 23. Structure of five Eu^{+3} sites (after Ref. [19]).

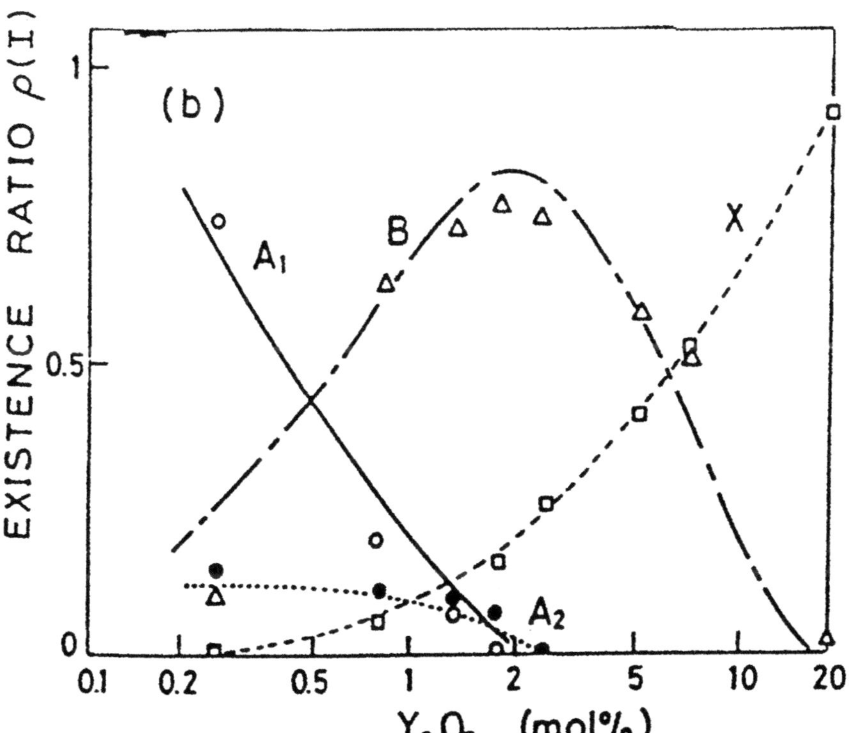

Figure 24. Concentration of the different Eu^{+3} sites as a function of Y_2O_3 concentration (after Ref. [19]).

3.4. Conclusions

We've examined doped YAG and ceria crystals and found that, at liquid nitrogen temperatures, multi-sites exist reflecting subtle changes in the crystal field affecting the dopants. Similar results can be found in neodymium-chromium-doped yttrium aluminum perovskite [20] and other similarly doped garnets [21]-[23].

We've taken the opportunity to present a new model for energy transfer between such multi-sites. We note that such studies may be crucial to understanding laser action in doubly-doped laser materials; a recent paper by Han et al.[24] shows that laser action in gadolinum scandium gallium garnet (GSGG) when pumped with narrow spectral lines may be dependent solely on energy transfer to and from multi-sites. This is an important area for future research.

4. Defect-Dopant Competition

4.1. Introduction

In previous sections, we have examined how the "isolated ion in the crystal matrix" model can break down, both by the long-range effect of defects and by the proximity of other dopants. In this section, we will examine how competition for energy between defects and dopants can strongly influence the luminescence properties of these materials. We will use as examples of the possible proceses that can occur the cathodoluminescent response of yttrium aluminum garnet (YAG) and lanthanum aluminum hexaaluminate (LMA).

4.2. Competition Between Defects and Dopants in Yttrium Aluminum Garnet (YAG)

YAG has been considered as a possible cathodoluinescent source, in particular when doped with cerium (Ce^{+3}) [28], [29]. In a set of papers in 1979, Robbins et al. [30]-[34] have carefully measured the luminescent properties of YAG. Based on his experimental results, Robbins proposed a model explaining the emission of luminescence from the different centers, as well as the process of energy transfer between defect centers of the YAG lattice and the cerium activator ions. We will detail the main results obtained and the models proposed by Robbins and other researchers.

4.2.1. Experimental Results from Cerium-doped YAG (Ce:YAG)

The cathodoluminescence spectra of cerium-doped YAG as a function of temperature is shown in Fig. 25. Also shown in this figure is the absorption peak of cerium. A broad peak is obtained with a strong temperature dependence with a peak of approximately 270 nm at 12°K and 310 nm at room

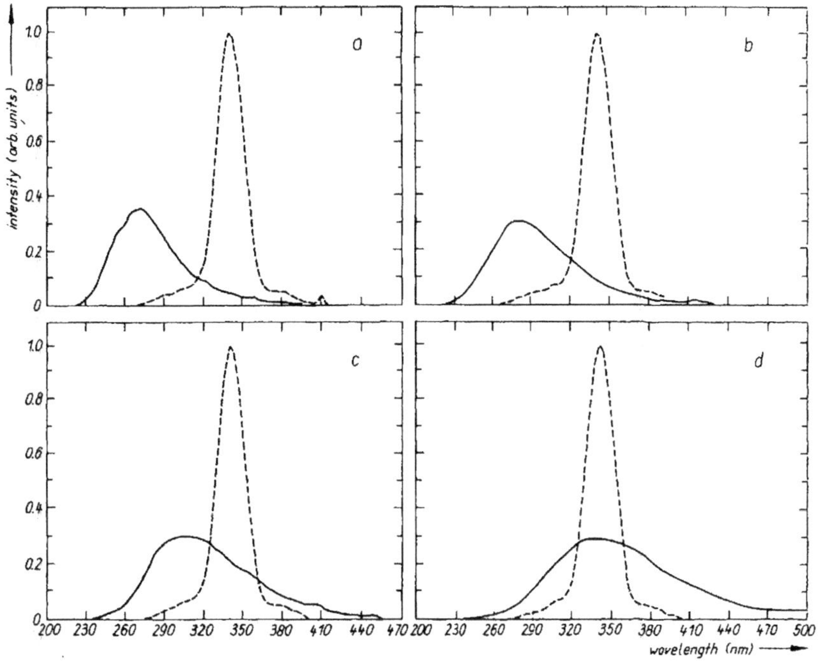

Figure 25. Cerium absorption (– – –) and defect emission (..............) in Ce:YAG for a) 12, b) 98, c(296 and d) 445°K (after Ref. [37]).

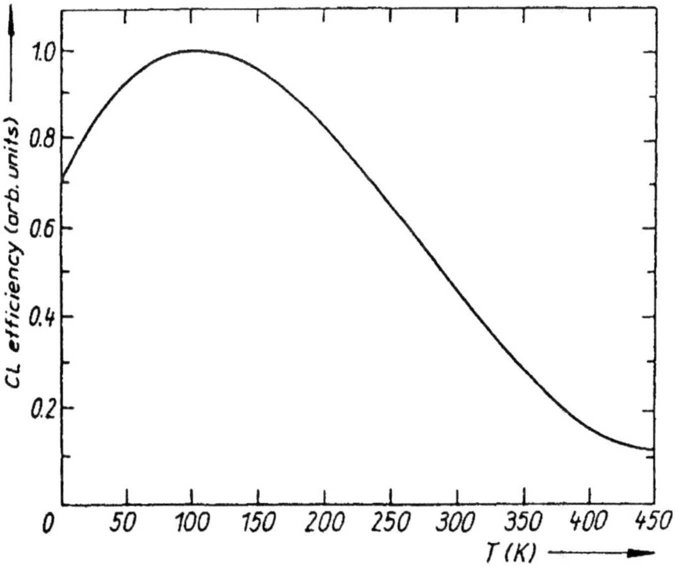

Figure 26. Total cathodoluminescence intensity for cerium ions and defect ions in Ce:YAG (after Ref. [37]).

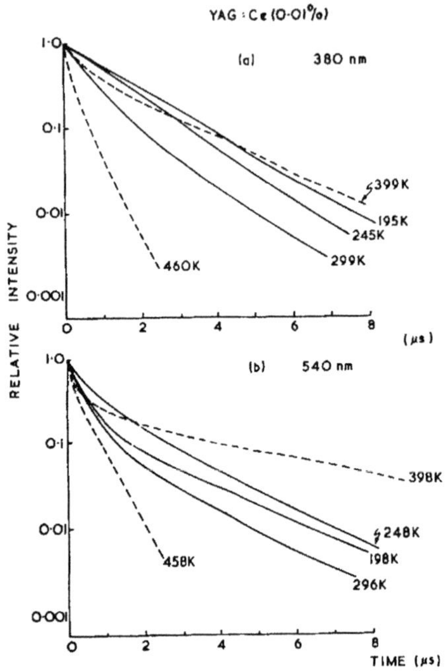

Figure 27. Cerium (540 nm) and defect (380 nm) cathodoluminescent emission decay in Ce:YAG at several temperatures (after Ref. [31]).

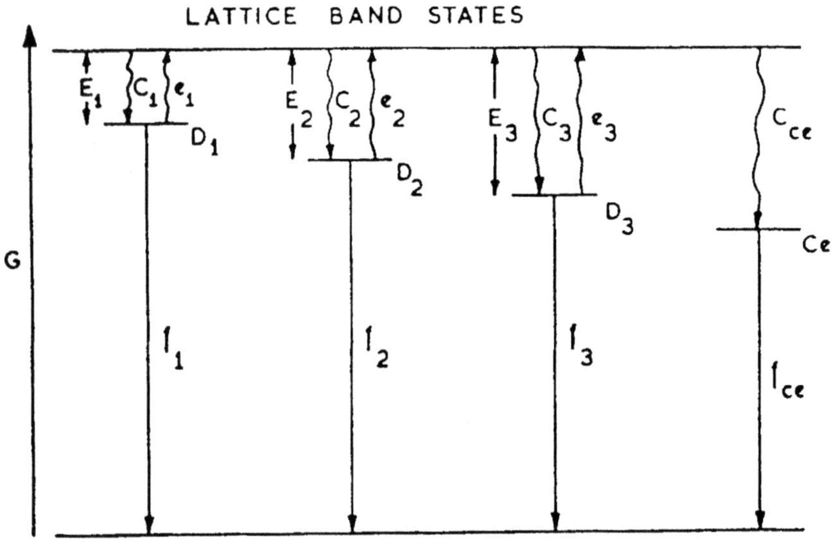

Figure 28. Defect model for Ce:YAG with three defect centers (D) and cerium (Ce) emission. Pump (G) is also shown. (after Ref. [31]).

temperature. When cerium is present, two peaks are observed: the defect at its characteristic wavelength and the cerium emission at 550 nm.

The ratio between the defect and cerium emissions is greatly dependent on the temperature. As the temperature rises, the defect intensity decreases while the cerium intensity increases (Fig. 26). At all temperatures, the theoretical cerium emission time constant (70 nsecs) is not observed; instead the cerium emission tends to follow the time constant of the defects (> 1 μsec) (Fig. 27). Moreover, as the cerium concentration in the crystal increases, the defect emission decreases.

Annealing the Ce:YAG in oxidizing atmospheres in high temperatures has a profound effect on the cathodoluminescent emission. When oxidized, the cathodoluminescent emission of both the defect and the cerium is decreased; moreover, the cerium exhibits its natural 70 nsecs time constant.

The connection of the defects in the crystal and the cerium ions is clear. The defects appear to compete with the cerium ions for energy; conditions yielding low cerium emission (low temperatures and low cerium emissions) also cause high defect intensity. On the other hand, the cerium ions time constant mirrors that of the defects. Clearly, an interactive model is necessary.

4.2.2. Models for Cerium-doped YAG Cathodolumeinescence

4.2.2.1. Robbins Model

To explain the experimental results, Robbins et al. [31] have proposed the model seen in Fig. 28. Three defect centers are proposed each with separate thermal characteristics. Almost all the excited carriers from the electron pumping arrive at the defects; these carriers can either be deactivated with resulting defect emission or rethermalized to the conduction band. Energy from the conduction band can also be trapped at the cerium ion and emitted as cerium emission at 550 nm.

The experimental data can now be explained. The greater the cerium concentration, the greater the competition for energy with the defect, and hence the decrease in the defect intensity. At low temperatures, energy trapped at defects does not rethermalize, and hence is emitted by the defect rather than the cerium ions. At higher temperatures, the energy is available at the cerium ions. The shift of the defect emission as the temperature rises is due to three different defects rethermalizing at different temperatures and emitting at different wavelengths.

To explain the decrease of both cerium and defect emission for oxidized crystals, Robbins et al. [31] proposed an unidentified shunt pathway which increases with oxidizing. This shunt both drains energy from the crystal and prevents "communication" between defects and dopants. The actual nature of the luminescent dopant itself has been associated by Robbins et al. [35] as cation vacancies or disorders.

4.2.2.2. Other models

It is interesting to note that many of the assumptions of the Robbins model have been questioned. Rotman et al. [36] have noted that if the defect were oxygen vacancies rather than cation irregularities, then the effect of oxygen annealing can be understood. The annealing at high temperatures eliminates the defect, lowering the defect emission. The entire shunt pathway can be eliminated if we assume that the cerium receives its energy directly from the defect through Förster-Dexter transfer rather than through the conduction band. We have shown [37] that the temperature dependence of the cerium emission can be explained by considering the shift of the defect emission in wavelength as a function of temperature. The shift in defect emission with temperature is alternatively explainable either by the Robbins' model of three defect centers or several different energy levels of an electron trapped at an oxygen vacancy [38].

We will not attempt to actually choose the appropriate model for the cathodoluminescent emission from Ce:YAG. There are two points to be learned from our example of Ce:YAG emission: 1. defects in crystals can have a major effect on the desired dopant emission (time constant, temperature dependence, reaction to oxygen annealing), in particular under cathodoluminescence, and 2. the interpretation of such data can be very tricky.

4.3. Competition Between Defects and Dopants in Lanthanum Magnesium Hexaaluminate (LMA)

LMA has been considered as a potential phosphor and cathodoluminescent source [39]-[44] when doped with europium (Eu^{+2}) and manganese (Mn^{+2}) (Eu,Mn:LMA). Based on our previous analysis on Ce:YAG, we performed similar cathodoluminescent measurements on Eu,Mn:LMA. Once again, a complicated set of data emerged, with a similar set of possibilities for modelling.

4.3.1. Experimental Results from Europium- and Manganese-Doped LMA [44]

Fig. 30 shows the cathodoluminescent spectrum from 0.1% europium-doped as-grown LMA. (All crystals had a 1% concentration of Mn^{+2}.) At liquid nitrogen temperature, three peaks are observed. The manganese in tetrahedral sites (replacing the magnesium ion) emits at approximately 520 nm. The europium ion emits at approximately 440 nm, while an unknown defect emits at 390 nm. (As expected, only the defect and manganese peaks appear in the 0% europium-doped as-grown crystal.)

At room temperature, the spectrum considerably changes. The europium and manganese peaks increase considerably over the peaks at liquid nitrogen temperatures while the defect emission is almost eliminated. At higher

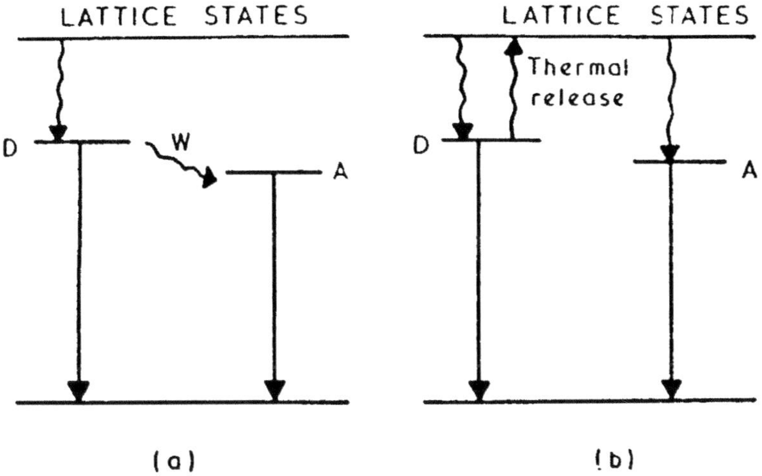

Figure 29. Comparison of two possible pathways for energy transfer between the defect (D) and the cerium (A), either through Förster-Dexter energy transfer (left) or thermalization through the conduction band (right) (after Ref. [30]).

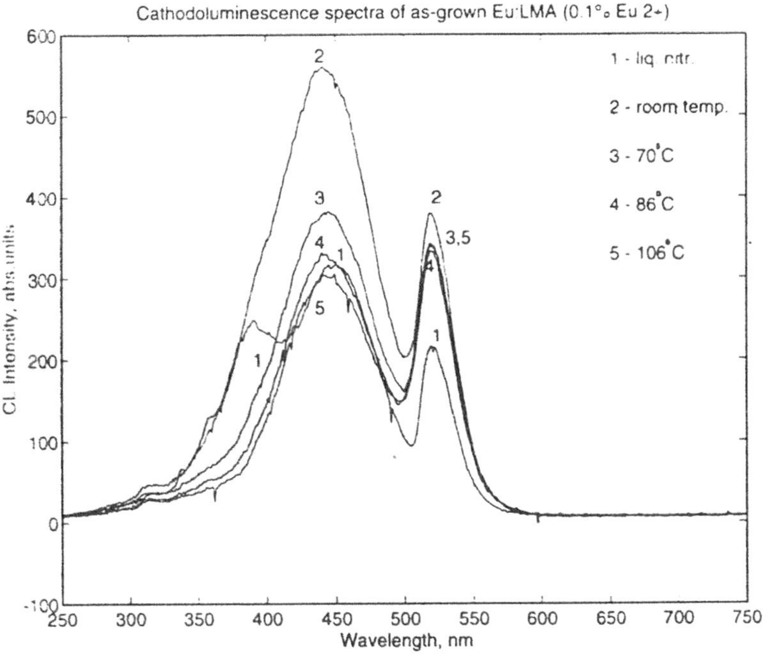

Figure 30. Cathodoluminescent spectra of 10% Eu:LMA at various temperatures (after Ref. [44]).

temperatures, the europium peak (and manganese peak, to a lesser degree) decrease (Fig. 31).

Fig. 32 shows the emission for the 10% Eu:LMA crystal. As the europium concentration increases, the ratio of the europium peak to the manganese peak increases. The defect emission is so small as to be almost insignificant. As in the 0.1% Eu:LMA crystal, both the europium and manganese peaks rise to a maximum and then decrease as the temperature increases.Oxidation had a major effect on the emission spectrum. The defect emission is considerably decreased relative to the europium and manganese peaks for the oxidized 0.1% Eu:LMA crystal (Fig. 32). In addition, the ratio of the europium to manganese emission decreases. The temperature dependence is similar to that described for the as-grown crystals.

4.3.2. Models for the Cathodolumiunescence of Eu,Mn:LMA

Similar to Ce:YAG, a multiple center model is necessary to understand the luminescence. One possibility can be seen in Fig. 33. Europium and manganese compete for energy from the conduction band. An increase in europium ions compared to the manganese concentration increases the ratio of the europium to manganese defect. Previous photoluminescence measurements have shown that energy does not efficiently pass between Eu^{+2} and Mn^{+2} in LMA [42], [43].

The role of the defect emission is still difficult to determine. Oxidation lowers both the defect and europium peaks. Raising the temperature in all the crystals eliminates the defect emission, and initially raises the europium and manganese emissions (although the emissions decrease with temperature at the highest temperatrues - this decrease appears to be the standard non-radiative relaxation that occurs at high temperatures for many rare-earth ions).

We note that both the models shown in Fig. 33 or in Fig. 34 can explain the data. The defect emission may decrease and the europium emission may increase with increasing temperature either due to the thermalization of the defect emission (Fig. 33) or increased energy due to phonon assistance (Fig. 34).

The lowering of the defect emission with oxygen annealing could be due to (1) an alternative deactivation pathway caused by oxidation (Fig. 33) or (2) the elimination of the defect itself under oxygen annealing (e.g. oxygen vacancies). Similarly, the lowering of the europium emission with oxygen annealing could be due to (1) an alternative deactivation pathway, (2) the elimination of the defect which feeds the europium or (3) the conversion of Eu^{+2} to Eu^{+3}.

Much of the speculation concerning the interaction of the defect with the Eu^{2+} ion can be resolved by measuring the time-dependence of the defect, Eu^{2+} and Mn^{2+} emissions to a pulsed excitation. Further measurements are necessary,

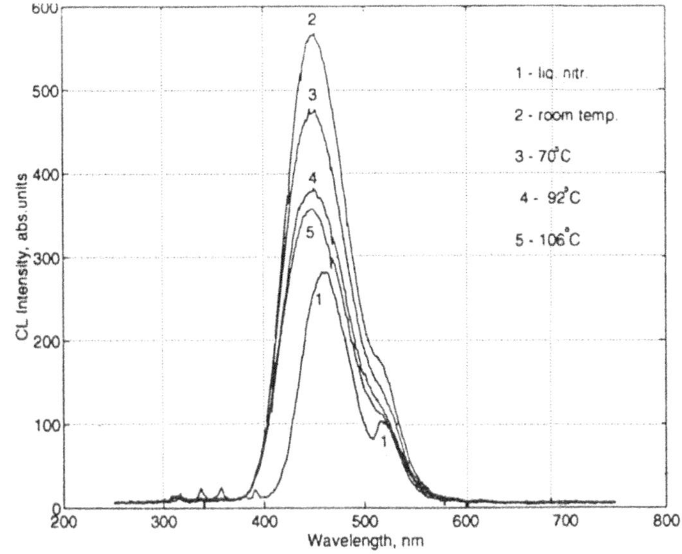

Figure 31. Cathodoluminescent spectra of 10% Eu:LMA at various temperatures (after Ref. [44]).

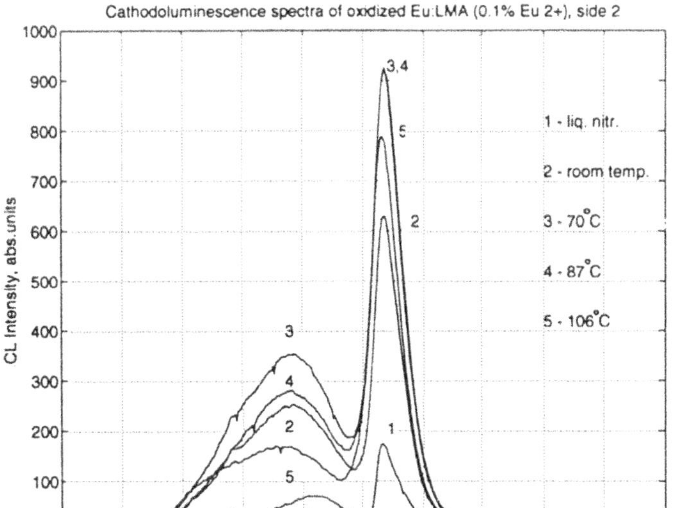

Figure 32. Cathodoluminescent spectra of oxygen-annealed Eu:LMA at various temperatures (after Ref. [44]).

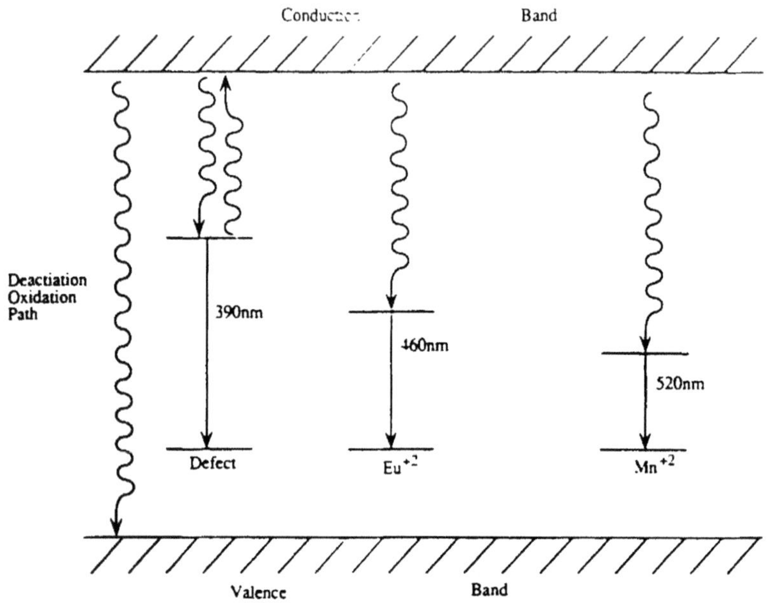

Figure 33. First possible model for cathodoluminecence in LMA (after Ref. [44]).

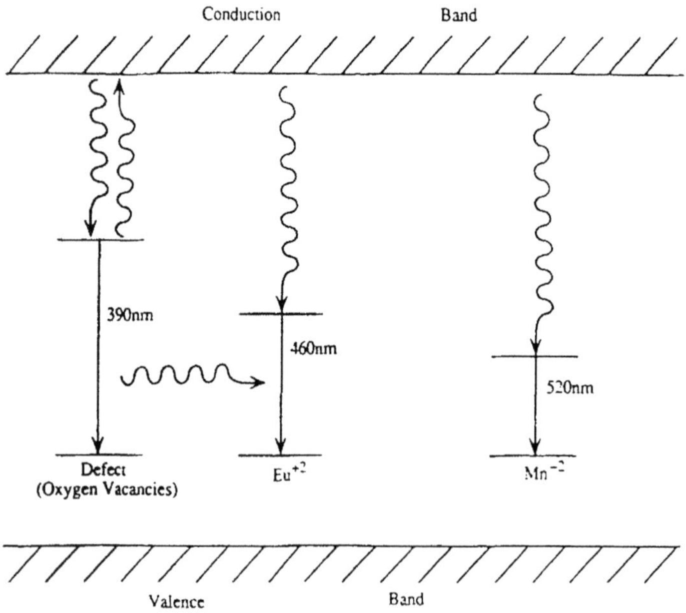

Figure 34. Second possible model for cathodoluminescence in LMA (after Ref. [44]).

4.4. Conclusions

The purpose of this analysis has been to show how defects and other dopants can affect the luminescent properties of potential phosphors. The time respone, intensity, and temperature dependence of these materials can be greatly affected by the defects and dopant interactions. Modelling these materials can be quite tricky, due to the need to fully understand these interactions.

5. Summary

Defects can have profound effects on the luminescent properties of the potential laser materials described in this chapter. The oxidation state of the dopants is controlled by the stochiometry of the crystal; we have outlined how to understand theoretically and measure experimentally these effects, using a combination of electrical conductivity and optical spectroscopy measurements. The defects can change the dynamic behavior and energy transfer properties of the dopants; site-selection spectroscopy can help to clarify the behavior of the constituents. Defects can compete with dopants for energy; cathodoluminescent measurements can help us propose models for these phenomena.

We have only touched the surface of the various experimental techniques; we have by design chosen those with which we have personal experience. Hopefully, these examples will illustrate both the need for and the advantage of careful analysis of defects in the crystal structure.

Acknowledgments

This work has covered fifteen years of work and could not have been written without the collaboration of many distinguished scientists. In particular, I'd like to thank Profs. Harry Tuller and Cardinal Warde of the Massachusetts Institute of Technology, Prof. Johan Prins of the Univ. of Witwatersrand, Dr. Frank Hartmann of the Brookhaven National Laboratory, Dr. Aryeh Weiss of the Jerusalem College of Technology, Dr. Jiri Mares of the Institute of Physics of the Czech National Academy of Sciences, Prof. Renata Reisfeld of the Hebrew University, Dr. Y. Kalisky of the Nuclear Research Centre in the Negev and Prof. Georges Boulon of the Univ. of Lyon. This work was partially supported by the Israel National Council for Research and Development, Jerusalem, Israel and by internal grants of Ben-Gurion University of the Negev, Beer-Sheva, Israel.

Appendix A: Defect Chemistry for an Oxide Doped with Isovalent Dopants

Once one knows qualitatively the main defect, one must consider quantitatively how it is determined by the growth and annealing conditions. Assuming the oxygen vacancy model, three major quasi-chemical reactions need to be considered.

In this appendix, we wish to consider the mass-action relations which govern the crystal chemistry. First, we have the generation and recombination of electrons e' (in the conduction band) and holes h (in the valence band). Using Kroger-Vink notation [8], this can be expressed as

$$O \rightarrow e' + h^{\bullet}, \tag{A1a}$$

with the mass-action relation yielding an np product of the form ($n \equiv$ concentration of electrons, $p \equiv$ concentration of holes)

$$np = K_i(T) = K_i^0 \exp(-E_g / kT), \tag{A1b}$$

where E_g is the thermal energy gap of the crystal, k is Boltzmann's constant and T is the temperature in degrees Kelvin.

Second, we consider the production of oxygen vacancies $V_0^{\cdot\cdot}$ by the removal of oxygen ions from the crystal into the gas phase.

$$O_0 \leftrightarrow V_0^{\cdot\cdot} + 2e' + \tfrac{1}{2}O_2. \tag{A2a}$$

The corresponding mass-action relation is given by

$$n^2[V_0^{\cdot\cdot}] = P_{O_2}^{-1/2} K_1[T] = K_1^0 P_{O_2}^{-1/2} \exp(-E_1 / kT), \tag{A2b}$$

in which P_{O_2} is the partial pressure of oxygen and E_1 the energy characterizing this redox process.

Finally, we consider the requirement of charge neutrality. This is described by the relation

$$n + [A'] = p + 2[V_0^{\cdot\cdot}], \tag{A3}$$

in which we have assumed impurity domination by acceptors $[A']$. As before, the predominant lattice defect is assumed to be a doubly-ionized oxygen vacancy.

Following conventional practice we simplify the analysis by examining the solutions for the unknown defect species under conditions for which the neutrality equation can be simplified to only two terms.

Under highly reducing conditions, large concentrations of $V_0^{\cdot\cdot}$ are generated which are balanced by the electrons released into the conduction band by these donors and so Eq. (A3) is approximated by

$$n = 2[V_0^{\cdot\cdot}] \; . \qquad \text{(A4a)}$$

As the degree of reduction is decreased, a condition of nearly complete compensation is ultimately achieved between the $V_0^{\cdot\cdot}$ and the acceptor impurities A'. This condition is described by

$$[A'] = 2[V_0^{\cdot\cdot}] \, . \qquad \text{(A4b)}$$

Finally, under oxidizing conditions, Eq. (A3) reverts to

$$[A'] = p \, . \qquad \text{(A4c)}$$

The resulting concentrations can be seen in Table 1.

Appendix B: Defect Chemistry for an Oxide Doped with Aliovalent Dopants

Eqs. A1 - A2 still describe the basic crystal chemistry of the host matrix. Ionization of Ni^{2+} to Ni^{3+} by the ionization of an electron to the conduction band is represented by

$$\mathrm{Ni}_{\mathrm{Al}}' \rightarrow \mathrm{Ni}_{\mathrm{Al}}^{x} + \mathrm{e}' . \qquad \text{(B1)}$$

where $\mathrm{Ni}_{\mathrm{Al}}'$ and $\mathrm{Ni}_{\mathrm{Al}}^{x}$ correspond to Ni^{+2} and Ni^{+3}, respectively. This yields a mass action relation of the form

$$[\mathrm{Ni}_{Al}^{x}]n / [\mathrm{Ni}_{Al}'] = K_2^0 \exp(-E_2 / kT), \qquad \text{(B2)}$$

where E_2 corresponds to the ionization energy.

The condition that requires that the crystal remain electrically neutral is satisfied by

$$p+2[2V_0^{\cdot\cdot}]=n+[A']+[\mathrm{Ni}'_{Al}], \tag{B3}$$

If we define the total concentration of nickel in the crystal to be $[\mathrm{Ni}_{tot}]$, then

$$[\mathrm{Ni}'_{\mathrm{Al}}]+[\mathrm{Ni}^{x}_{\mathrm{Al}}]=[\mathrm{Ni}_{\mathrm{tot}}]. \tag{B4}$$

With the five Eqs. (A1b), (A2b), (B2 - B4), it is now possible, in principle, to solve for the given unknown defect concentrations $[V_{\mathrm{O}}^{\cdot\cdot}]$, $[\mathrm{Ni}'_{\mathrm{Al}}]$, $[\mathrm{Ni}^{x}_{\mathrm{Al}}]$, n, and p as a function of temperature, P_{O_2}, and the appropriate equilibrium constants. However, it is often more convenient to solve such problems piecewise by simplifying expressions, such as Eqs. (B3) and (B4), to include only the term on each side of the equality sign which predominates under specified conditions of temperature and amtosphere.

Under highly oxidizing environments one would expect $[V_{\mathrm{O}}^{\cdot\cdot}]$, n and $\mathrm{Ni}'_{\mathrm{Al}}$ to be depressed relative to p and $[A']$ (see Eqs. A3) and (B2)). Consequently, Eqs. (B3) and (B4) could be simplified to read

$$p=[A'] \tag{B5}$$

and

$$[\mathrm{Ni}^{x}_{\mathrm{Al}}]=[\mathrm{Ni}_{\mathrm{tot}}], \tag{B6}$$

respectively. We identify this defect regime as region I. Similarly, one can demonstrate [9] in a straightforward way that the following defect regimes follow successively with decreasing P_{O_2}:

$$\text{region II} \quad 2[\mathrm{V}_{\mathrm{O}}^{\cdot\cdot}]=[\mathrm{A}'], \quad [\mathrm{Ni}^{x}_{\mathrm{Al}}]=[\mathrm{Ni}_{\mathrm{tot}}]; \tag{B7}$$

$$\text{region III} \quad 2[\mathrm{V}_{\mathrm{O}}^{\cdot\cdot}]=[\mathrm{Ni}'_{\mathrm{Al}}], \quad [\mathrm{Ni}^{x}_{\mathrm{Al}}]=[\mathrm{Ni}_{\mathrm{tot}}]; \tag{B8}$$

$$\text{region IV} \quad 2[\mathrm{V}_{\mathrm{O}}^{\cdot\cdot}]=[\mathrm{Ni}'_{\mathrm{Al}}], \quad [\mathrm{Ni}'_{\mathrm{Al}}]=[\mathrm{Ni}_{\mathrm{tot}}]; \tag{B9}$$

$$\text{region V} \quad 2[\mathrm{V}_{\mathrm{O}}^{\cdot\cdot}]=\mathrm{n}, \quad [\mathrm{Ni}'_{\mathrm{Al}}]=[\mathrm{Ni}_{\mathrm{tot}}]. \tag{B10}$$

By substituting the appropriate simplified neutrality and nickel conservation relations into Eqs. (A1b) (A2b) and (B2), one can easily solve for each of the five defect species. The solutions for each region are summarized in Table 2 and are shown plotted in Fig. 4. It is immediately obvious that the P_{O_2} dependence of each constituent is different in each defect regime. Note, in particular, that the oxygen vacancy density, $[V_O^{\cdot\cdot}]$, exhibits two P_{O_2}-independent plateaus in regions II and IV which are separated by region III in which $[V_O^{\cdot\cdot}] \propto P_{O_2}^{-1/6}$.

The actual widths of the defect regimes depend upon the concentrations of acceptors in the crystal and the relative magnitudes of the equilibrium constants. In particular, in region III we have implicitly assumed that $[\mathrm{Ni}'_{\mathrm{Al}}]$ becomes greater than $[A']$ at reduced P_{O_2}'s. However, if $[\mathrm{Ni}_{\mathrm{tot}}] \cong [A']$, a more complicated situation, as we believe is the case with our crystal, ensues. Instead of assuming that $[V_O^{\cdot\cdot}]$ is fixed by one type of acceptor or the other, we must consider the situation which occurs at the transition between regions II and III, i.e.,

$$2[\mathrm{V_O^{\cdot\cdot}}] = [\mathrm{A}'] + [\mathrm{Ni}'_{\mathrm{Al}}], \qquad \text{(B11)}$$

where

$$[\mathrm{Ni}^{\times}_{\mathrm{Al}}] \cong [\mathrm{Ni}_{\mathrm{tot}}]. \qquad \text{(B12)}$$

A cubic equation of the form

$$x^2\left(\frac{[A']}{[\mathrm{Ni}_{\mathrm{tot}}]} + x\right) = C(1-x)^2 \times \exp[-(E_1 - 2E_2)/kT] P_{\mathrm{O}_2}^{-1/2}, \qquad \text{(B13)}$$

in which

$$\mathrm{x} \equiv [\mathrm{Ni}'_{\mathrm{Al}}]/[\mathrm{Ni}_{\mathrm{tot}}], \qquad \text{(B14)}$$

and

$$C = K_1^0 / 2(K_2^0)^2 \qquad \text{(B15)}$$

is obtained when Eqs. (A1b), (B2), (B11) and (B12) are solved simultaneously. The values of K_1^0, K_2^0, E_1, and E_2 determine the P_{O_2} range at a given temperature at which the nickel makes the transition from trivalence to divalance. A plot of x vs P_{O_2} at constant $T, [A']/[\mathrm{Ni}_{\mathrm{tot}}] = 2$ (an assignment explained later below), and a judicious choice for the constant C $\{C \exp[-(E_1 - 2E_2)/kT] = 0.001$ are shown in Fig. 5. For this value of C, the nickel ions convert from trivalence to divalence at partial pressures of oxygen readily attainable under laboratory conditions. It can be shown that x has a $P_{O_2}^{-1/4}$ dependence when $x << 1$; the activation energy in this regime is $(E_1 - 2E_2)/2$. As x becomes larger and begins to saturate, the P_{O_2} dependence lessens, and the activation energy is no longer a constant over a range of temperatures at a given P_{O_2}. Also shown in Fig. 5 is a plot of $[V_O^{\cdot\cdot}]$ as a function of P_{O_2} derived from Eq. B11.

Appendix C: Energy Transfer Modeling

We will present a general theory for energy transfer between donors and acceptors, first when randomly scattered through the crystal followed by a model for correlated donors and acceptors.

C.1. Standard Energy Transfer Model

Although the mathematics of energy transfer looks complicated, the fundamental underlying physics is easily expressed in several basic equations [18], [25]-[27]. Consider the case where a single excited donor at time $t = 0$ is located in space surrounded by N_A acceptors, where the k th acceptor is located at distance r_k from the donor. In this case, the probability as a function of time $\rho(t)$ that the donor is still excited is [27]

$$\rho(t) = \exp(-t / \tau_0^D) \prod_{k=1}^{N_A} \exp[-tn(r_k)], \tag{C1}$$

where τ_0^D is the spontaneous decay time of the donors and $n(r_k)$ is the inverse of the average time for energy transfer between a donor and an acceptor separated by distance r_k.

The situation described here essentially allows the donor to decay in two ways. It can decay via spontaneous emission (given by the $\exp(-t/\tau_0^D)$ term) and it can decay while transferring its energy to one of the acceptors (the $\prod_{k=1}^{N_A}[-tn(r_k)]$ term). Unfortunately, to expand the model from a single donor to many donors, we must face the fact that each donor faces a different set of acceptor positions. Moreover, in general, we do not know the positions r_k of the N_A acceptors. If we are willing to consider the case where the position of one acceptor does not affect the position of the other acceptors, then further progress is possible.

Consider a single donor and a single acceptor whose radial distance is unknown but associated with the probability distribution $w(r)$. In that case the probability $\rho(t)$ that a single excited donor at time 0 will remain in that state by time t is given by [27].

$$\rho(t) = \exp(-t / \tau_0^D) \int_V \exp[-tn(r)] w(r) dV, \qquad \text{(C2)}$$

where the integral is over the volume V. If we assume that the geometry of the crystal is such that the donor is located in the center of a sphere with radius r_V and that the position of the acceptor is uanaffected by the presence of the donor, then

$$w(r)dV = (4\pi r^2 dr) / (4\pi r_V^3 / 3). \qquad \text{(C3)}$$

If the position of one acceptor does not affect the position of a second acceptor (practical in lightly doped crystals where the probability of two acceptors being at the same site is negligible), we obtain for several acceptors

$$\rho(t) = \exp(-t / \tau_0^D) \lim_{\substack{N_A \to \infty \\ V \to \infty}} (\int_V \exp[-tn(r) w(r) dV])^{N_A}. \qquad \text{(C4)}$$

The N_A acceptors have been introduced by raising the term representing the energy transfer probability from donor to acceptor $(\int \exp[-tn(r)] w(r) dV)$ to the N_A power. The volume of the crystal and the number of acceptors are allowed to go to infinity in such a way that the concentration of acceptors, c, is held fixed at $N_A / [(4/3)\pi r_V^3]$.

We will choose the multipole interaction where

$$n(r) = (r_0 / r)^s (1 / \tau_0^D), \qquad \text{(C5)}$$

where $s = 6$ for dipole-dipole interactions, $s = 8$ for dipole-quadrupole interactions, etc... r_0 is the critical distance for which the rate of transfer of a donor and acceptor separated at that distance equals the rate of spontaneous decay for the donor. If we multiply both sides of Eq. (C4) by $N_D(0)$, the number of donors excited at time $t = 0$, then the left hand side of the equation will become $N_D(t)$, the number of donors excited at time t. In such a case, an analytical solution can be obtained:

$$N_D(t) = N_D(0)\exp[-t/\tau_0^D - \Gamma(1-3/s)Z_V^{3/s}]. \qquad \text{(C6)}$$

$\Gamma(x)$ is the well-known gamma function [18] and the factor Z_V, incorporating the "spatially dependent" rate, is given by

$$Z_V^{3/s} = (r_0/r_V)^3\,(t/\tau_0^D)^{3/s}. \qquad \text{(C7)}$$

Eq. (C6) can be simplified to

$$N_D(t) = N_D(0)\exp[-t/\tau_0^D - (c/c_0)\Gamma(1-3/s)(t/\tau_0^D)^{3/s}], \qquad \text{(C8)}$$

where c_0 is the critical concentration defined by

$$c_0 = 3/(4\pi r_0^3). \qquad \text{(C9)}$$

If we now consider the number of excited acceptors, it is apparent that the number increases with increased energy transfer from the donors and decreases due to the spontaneous decay of the acceptors with characteristic decay time τ_0^A. Let us start with the energy transfer from excited donors: the change in the number of donors in time $\Delta t'$ at time t' is $(dN_D(t')/dt')\Delta t'$. The number that decayed spontaneously in that period is $-N_D(t')\Delta t'/\tau_D^0$. Thus the difference between these two is the number of excited donors which returned to the ground-state due to the transfer of energy to the acceptors; correspondingly, the number of excited acceptors has increased. Those acceptors which have now been excited at time t' will now themselves sponaneously decay at a rate $1/\tau_0^A$. Thus, by integrating over time t', the overall concentration of excited acceptors can be given by

$$N_A(t) = \int_0^t \exp[-(t-t')/\tau_0^A]\left(-\frac{N_D(t')}{\tau_0^D} - \frac{dN_D(t')}{dt'}\right)dt'. \qquad \text{(C10)}$$

Analytical solutions of this expression are possible for different $N_D(t)$.

C.2. Enhanced Energy Transfer - Correlated Placement [27]

The assumption that the donor never affects the position of the acceptor is questionable. It is obvious that there is a minimum distance within which the acceptor could not colocate with the donor. There are physical reasons why "correlation" would occur. ("Enhanced correlation" will indicate that the presence of a donor enhances the probability that an acceptor will locate nearby;

"excluded correlation" indicates that the donor tends to inhibit the acceptor from colocating.) Moreover, in the donor/acceptor subsite created by the presence of a neighboring acceptor/donor will certainly correspond to a non-uniform distribution.

The implementation of a model with correlation is as follows: Eq. (C8) is based on the assumption of a uniform probability for the distribution of acceptor locations. We will apply the Förster-Dexter approach to the treatment of a simple non-uniform distribution. The first distribution considered here describes an excluded volume V_1 of the acceptors around each donor ion to a radius of r_1:

$$w(r)dV = 0 \qquad \text{for } r < r_1, \\ = \frac{4\pi r^2 dr}{V - V_1} \qquad \text{for } r \geq r_1 \qquad \text{(C11)}$$

We must now solve Eq. (C4) for the case in which we integrate from some fixed intermediate radius r_1 instead of from the origin. The quantity $V - V_1$ maintains the overall normalization where V_1 is the excluded volume $4\pi r_1^3/3$. This yields

$$N_D(t) = N_D(0)\exp(-t/\tau_0) \\ \lim_{\substack{N_A\to\infty \\ V\to\infty}} [(1 - V_1/V)^{-1}(3/s)Z_V^{3/s} \int_{Z_V}^{Z_1} Z^{-1-3/s}\exp(-Z)dZ]^{N_A}, \qquad \text{(C12)}$$

where the r-dependent Z is

$$Z = (r_0/r)^s t/\tau_0^D. \qquad \text{(C13)}$$

As shown in Ref. 27, this can be solved and approximated as

$$N_D(t) = N_D(0)\exp(-t/\tau_0^D)\exp\{cV_1[1 - \Phi(Z_1)/\exp(Z_1)]\}. \qquad \text{(C14)}$$

where $\Phi(Z_1)$ is the hypergeometric function defined in Ref. 27 and Z_1 is Z evaluated at $r = r_1$.

Eq. (C14) gives the time-dependent excited donor ion concentration for situations where, on the average, an excluded volume out to radius r_1 exists around the donor ions. $N_A(t)$ is obtained numerically from Eq. (C10).

The approach used above essentially excludes the acceptor from a sphere around the donor and hence slows down the energy transfer rate from the donor. The standard donor decay represents an average of different donors, some of which happen to have acceptors nearby, some of which do not. By eliminating acceptors from the surroundings of the donor, we limit the averaging of the donors to those with "far-away" acceptors.

While this distribution is important theoretically, in a crystal with many donor sites distributed either systematically or randomly throughout the lattice, it is unreasonable that a donor would affect the position of acceptors deep within the lattice. Thus, we propose a three stage model.

The approach shown above is now extended to a three-region acceptor distribution. The same basic steps used in deriving Eq. (C14) apply. The normalized radial distribution has an enhanced concentration of A/V out to radius r_1 and a compensating depleted region of concentration B/V between r_1 and r_2 in the volume between V_1 and V_2 as seen in Fig. C1 or the reverse case seen in Fig. C2. The radius r_2 must not be larger than half the average distance between donors (for an even distribution of donors). The following constraint then applies:

$$Ar_1^3 + B(r_2^3 - r_1^3) = r_2^3, \tag{C15}$$

where the "donor volume" $V_2 = V/N_D$. Following similar steps as used in the derivation of Eq. (C14), we find

$$\begin{aligned} N_D(T) = N_D(0)\exp[-t/\tau_0^D - A(c/c_0)(\pi t/\tau_0^D)^{1/2} \\ - cV_1(B-A)\Phi(Z_1)\exp(-Z_1) - cV_2(1-B)\Phi(Z_2)\exp(-Z_2)]. \end{aligned} \tag{C16}$$

For $A < 1$ we speak of an "excluded placement" and for $A > 1$ we speak of an "enhanced" placement. Eq. (C16) gives the uniform Förster-Dexter result when $A = B$; Eq. (C16) is consistent, as expected, with Eq. (C14) when $A = 0$ and r_2 gets large such that $V_2 >> V_1$.

The generalization of Eq. (C16) for an arbitrary distribution is now fairly straightforward. By extending the approach used in deriving Eq. (C16), we find

$$N_D(t) = N_D(0)\exp\left(-\frac{t}{\tau_0^D} - \int_0^{r_1} \frac{dw(r)}{dr} cV(r)\Phi(Z_r)\exp(-Z_r)dr\right), \tag{C17}$$

where $w(r)$ is the actual radial probability distribution, $V(r)$ is the volume out to radius r, and Z_r is found by evaluating Eq. (C13) at radius r. Eq. (C17) strictly applies provided $w(r)$ goes to 1 as r_v goes to infinity or, alternatively, that the variations in $w(r)$ are most important for smaller values of r, which is the case in practice. The effects on the fluorescent emission are generally seen in the initial time period following the excitation of the donors.

References

[1] Tissue, B. M., Jia, W., Lu, L. and Yen, W. M., 1991: "Coloration of Chromium-Doped Yttrium Aluminum Garnet Single Crystal Fibers Using a Divalent Codopant," *J. Appl. Phys.* **70**, pp. 3755-3777.

[2] Ostroumov, V. B., Privis, Yu. S., Smirnov, V. A. and Shcherbakov, I. A., 1986: "Sensitizing of Nd^{3+} Luminescence by Cr^{3+} in Gallium Garnets," *J. Opt. Soc. Am. B* **3**, pp. 81-101.

[3] Mears, R. J. and Baker, S. R., 1992: "Erbium Fibre Amplifiers and Lasers," *Opt. Quant. Elec.* **24**, pp. 5127-538.

[4] Vivien, D., 1994: "Chemical, Structural and Synthesis Effects in Inorganic Laser Materials," Opt. Mat. **4**, pp. 5-10.

[5] Collongues, R., 1992: "Nonstochiometry in Oxides," *Prog. Cryst. Growth and Charact.* **25**, pp. 203-240.

[6] Slack, G. A., Oliver, D. W., Chrenko, R. M. and Roberts, S., 1969: "Optical Absorption of Y_3Al_5O12 from 10- to 55000-cm^{-1} Wave Numbers," *Phys. Rev.* **177**, pp.1308-1314.

[7] Metselaar, R. and Larsen, P. K., 1976: "Diffusion of Oxygen Vacancies in Yttrium Iron Garnet Investigated by Dynamic Conductivity Measurements," *J. Phys. Chem.* **37**, pp. 599-605.

[8] Rotman, S. R., Tandon, R. P. and Tuller, H. L., 1985: "Defect-Property Correlations in Garnet Crystals: the Electrical Conductivity and Defect Structure of Luminescent Cerium-Doped Yttrium Aluminum Garnet," *J. Appl. Phys.* **57**, pp.1951-1955.

[9] Rotman, S. R. and Tuller, H. L., 1987: "Defect-Property Correlations in Garnet Crystals: III. Electrical Conductivity and Defect Structure of Luminescent Nickel-Doped Yttrium-Aluminum-Garnet," *J. Appl. Phys.* **62**, pp. 1305-1312.

[10] Rotman, S. R., Roth, M., Tuller, H. L. and Warde, C., 1989: "Defect-Property Correlations in Garnet Crystals: IV. The Optical Properties of Nickel-Doped Yttrium-Aluminum-Garnet," *J. Appl. Phys.* **66** (3), pp. 1366-1369.

[11] Nie, W., Boulon, G. and Mares, J., 1989: "Spectroscopy of Multisites Chromium (III) in Yttrium Garnet," *Chem. Phys. Letts.* **160**, pp. 597-601.

[12] Nie, W., Boulon, G. and Monteil, A., 1989: "Zero-Phonon Lines and Energy Transfer Between Chromium (III) and Neodymium (III) Multisites in Yttrium Aluminum Garnet (YAG)," *Chem. Phys. Letts.* **164**, pp. 106-112.

[13] Mares, J. A., Nie, W. and Boulon, G., 1990: "Multisites and Energy Transfer in Cr^{3+} - Nd^{3+} Codoped $Y_3Al_5O_{12}$ and $YAlO_3$ Crystals," *J. de Phys.* **51**, pp.1655-1669.

[14] Nie, W., Monteil, A. and Boulon, G., 1990: "Spectroscopy of Multisites Cr^{3+} and Nd^{3+} Multisites in $Y_3Al_5O_{12}$ Laser Crystals," *Opt. Quant. Elec.* **22**, pp. S227-S245 .

[15] Mares, J. A., Khas, Z., Nie, W. and Boulon, G., 1991: "Non-Radiative Energy Transfer Between Cr^{3+} and Nd^{3+} Multisites in $Y_3Al_5O_{12}$ Laser Crystals," *J. Phys. I*, **1**, pp. 881-899.

[16] Mares, J. A., Nie, W. and Boulon, G., 1991: "Energy Transfer Processes Between Various Cr^{3+} and Nd^{3+} Multisites in YAG:Nd,Cr," *J. Lum*, **48-49**, pp. 227-231.

[17] Nie, W., Kalisky, Y., Pedrini, C., Monteil, A. and Boulon, G., 1990: "Energy Transfer from Cr^{3+} Multisites to Tm^{3+} Multisites in Yttrium Aluminum Garnet," *Opt. Quant. Elec.* **22**, pp. S123-S131.

[18] Rotman, S. R., Eyal, A., Kalisky, Y., Brenier, A., Pedrini, C., Boulon, G., 1994: "Modelling Unusually Fast Energy Transfer in Crystals," Opt. Mat. **4**, pp. 31-33.

[19] Yugami, H., Nakajami, A., Ishigame, M. and Suemoto, T., 1991: "Local Structures in the Superionic Conductor Y^{3+}-doped CeO_2 Studied Using Site-Selective Spectroscopy," *Phys. Rev. B* **44**, pp. 4862-4871.

[20] Nelson, P. R., Brenier, A., Pedrini, C., Boulon, G. and Mares, J., 1991: "Chromium-Neodymium Energy Transfer in YAP," *J. de Phys. IV* (C7) **1**, pp. 375-378.

[21] Monteil, A., 1990: "Simulation and Application of Site-Selective Optical Spectroscop to chromium-Doped Gadolinium Garnets," *J. Phys. Cond. Mat.* **2**, pp. 9639-9651.

[22] Monteil, A., Nie, W., Madej, C. and Boulon, G., 1990: "Multisites Cr^{3+} in GGG and GSGG Garnets," *Op. Quant. Elec.* **22**, pp. S247-S257.

[23] Vergara, I., Monteil, A., Boulon, G., Madej, C. and Garcia Sole, J., 1990: "Influence of Multisites in the Energy Transfer of Chromium and Neodymium Doubly Doped GGG Crystals," *Mat. Chem. Phys.* **26**, pp.181-191.

[24] Han, T. P. J.. Scott, M. A., Jacque, F., Callagher, H. G. and Henderson, B., 1993: "Nd^{3+} - Cr^{3+} Pairs in $Gd_3Sc_2Ga_3O_{12}$ Garnet Crystals," *Chem. Phys. Letts.* **208** (1,2) pp. 63-67.

[25] Eyal, A. and Rotman, S. R., 1993: "The Determination of Acceptor Distribution from the Donor Decay Response in Codoped Crystals," *Chem. Phys. Lett.* **206** (1-4) pp. 113-118.

[26] Inokuti, M. and Hirayama, F., 1965: "Influence of Energy Transfer by the Exchange Mechanism on Donor Luminescence," *J. Chem. Phys.* **43**, pp. 1978-1989.

[27] Rotman, S. R. and Hartmann, F. X., 1988: "Non-Radiative Energy Transfer in Non-Uniform Codoped Laser Crystals," *Chem. Phys. Lett.* **152** (4,5), pp. 311-378.

[28] Blasse, G. and Bril, A., 1967: "A New Phosphor for Flying-Spot Cathode-Ray Tubes for Color Television: Yellow-Emitting $Y_3Al_5O_{12}$:Ce^{3+}," *Appl. Phys. Letts.* **11**, pp. 53-55.

[29] Wong, C. M., Rotman, S. R. and Warde, C., 1984: "Optical Studies of Cerium Doped Yttrium Aluminum Garnet Single Crystals," *Appl. Phys. Lett.* **44**, pp. 1038-1041.

[30] Robbins, D. J., Cockayne, B., Lent, B., Duckworth, C. N. and Glasper, J. L., 1979: "Investigation of Competitive Recombination Processes in Rare-Earth Activated Garnet Phosphors," *Phys. Rev. B* **19**, pp. 1254-1269.

[31] Robbins, D. J., Cockayne, B., Glasper, J. L. and Lent, B., 1979: "The Temperature Dependence of Rare-Earth Activated Garnet Phosphors. I. Intensity and Lifetime Measurements on Undoped and Ce-doped $Y_3Al_5O_{12}$," *J. Electrochem. Soc.* **126**, pp. 1213-1220.

[32] Robbins, D. J., Cockayne, B., Glasper, J. L. and Lent, B., 1979: "The Temperature Dependence of Rare-Earth Activated Garnet Phosphors. II. A Comparative Study of Ce^{3+}, Eu^{3+}, Tb^{3+} and Gd^{3+} in $Y_3Al_5O_{12}$," *J. Electrochem. Soc.* **126**, pp.1221-1228.

[33] Robbins, D. J., 1979: "The Effects of Crystal Field and Temperature on the Photoluminescence Excitation Efficiency of Ce^{3+} in YAG," *J. Electrochem. Soc.* **126**, pp. 1550-1555.

[34] Robbins, D. J., Cockayne, B., Lent, B. and Glasper, J. L., 1979: "The Relationship Between Concentration and Efficiency in Rare Earth Activated Phosphors," *J. Electrochem. Soc.* **126**, pp. 1556-1563.

[35] Hayes, W., Yamaga, M., Robbins, D. J. and Cockayne, B., 1980: "Optical Detection of EPR of Recombination Centres in YAG," *J. Phys. C* **13**, pp. L1085-L1089.

[36] Rotman, S. R. and Warde, C., 1985: "Defect Luminescence in Cerium-Doped Yttrium Aluminum Garnet,"*J. Appl. Phys.* **58**, pp. 522-525.

[37] Aizenberg, G. E. and Rotman, S. R., 1991: "A Non-Radiative Energy Transfer Model for Cerium-Doped Yttrium Aluminum Garnet (Ce:YAG)," *Phys. Stat. Sol. (a)* **126**, pp. 263-273.

[38] Rozenfeld, Y. and Rotman, S. R., 1993: "The Luminescence of Defects in Yttrium Aluminum Garnet," *Phys. Stat. Sol. (a)* **139**, pp.249-262.

[39] Stevels, A. L. N. and Schrama-De Pauw, A. D. M., 1976: "Effects of Defects on the quantum Efficiency of Eu^{2+}-Doped Aluminates with the Magnetoplumbite-Type Crystal Structure," *J. Lum.* **14**, pp. 147-153.

[40] Stevels, A. L. N. and Verstegen, J. M. P. J., 1976: "$Eu^{2+} \rightarrow Mn^{2+}$ Energy Transfer in Hexagonal Aluminates," *J. Lum.* **14**, pp. 207-218.

[41] Verstegen, J. M. P. J. and Stevels, A. L. N., 1974: "The Relation Between Crystal Structure and Luminescence in β-Alumina and Magnetoplumbite Phases," *J. Lum.* **9**, pp. 406-414.

[42] Luria, E. and Rotman, S. R., 1994: "Converting Eu^{+2} to Eu^{+3} in Europium-Doped LMA," *J. Lum.* **60-61**, pp. 67-69

[43] Luria, E. and Rotman, S. R., 1994: "Time decay of the Excited States of Eu^{+2} in Europium-Doped LMA," *J. de Phys.* IV **4**, pp. 435-438.

TABLE 1. Concentrations of electronic and ionic defect species under different neutrality conditions

	$n = 2[V_{O}^{\cdot\cdot}]$	$[A'] = 2[V_{O}^{\cdot\cdot}]$	$p = [A']$
n	$(2K_{1/3}^{0})^{1/3}\exp(-E_1/3kT)P_{O_2}^{-1/4}$	$\left(\frac{2K_1^0}{[A']}\right)^{1/2}\exp-(E_1/2kT)P_{O_2}^{-1/4}$	$\frac{K_i^0\exp(-E_g/kT)}{[A']}$
p	$\left(\frac{K_i^{0^3}}{2K_1^0}\right)^{1/3}\exp[-(E_g-E_1/3)/kT]P_{O_2}^{1/6}$	$\left(\frac{K_i^{0^2}[A']}{2K_1^0}\right)^{1/2}\exp[-(E_g-E_1/2)/kT]P_{O_2}^{1/4}$	$[A']$
$V_{O}^{\cdot\cdot}$	$\left(\frac{K_i^0}{4}\right)^{1/3}\exp(-E_1/3kT)P_{O_2}^{-1/6}$	$\left[\frac{A'}{2}\right]$	$\frac{K_1^0[A']^2}{(K_i^0)^2}\exp[-(E_1-2E_g)/kT]P_{O_2}^{-1/2}$

Table 2. The concentrations of the four major defects n, p $[V_O^{\cdot\cdot}]$*, and* $[Ni'_{Al}]$ *are given in terms of the equilibrium constants, the total nickel concentration* $[Ni_{tot}]$*, the acceptor density* $[A']$ *and the* P_{O_2} *for each of the defect regions I-V at the top of the table. The second row indicates the approximate valence of* Ni_{Al}*, and the third row gives the approximate neutrality relation.*

	I	II	III	IV	V
		$[Ni_{Al}^x]=[Ni_{tot}]$		$[Ni'_{Al}]=[Ni_{tot}]$	
	$p=[A']$	$2[V_O^{\cdot\cdot}]=[A']$	$2[V_O^{\cdot\cdot}]=[Ni'_{Al}]$	$2[V_O^{\cdot\cdot}]=[Ni'_{Al}]$	$2[V_O^{\cdot\cdot}]=n$
n	$\frac{K_i}{[A']}$	$\left(\frac{2K_1}{[A']}\right)^{1/2} P_{O_2}^{-1/4}$	$(2K_1K_2)^{1/3}[Ni_{tot}]^{-1/3} P_{O_2}^{-1/6}$	$\left(\frac{2K_1}{[Ni_{tot}]}\right)^{1/2} P_{O_2}^{-1/4}$	$(2K_1)^{1/3} P_{O_2}^{-1/6}$
p	$[A']$	$\left(\frac{[A']K_i^2}{2K_1}\right)^{1/2} P_{O_2}^{1/4}$	$\left(\frac{K_i^3}{2K_1K_2}\right)^{1/3} [Ni_{tot}]^{1/3} P_{O_2}^{1/6}$	$\left(\frac{[Ni_{tot}]}{2K_1}\right)^{1/2} K_i P_{O_2}^{1/4}$	$\frac{K_i}{(2K_1)^{1/3}} P_{O_2}^{1/6}$
$[V_O^{\cdot\cdot}]$	$\frac{K_1[A']^{1/2}}{K_i^2} P_{O_2}^{1/4}$	$\frac{[A']}{2}$	$[Ni_{tot}]^{2/3}\left(\frac{K_1}{4K_2^2}\right)^{1/3} P_{O_2}^{-1/6}$	$\frac{[Ni_{tot}]}{2}$	$\left(\frac{K_1}{4}\right)^{1/3} P_{O_2}^{-1/6}$
$[Ni'_{Al}]$	$\frac{[Ni_{tot}]K_i}{[A']K_2}$	$\left(\frac{2K_1}{[A']K_2^2}\right)^{1/2} [Ni_{tot}] P_{O_2}^{-1/4}$	$[Ni_{tot}]^{2/3}\left(\frac{2K_1}{K_2^2}\right)^{1/3} P_{O_2}^{-1/6}$	$[Ni_{tot}]$	$[Ni_{tot}]$

SOLID STATE LASERS

Yehoshua Kalisky

Laser Department, Nuclear Research Centre-Negev
P.O. Box 9001, Beer-Sheva 84190, Israel

Abstract

Solid state lasers, especially those based on oxide and fluoride hosts, are presently widely acceptable hosts for various lasing ions. The oxide and fluoride crystals have various optical, physical and mechanical properties which can be utilized for specific applications.

The chapter starts with the properties of yttrium aluminum garnet (YAG) as a host for the Nd^{3+} *ion with lasing at 1.06 mm, and the effects of changes in the garnet composition on the laser performance. It then proceeds to 2 micron lasers, i.e., thulium and holmium doped crystals, emphasizing the various pumping modes such as diode pumping.*

The last section describes various fluoride laser hosts and their unique properties.

SOLID STATE LASERS

Yehoshua Kalisky

Contents

Solid State Lasers

Yehoshua Kalisky
Laser Department, Nuclear Research Centre-Negev
P.O. Box 9001, Beer-Sheva 84190, Israel

1. Introduction

The solid-state laser, initially developed in the 60's, has recently seen a tremendous burst of new research. Developments include 1) tunable lasers, 2) codoped lasers (with separate ions performing the absorption of the pump energy and the emission of the lasing emission), and 3) upconversion lasers in which the lasing output is at a shorter wavelength than the pump.

In this chapter, we will review the latest results for solid-state lasers based on garnets and fluorides. While many other materials have been tried successfully, these will demonstrate much of the research which has gone on in the laser development area.

2. Garnets

2.1 PHYSICAL CHARACTERISTICS OF GARNETS AND MIXED GARNETS

There are several examples of garnets used as laser materials. The most commercially common material is yttrium aluminum garnet (YAG), $Y_3Al_5O_{12}$. Other examples of garnets are: gadolinium gallium garnet (GGG)-$Gd_3Ga_5O_{12}$, gadolinium scandium gallium garnet (GSGG)-$Gd_3Sc_2Ga_3O_{12}$, gadolinium scandium aluminum garnet (GSAG)-$Gd_3Sc_2Al_3O_{12}$ and yttrium scandium aluminum garnet (YSAG)-$Y_3Sc_5O_{12}$ [1]. These crystals have good optical, thermal and mechanical properties which pertain to laser materials, as shown in Tables 1 and 2 of the first chapter of this book. The tables also contain for comparative information the optical and mechanical properties of YLF. The properties of fluorides as laser matrices will be discussed in more detail in Section 4 of this chapter.

Crystal growth technique and fabrication of YAG are today well established procedures, and are used by several companies as their main laser host product. YAG and other garnets serve as laser medium for Nd^{3+} as well as for mid IR lasers such as Ho^{3+}, Tm^{3+}, and Er^{3+}. Substituting Y^{3+} and Al^{3+} with larger cations (Sc^{3+}, Gd^{3+}, or Ga^{3+}) yields a larger crystalline cell with a lower crystal field. The general composition of such a garnet is $A_3B_2C_3O_{12}$. The garnet structure is composed of a dodecahedral site (A) with a coordination number 8, an octahedral site (B) with a coordination number 6, and a tetrahedral site (C) of coordination number 4.

Garnets based on scandium, gallium or gadolinium like GSAG, GSGG, GGG exhibit a better distribution coefficient for Nd^{3+} since the main cation in YAG is replaced by Gd, Sc or Ga which have larger ionic radii, close to that of a Nd^{3+} ion [1]. The distribution coefficient k of a crystal is another important parameter in crystal growth technique. It is a measure of the segregation of the ion to be doped into the crystal and is defined as:

$$C(s)= kC(m)*(1-g(s))^{k-1} \tag{1}$$

where C(m) is the concentration in the melt, C(s) is the concentration of the ion in the solid sample at a certain distance along the crystal axis, g(s) is the crystallized fraction of the melt, and k is the distribution coefficient.

The method of crystal growth for garnets is the pulling technique also known as the Czochralski method. By using this method, relatively large crystals of high quality can be obtained. A detailed description of the method can be found in a recent review by Kokta [2]. Undoped YAG is grown by this method easily, with a rate of 1 cm/hour, from iridium crucibles, under nitrogen atmosphere containing 300-1000 ppm oxygen. Doping YAG with Nd^{3+} creates difficulties as a result of its low distribution coefficient - 0.16-0.18. The low distribution coefficient result in a very slow growth rate, on the order of 0.5 mm/hour.

Ions such as Gd^{3+}, Sc^{3+}, or Ga^{3+} are bigger than Y^{3+} or Al^{3+}; therefore, their crystalline cell is larger, and the doping of rare earth ions (like Nd^{3+}, Ho^{3+}, Er^{3+}, etc) is easier than in YAG. This clear advantage has led in the 80's and 90's to extensive scientific research. Several attempts were made to synthetize Al-substituted scandium garnets [3] and scandium-substituted rare-earth aluminum garnets [4]. All the crystals were grown along <111> orientation, with pull rates of about 0.015"/hour or less and 15 RPM rotation rate. As an example, the distribution coefficient of Nd is increased from 0.18 in YAG [5] to 0.6 in GSAG [6].

Mixtures of transition metal ions (Cr^{3+}) and rare earth ions (like Nd^{3+}) are used to improve the lasing properties of Nd doped garnets. An efficient energy transfer Cr → Nd:YAG was studied years ago by Kiss et al. [7] with substantial improvement in laser efficiency (relative to Nd-only doped YAG). Efficient energy transfer in $Cr^{3+} \rightarrow Nd^{3+}$ systems is typical in other solid state hosts like glasses. As an example, there exists a strong dipole-dipole coupling between Cr^{3+} to Nd^{3+} in fluorophosphate glass [8], with a linear dependence on Nd^{3+} concentration and high transfer efficiency (≈60%) for concentrations up to 2.0 wt % of Nd^{3+}. The study of Cr,Nd:YAG, and rare earth doped scandium garnets (defined as RESG) like GSGG, GSAG,YSAG, YSGG is a well established subject summarized in an excellent review by Smirnov and Shcherbakov in 1988 [9]. The following discussion will serve only as an introductory background to the broader subject of Cr^{3+} and other rare earth codoped garnets.

2.2 CHROMIUM-DOPED AND NEODYMIUM-DOPED GARNETS

Cr^{3+} occupies sites with octahedral symmetry and therefore the optical transitions within its 3d levels are Laporte forbidden. The configurational diagram of Cr^{3+} in an intermediate crystal field is shown schematically in Figure 1 and the energy levels' dependence on the ligand field strength in d^3 system in octahedral symmetry, based on Tanabe-Sugano work [10,11], is presented in Figure 2. The $^2E \rightarrow ^4A_2$ transition is a narrow, spin forbidden transition, while the $^4T_2 \rightarrow ^4A_2$ is a spin allowed transition, strongly coupled to the vibrational states. Therefore, this transition is broad enough to overlap with the useful absorption lines of rare-earth ions like Nd^{3+}, Tm^{3+}, or Ho^{3+}. This fact is important in improving the pumping efficiency of Nd^{3+} as well as other rare earth lasers; in order to improve pumping efficiency, energy sensitizers which have broad and allowed transitions are being used. The sensitizer absorbs the useful energy and transfers it resonantly or non-resonantly to the higher levels of an energy acceptor (activator) which emits it as fluorescence or laser radiation. This subject, which is of fundamental importance to a proper design of solid state laser materials, was reviewed in the chapter in this book by Zharikov et al.

From the Sugano-Tanabe diagram presented in Figure 2, it is clear that the position of the $^4T_2$ level depends on the strength of the crystal field and hence on the crystal structure. Introducing Cr^{3+} into rare-earth scandium garnets like GSGG, YSGG, GGG, GSAG or LLGG - $(La_3Lu_2)Ga_3O_{12}$ has two main advantages: first, the distribution coefficient of the doped Cr^{3+} ions is close to unity, and second, Cr^{3+} is subjected to lower crystal fields in these garnets due to their larger lattice constants relative to YAG. The lower crystal field leads to a decrease in the energy gap ΔE between $^4T_2$ and ^2E as depicted in Figure 2. As a result of lowering $^4T_2$, Cr^{3+} becomes a better sensitizer since the $^4T_2$ level is closer in its energy to the energy levels of the activator (energy acceptor) ion. Table 3 in the first chapter of this book, based mainly on the data presented in the manuscript of Smirnov and Shcherbakov [9], gives the values of some ΔE related to various hosts.

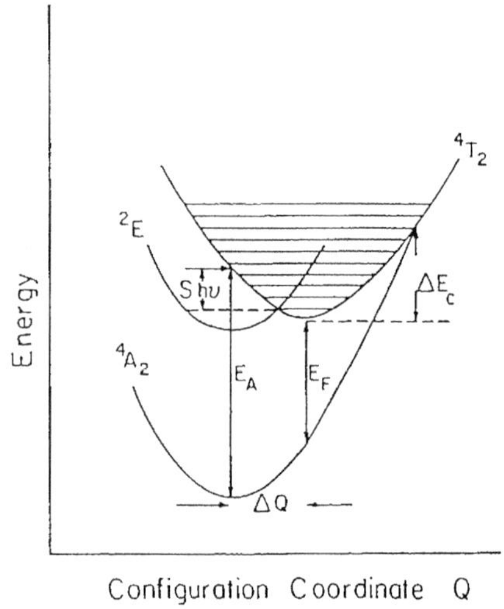

Figure 1. A schematic representation of the configuration coordinate diagram for Cr^{3+} in octahedral site. (Taken from Ref. [10]).

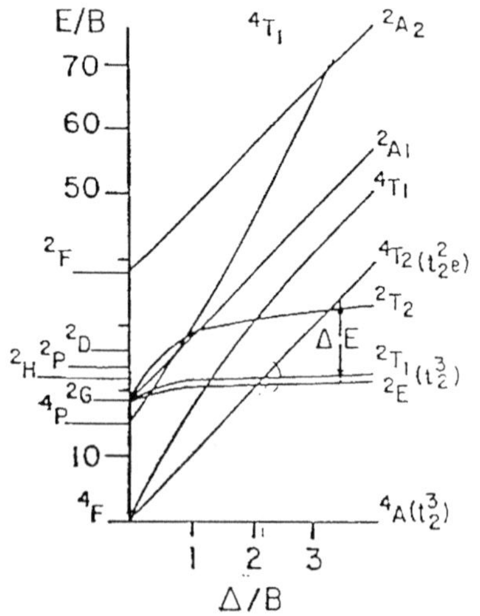

Figure 2. Energy levels' dependence on the ligand field strength in d^3 electronic configuration located in an octahedrally coordinated symmetric site. (Taken from Ref. [10].)

As a result of these close to unity distribution coefficients, large homogeneous GSGG crystals, codoped with Cr^{3+} and Nd^{3+} can be fabricated [12]. Both Zharikov [12,13] and Pruss et al. [14] conducted spectroscopic and luminescence studies in GSGG codoped with Cr^{3+} and Nd^{3+} and demonstrated an efficient energy transfer $Cr^{3+} \rightarrow Nd^{3+}$. Pruss and Huber reported transfer efficiency of 86% within a duration of 17 msec. A proof of efficient energy transfer Cr → Nd is given in Figure 3 where the emission spectrum of Cr^{3+} and Nd^{3+} is observed by 647.1 nm excitation. In particular, Nd^{3+} emission of both $^4F_{3/2} \rightarrow {}^4I_{9/2}$ as well as $^4F_{3/2} \rightarrow {}^4I_{11/2}$ (peaking at 1.06 mm) were recorded by direct excitation into the Cr^{3+} ($^4A_2 \rightarrow {}^2E$) band, by using the 647.1 nm Kr-line laser [14]. The same authors also reported CW operation of Cr,Nd:GSGG with a slope efficiency of 41% while pumping into the Cr^{3+} absorption bands. The performance curve of this laser pumped by Kr- line is shown in Figure 4.

At the same time, Zharikov et al.[12] compared the free running laser performance of Cr:Nd:GSGG and Nd:GSGG. They found that the former was much more efficient than the latter. The improved performance of the codoped crystal can be seen in Figure 5 [12]. The Figure compares (under the same experimental conditions) the laser performance of Nd:GSGG, (curve 1), with the performance of Cr:Nd:GSGG (curve 2), as well as the performance of Nd:YAG (curve 3). They further characterized the energy transfer in the Cr → Nd system and found that the probability for energy transfer experiments on Cr → Nd is higher in GSGG by a factor of ≅20 compared to YAG. Zharikov et al. [13] also compared the performance of the GSGG codoped host to Nd:YAG. They found that Cr,Nd:GSGG performed better than Nd:YAG both in a free-running pulsed mode and Q-switched mode. The slope efficiency for Cr,Nd:GSGG in a free running mode is reported to be higher by a factor of ≅3.5 than in YAG [13].

For a Cr:Nd:YSGG laser, free running, pulsed operation with a slope efficiency of η=11.5% was reported. Also, a maximum output energy of 20 J with equal concentrations of 3×10^{20} cm^{-3} for both Cr^{3+} and Nd^{3+} and a total efficiency of 10.5% were obtained. The Q-switched operation @50Hz repetition rate yielded slope efficiency of 5.4% with a total efficiency of 3.6%. [Results for lasing performance of Cr,Nd:RESG were summarized by Smirnov and Shcherbakov [9].]

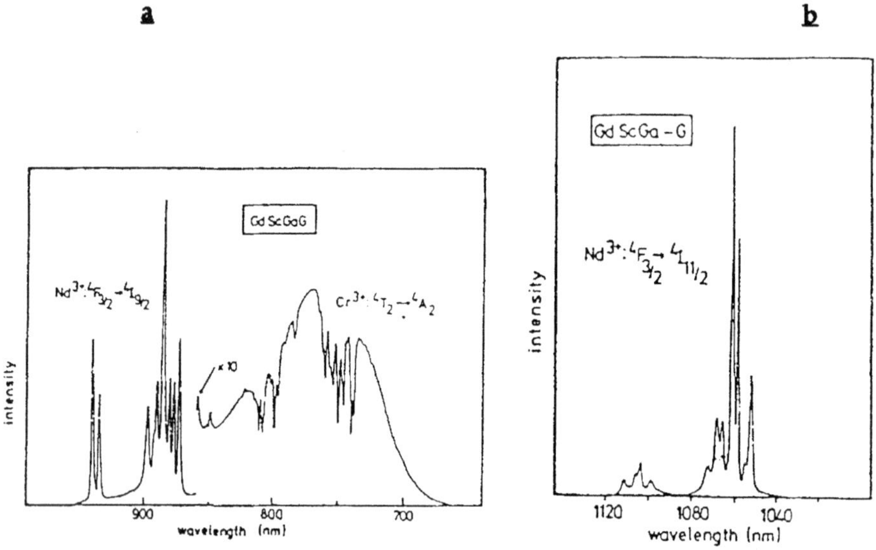

Figure 3. Emission spectrum of C,Nd:GSGG at ambient temperature excited into Cr^{3+} band at 647.1nm. (Taken from Ref. [14]). a: 700-950 nm; b: 1040-1120 nm

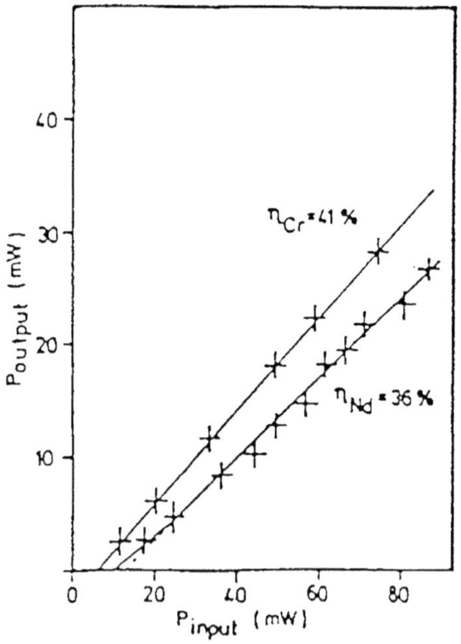

Figure 4. Performance of the Cr,Nd:GSGG laser pumped into Cr^{3+} absorption band $^4A_2 \rightarrow {}^4T_2$ and directly into Nd^{3+} levels. (Taken from Ref. [14].)

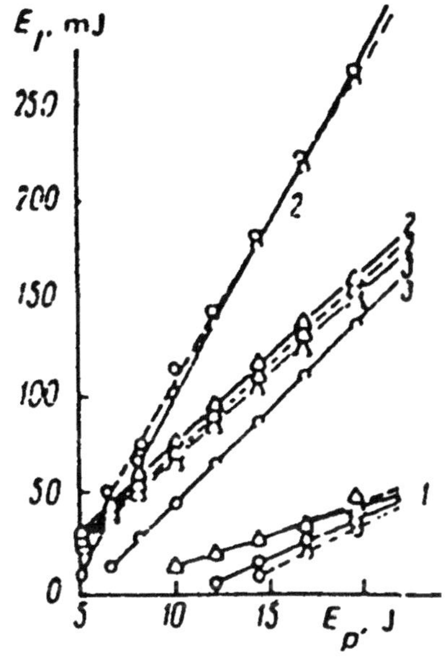

Figure 5. Dependence of the free-running output energy on input energy of Nd:GSGG (curve 1), Cr:Nd:GSGG (curve 2) and Nd:YAG (curve 3), for two different output couplers: R=83%-Δ; R=48%-o. (Taken from Ref. [12].)

The results described in this section demonstrate that with proper choice of host and sensitizer, the pumping efficiency of a laser system can be improved. However, there might be some financial limitations on the use of scandium-based garnets. In the following section we will describe alternatives to the rare earth scandium garnets.

2.3 DISORDERED GARNETS

GSGG, as was noted earlie,r has a lower crystal field and therefore will lower the energy gap ΔE between $^4T_2$ and ^2E to a value of 50 cm^{-1}. The presence of expensive scandium is the biggest disadvantage of this material. In order to get a garnet with a lower crystal field, it was proposed by Brenier et al. [15] to use Ca,Zr - substituted GGG.

The garnet structure has three different symmetry sites, and this allows for multiple substitution. For example, in the garnet GGG ($Gd_3Ga_5O_{12}$), which has the following symmetry-sites: Gd_3- Dodecahedral, Ga_2 - Octahedral, Ga_3 - Tetrahedral, there is a possibility that some Gd^{3+} will occupy the Ga octahedral sites. The garnet will have the following chemical formula:

$[Gd_3]\{Ga_{2-x}Gd_x\}(Ga_3)O_{12}$. The difference in the ionic radii of Ga and Gd is a source for defects in the crystal. The disorder in the GGG crystal can be increased by multisubstitution of the garnet with Ca and Zr ions. The Ca ions substitute for the Gd ions in the dodecahedral sites and the Zr ions substitute for the Ga ions in the octahedral sites. The increase in the disorder is accompanied by an increase in the average lattice constant with a consequent decrease in the crystal field as well as a decrease in the energy gap ($^4T_2$ - ^2E) to 10 cm^{-1}. The (Ca,Zr) - substituted GGG exhibits an intermediate disordered structure between pure crystalline and pure glassy structure. Table 3 in Chapter 1 presents comparative values of the energy gap for different garnets which also include substituted garnets. The lowering in ΔE has an effect on the performance of Cr^{3+} as a tunable laser and as an efficient sensitizer for rare earth lasing ions. Cr^{3+} doped GGG emission usually has an exponential profile at all temperatures. However, in (Ca,Zr)-GGG, the decay is non-exponential at low temperatures. The defects introduced into the crystal by Ca and Zr, which substitute Gd and Ga, respectively, create a non-homogeneous structure in the crystal. This will yield wide distribution of the lattice constants as well as a distribution of the local crystal fields.

Theoretical treatment of the spectral and temporal characteristics of Cr^{3+} in Ca,Zr substituted GGG was presented by Grinberg et al. [16]. In the case of intermediate crystal fields (as the case of mixed garnets), the first electronic excited state in octahedrally coordinated Cr^{3+}, is a mixture of the ^2E state and components of the $^4T_2$ state. The emission spectrum at low temperatures will have spectral characteristics of both electronic states [16]. By increasing the temperature, the population of $^4T_2$ level will increase and the decay curve will have the typical characteristics of this state. The decay of $^4T_2$ will also be faster as a result of the spin-allowed transitions from $^4T_2$ to the ground level $^4A_2$.

Another kind of mixed garnets is based on chromium doped $Y_3(In,Ga)Ga_3O_{12}$ (YIGG) [17]. This material is close in its crystal field strength to GSGG. The values of ΔE in Table 3 of Chapter 1 are proportional to the strength of the crystal field. The crystal field is lower in YIGG than in YAG, and comparable to other garnets like GSGG and GGG. The emission of Cr^{3+} -in this crystal exhibits a large Stokes shift, of 2490 cm^{-1}, higher than other garnets like YAG, GSGG, and GIGG. This fact has several advantages for YIGG:

1. High luminescence efficiencies.
2. Small probability of self absorption of the emission from the $^4T_2$ level by a ground-state absorption via the $^4A_2 \rightarrow {}^4T_2$ transition.

3. Broad band emission—a potential candidate for a tunable laser and a good sensitizer.
4. Multisite structure reflected by its large inhomogeneous width as a result of the replacement of In sites with the Ga octahedral sites and the difference in the ionic radii between In and Ga.

3. Two Micron lasers: Holmium and Thulium Doped Crystals

3.1 INTRODUCTION

Solid state lasers doped with Ho^{3+}, Tm^{3+} or Er^{3+} have several important potential uses: medical applications [18], eye-safe range finders, pump sources for OPO, light detection and ranging (LIDAR) applications [19], wind velocity measurements [20], and atmospheric measurements to monitor the health of our planet [21].

Holmium lasers emit radiation at a wavelength range peaking at ~2.1 μm, resulting from the $^5I_7 \rightarrow {}^5I_8$ electronic transition. Stimulated emission in Ho^{3+} doped $CaWO_4$ was first reported by Johnson et al. [22]. Later, stimulated emission was obtained in singly-doped Ho^{3+} glasses [23], (lithium-magnesium-aluminum-slicate), with emission peaking at 2.08 um and operating at 77 K. In the latter case, the output pulses were associated with internal Q-switching, which depends on small amounts of added iron (Fe^{2+}). Also, laser emission at ~2.1 μm in Ho^{3+} doped single crystal of HoF_3 at 77 K was obtained [24]. The cavity in this case was an elliptical cylinder with threshold energy of ~1 J/pulse. The laser transition terminal level was assigned to the highest energy levels of the $^5I_8$ ground state, which is at 532 cm^{-1}. It is worthwhile to note that this system is relatively efficient despite the 3-level nature of the holmium laser, due to efficient cooling of the crystal. By cooling, the high energy lying Stark components are depopulated and population inversion can be easily achieved.

Another way to increase the population inversion in the holmium lasing system and hence lasing efficiency, is to utilize energy transfer from efficient absorbers to the $^5I_7$ lasing level of Ho^{3+}. This will be discussed further in Section 3.2. The first experiments which utilized energy transfer were conducted by Johnson et al. [25] by codoping Ho^{3+} with Yb^{3+}, Tm^{3+}, Er^{3+} ions, in YAG. The laser operated at 77 K to enhance its efficiency. Output power obtained was 7.6 W at a wavelength of 2.12 μm, while the input power was 550 W. The optical transitions of Ho^{3+} cover only 11% of the spectral region of 0.6–2 μm. Codoping with Er, Yb and Tm provides an extension up to 89% of this spectral range. The CW pumping source is a black body source. At the temperature of 3000 K its spectral distribution lies in the 0.6–2 μm region and thus results in a pumping efficiency of 62%. If one assumes non-radiative losses, then the maximum expected efficiency of the system is 30% [25]. Further improvement in laser holmium performance was achieved when holmium was doped in a crystal with low non-radiative losses. An example for such a crystal is $YLiF_4$ (YLF) discussed in section 4 of this chapter. Laser experiments in Er:Tm:Ho:$YLiF_4$ (denoted here as αβHo:YLF) were performed by Chicklis et al. [26]. This system had several fundamental advantages which led to efficient pumping: holmium was codoped with Tm^{3+} and Er^{3+} , which were used as efficient energy sensitizers. The crystalline host had lower non-radiative losses compared to oxide crystals; the laser was operated in a pulsed mode. Pulsed mode operation prevented the accumulation of heat in the laser rod by allowing thermal relaxation (depending on the repetition rate) and therefore minimizing the thermal load on the crystal. As a result of the above advantages, the 2.1 μm holmium laser emission was observed at room temperature. Chicklis et al. [26] reported a maximum output energy of 400 mJ/pulse with 1.3% slope efficiency in a free running mode, and 2% slope efficiency in a Q-switched mode. Another important fluoride crystal is Ho^{3+} doped in a $BaYb_2F_8$[27a,27b] crystal. Antipenko et al. [27a] calculated the oscillator strengths and radiative transitions among the elctronic levels of Ho:$BaYb_2F_8$ for the $^5I_7 \rightarrow {}^5I_8$ transition of Ho^{3+}. Table 1 presents spectroscopic data relevant to the $^5I_7 \rightarrow {}^5I_8$ lasing transition at $\lambda \cong 2$ μm for $BaYb_2F_8$, $YLiF_4$ [28], and YAG laser crystals.

Table 1: Oscillator strength (f), radiative transition rates (A), and stimulated emission cross-section (σ_{em}), in holmium doped fluorides for the $^5I_7 \rightarrow {}^5I_8$ transition (the data relates to electric-dipole transitions only).

Host	f x 10^6	A, sec^{-1}	σ_{em}, cm^2 (@295 K)
$BaYb_2F_8$	1.67	73	0.78×10^{-19}
YLF	1.12	59	1.12×10^{-19} [28] 1.4×10^{-19} [36]

As was noted by Johnson et al. [27b], $BaYb_2F_8$, like other fluorides such as YLF, or LaF_3 has low phonon energies, and the orbital coupling of the ion to the lattice is relatively weak. These characteristics permit longer fluorescence lifetimes and higher stimulated transitions between a variety of exited states.

An overall historical summary of the laser system based on Ho^{3+} codoped with Tm^{3+} and Er^{3+} ions is presented briefly in References [29] and [30].

3.2 ADVANTAGES OF THE HOLMIUM LASER

A significant advantage of the holmium laser is its high gain cross-section and the long lifetime of the $^5I_7$ (7–14 msec) upper laser level, which results in high energy storage capability and efficient Q-switched operation. The special characteristics of an Ho^{3+} doped solid matrix, especially when the holmium is codoped with Er^{3+} and Tm^{3+}, have motivated extensive research on CW and pulsed holmium laser operation as well as spectroscopic studies of several holmium-doped host materials, with an emphasis on YLF and YAG. A summary of the subject can be found in reference [31]. Holmium is a quasi–three–level laser system at room temperature. The lower $^5I_8$ ground state has Stark components over a 300–500 cm^{-1} energy domain. When cooled, the lower $^5I_8$ Stark levels are populated and the molecular system becomes a quasi–four–level system. Therefore, it is expected to obtain an efficient laser emission of the holmium ion at low temperatures.

3.3 UTILIZING ENERGY TRANSFER

Codoping Ho^{3+} with Tm^{3+} and Er^{3+} in YAG or YLF (αβHo:YAG or αβHo:YLF, respectively) leads to the efficient laser operation [32]. This is explained as a result of energy transfer from high energy levels of Er^{3+} to the $^3H_4$ level of Tm^{3+} and then, by a triple cross-relaxation, to the $^5I_7$ electronic level of Ho^{3+}. An alternate way to improve the holmium laser efficiency at 2.1 μm was suggested by Antipenko et al. [33,34] They suggested codoping Ho^{3+} with Tm^{3+} and Cr^{3+}. As a result of this ionic combination, a significant part of the flash lamp pumping light is absorbed efficiently by the broad band, allowing electronic transitions of Cr^{3+} (the $^4T_2$ and $^4T_1$ levels), which then is transferred efficiently to the $^3H_4$ level of Tm^{3+} and then, by double-cross-relaxation, to the $^5I_7$ electronic level of holmium. The process can be schematically described in Figure 6 below.

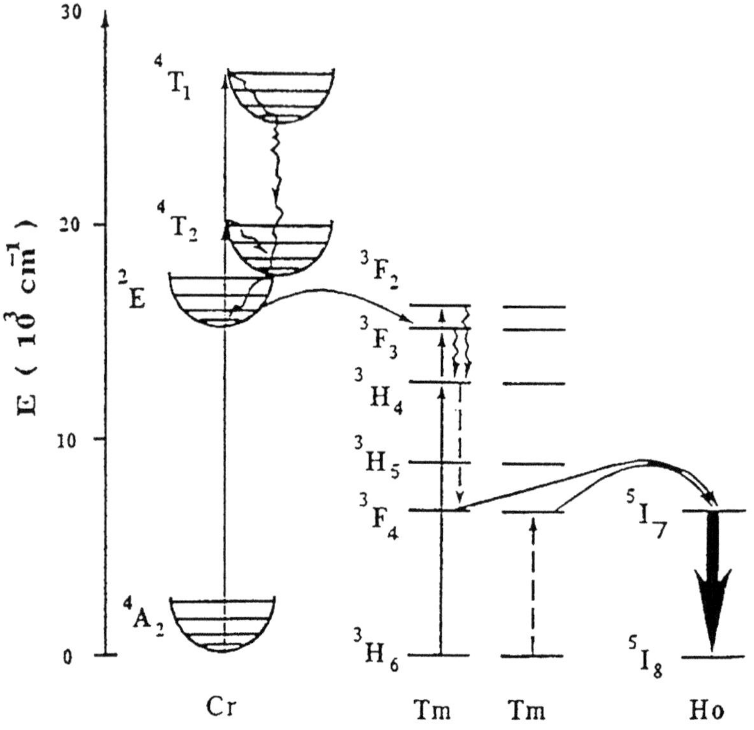

Figure 6: Schematic sensitization of 2.1 μm Ho^{3+} laser emission with Cr-Tm energy transfer, Tm-Tm cross relaxation and Tm-Ho energy transfer. (Taken from reference [40].)

When higher repetition rates are required, a cooling system has to be used for eliminating thermal effects in the lasing crystal and maintaining a good beam quality. In these cases, the use of the YLF crystal as a host is favorable due to its negative change of refraction index with temperature. YLF doped crystals have additional advantages which make the holmium doped YLF a favorable laser host. YLF is characterized by low phonon energies that result in low non-radiative rates and low UV absorption, which minimize the formation of color centers.

Holmium laser emission at ~ 2.1 μm can be achieved by two methods:

1. Conventional lamp pumping – using a flashlamp or CW source to produce pulsed or CW laser operation.
2. Diode pumping scheme.

3.4 CONVENTIONAL PUMPING

3.4.1 CW Laser Operation

The nature of the 2 μm Ho^{3+} laser level requires cryogenic cooling in order to obtain population inversion and efficient CW laser operation. Beck and Gurs [32] operated successfully a Ho:YAG laser, cryogenically cooled with 50 W output power and 6.5% slope efficiency; a 60 W output power with 4.7% slope efficiency was reported later by Lotem et al. [35], using αβHo:YLF with a nominal concentration of holmium of 0.1% at.

A systematic comparison between YAG and YLF [31,36] reveals that for systems designed for a high average output power (≈40 W), the two host crystals may perform similarly. In such systems, if the maximum laser power is not the important issue, the YAG is a better choice due to its availability and superior mechanical properties. In cases where linearly polarized laser radiation is required, YLF would be the appropriate choice.

Laser parameters such as the stimulated emission cross-section, round trip loss and population inversion at threshold were calculated [36] using Findley-Clay analysis for a CW holmium laser. The value of the stimulated emission cross-section for Ho:YLF is $\sigma=1.4\times10^{-19}$ cm^2 while that for YAG is $\sigma=1.2\times10^{-19}$ cm^2. The population inversion densities at threshold are very similar for αβHo:YAG and αβHo:YLF, namely, $(N_2 - N_1)_{th} \cong 1.6\times10^{17}\,\mathrm{cm}^{-3}$.

The thermal effect on the laser rod is an important parameter, which may cause beam quality degradation and laser rod fracture. It was found that the net heat dissipation while lasing at a pumping level of 1500 W (lamp input power) was 17.4 W/cm and 19.4 W/cm for αβHo:YLF and αβHo:YAG laser rods, respectively. This thermal load is within the theoretical fracture ranges for YAG and YLF, which are, at room temperature, 60 W/cm and 11 W/cm, respectively.

The laser setup for CW operation of a cryogenically cooled holmium laser is described by several authors and summarized in Refs. [31], [32] and [35]. As stated previously, the $^5I_7 \rightarrow {}^5I_8$ transition of holmium laser is a 3-level system at room temperature. This requires cryogenic cooling in order to obtain efficient CW laser operation. The laser rod is usually cooled by a massive fast flow of liquid nitrogen (7 liter/min) at a pressure range of 2–3 atm. The laser head used in various experiments is a water cooled gold-plated elliptical reflector. The elliptical cavity is almost circular to promote low order modes. See Figure 7 for a schematic description of the water-cooled elliptical laser head. The CW pumping is achieved by using a linear, air-cooled tungsten-halogen lamp with a nominal electrical power of 1000 W or 1500 W. In order to achieve thermal insulation, the laser head chamber is evacuated.

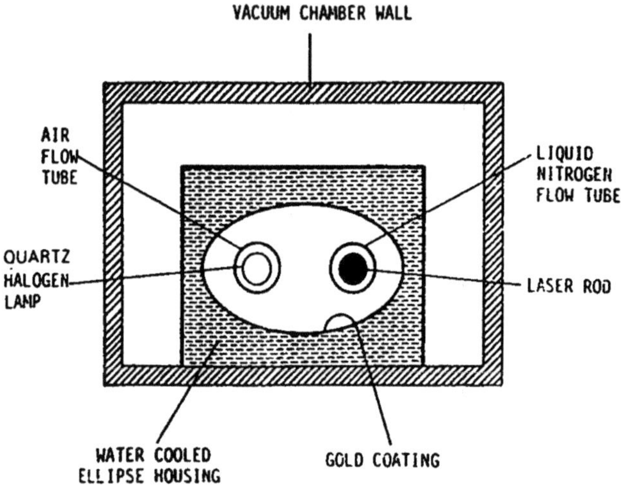

Figure 7: Schematic description of the water cooled elliptical laser head. (Taken from Ref. [31].)

3.4.2 Pulsed Operation of Holmium Laser

Pulsed operation of the holmium laser is an essential requirement for reducing the thermal load on the laser rod. It was found [31] that the optimal performance of αβHo:YLF or αβHo:YAG is around 150–160 K, where maximum slope efficiency is obtained. However, the threshold energy increased with an increase in temperature.

Laser crystals αβHo:YAG or αβHo:YLF operate successfully at a temperature of T≈290 K with an output energy of 2.5 J/pulse, an efficiency of 0.8%, and a threshold energy of ≈100 J. Crystals of Ho^{3+} codoped with Tm^{3+} and Er^{3+} have strong absorption lines in the visible and UV part of the spectrum. Therefore it is expected that the laser performance would improve by increasing the UV or visible lines in the emission spectrum of the flash tube. This is achieved by increasing the current density in the flashtube; the effect of current density inside the flashlamp on the laser performance is measured by controlling the pulse width and maintaining the same electrical energy. This is displayed in Figure 8, where, as expected, an increase in slope efficiency is obtained by an increase in the current density.

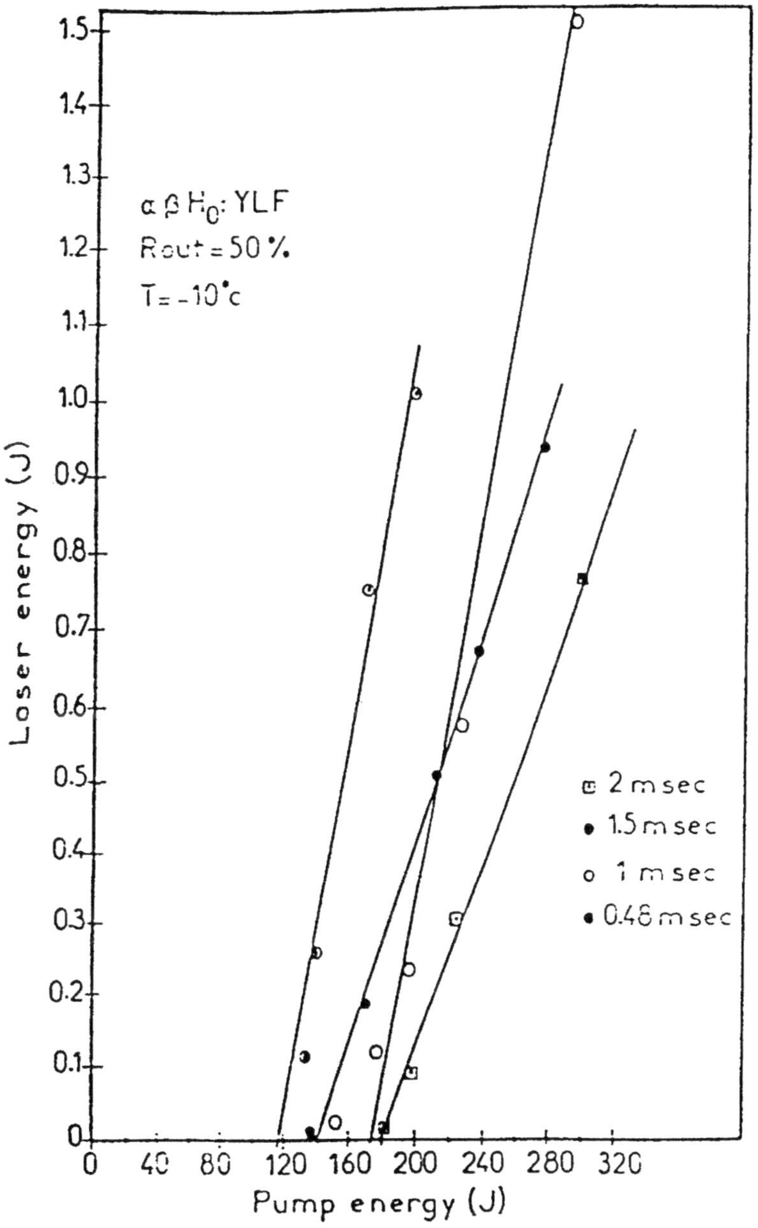

Figure 8: Laser output energy as a function of electrical energy deposited in the flashlamp for αβHo:YLF at -10°C, for different pulse widths. (Taken from reference [37].)

3.5 DIODE PUMPING

Diode pumping of 2 μm lasers is achieved in the codoped Tm:Ho:YAG or YLF system with efficiency as high as 70%. Holmium 2.1 μm laser emission from diode-pumped Ho:YLF crystals was first reported at 77 K. However, with the advent of diode arrays, it has become possible to pump efficiently the Tm^{3+} and the Ho^{3+} ions even at ambient temperatures. The excited Tm^{3+} ions sensitize the holmium ions to operate at the $^5I_7 \rightarrow {}^5I_7$ transition. The spectroscopic scheme in Tm:Ho:YAG is illustrated in Figure 9.

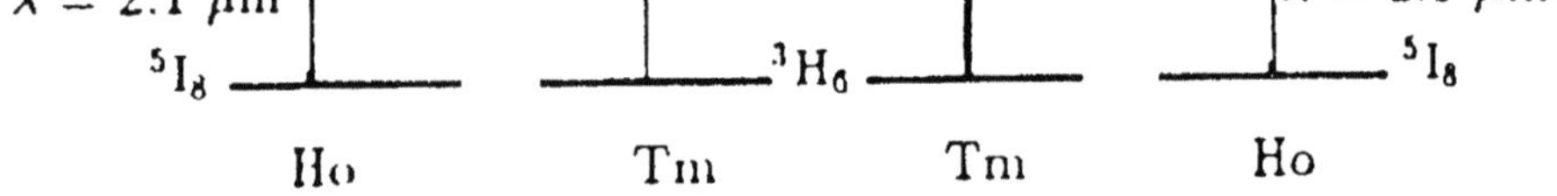

Figure 9: Pumping scheme of Tm, Ho:YAG. In some descriptions the $^3H_4$ and $^3F_4$ energy positions are reversed. It is by now widely accepted that $^3F_4$ is lower than the $^3H_4$ level. (Taken from Ref. [38].)

At high concentrations, usually above 3% at. of Tm^{3+}, cross-relaxation between two Tm^{3+} ions is responsible for the depletion of the $^3H_4$ level. The cross-relaxation process provides two excited Tm^{3+} in the $^3F_4$ state for every 785 nm pump photon. Theoretically, the quantum efficiency of such a process is two. Schematically, the process can be described as the following cross-relaxation process:

$$Tm(^3H_4 \rightarrow {}^3F_4) \rightarrow Tm(^3H_6 \rightarrow {}^3F_4)$$

or

$$Tm(^3H_4, {}^3H_6) \rightarrow Tm(^3F_4, {}^3F_4)$$

The double cross-relaxation is efficient and is associated with a fast spatial energy migration among Tm^{3+} ions, subsequent energy transfer to the $^5I_7$ level of Ho^{3+}, and, finally, laser emission via the Ho^{3+} $^5I_7 \rightarrow {}^5I_8$ transition.

The $^3H_4$ manifold of Tm^{3+} is pumped by the AlGaAs diode laser at 780–785 nm. The ground state $^3H_6$ absorbs one photon at 785 nm into the $^3H_4$ level (Figure 10). At low Tm^{3+} ion density, where ion-ion interactions are small, the $^3H_4$ level decays to the $^3F_4$ level radiatively or non-radiatively, via a multiphonon process, which then subsequently transfers its energy from the $^3F_4$ state into the $^5I_7$ level of holmium. At a low pumping rate into the $^3H_4$ level, the dominant process is a non-radiative multiphonon decay into the $^3F_4$ state of Tm^{3+} ion, with subsequent energy transfer to the Ho^{3+} $^5I_7$ level. At high pumping rates, laser action of the $^3H_4 \rightarrow {}^3H_5$ ($\lambda = 2.31$ μm) transition in Tm^{3+} is obtained and cascade emission, which involves two different ions, occurs [38].

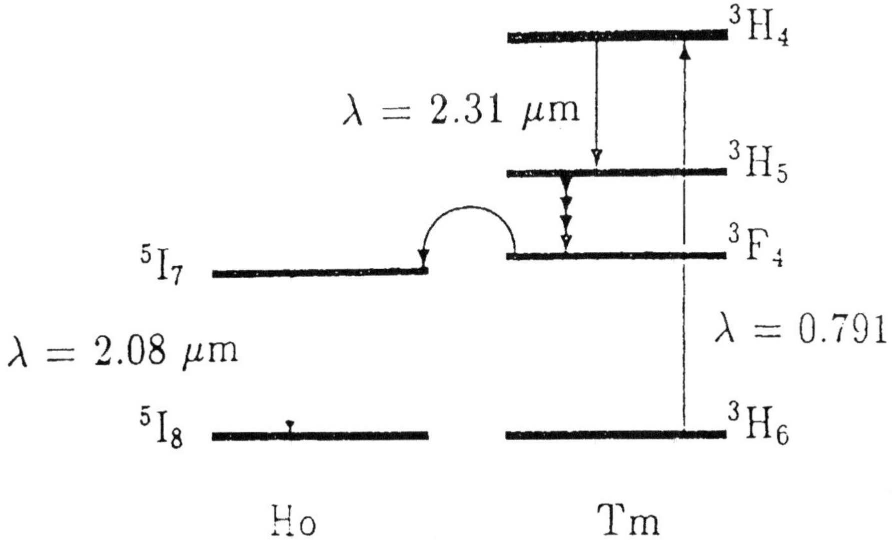

Figure 10: Pumping diagram for cascade laser emission in Tm:Ho:YLF. (Taken from Ref. [38].)

From this figure it can be seen that the terminal $^3H_5$ transition decays non-radiatively to the $^3F_4$ state, which then transfers its energy to the neighboring $^5I_7$ level of holmium. This cascade operation produces a change in the slope efficiency when Tm^{3+} laser is operated, due to the change of the emission wavelength from 2.31 μm to 2.08 μm.

From the mechanism presented here we observe that two infrared photons are produced as a result of one absorbed photon. The output power of the system is plotted in Figure 11. The initial slope efficiency for the holmium emission $^5I_7 \rightarrow {}^5I_8$ is 26%. The slope efficiency changes at the onset of the 2.31 μm emission; above that threshold, for both laser wavelengths, the slope efficiency is 42%.

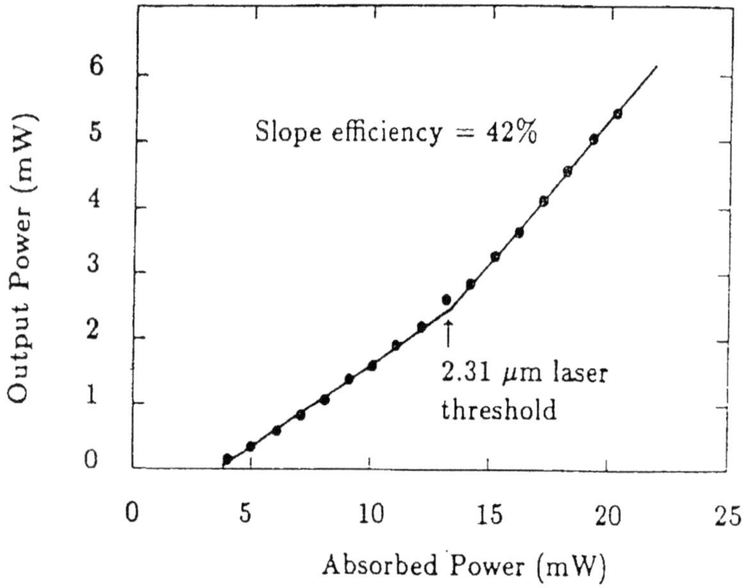

Figure 11: Slope efficiency for the Tm, Ho:YLF laser system. (Taken from Ref. [38].)

Note here that the Cr^{3+} ion which is codoped with Tm^{3+} and Ho^{3+} ions does not have an active role in diode pumping. However, its presence allows us to obtain 2 μm lasing with pump sources that have broad bands in the visible or UV spectral range. It also allows the use of a visible laser source, such as Kr^+ laser, or a dye laser, such as rhodamine 6G. The pumping scheme using a Kr^+ laser is performed via Cr^{3+} ($^4T_2$ or $^4T_1$ levels) and is presented in Figure 12 [39].

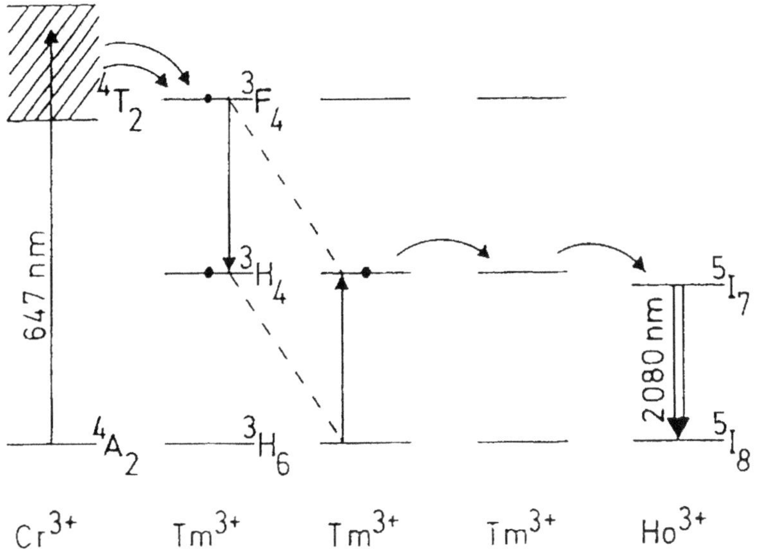

Figure 12: A pump scheme of the 2.1 µm Ho^{3+} laser with Cr-Tm energy transfer. The holmium was pumped with a Krypton ion laser.

Tables 2 and 3 summarize some of the significant results of diode pumped Tm:Ho-doped crystals [40].

Table 2: A brief summary of pioneering results of diode pumped holmium lasers.

Author/ Material	Pumping Wavelength (nm)	Slope Efficiency (%)	Threshold Power (mW)	Max. Output Power (mW)	Operating Temp. (K)
Duczynski et al. (1986) Cr:Tm:Ho:YSGG Cr:Tm:Ho:YSAG	647.1 nm	13%	25 mW	@60 mW	300
Fan et al. (1987) Tm:Ho:YAG	781 nm	19%	4.4 mW	1.2 mW	300
Kintz et al. (1987) Tm:Ho:YAG	785 nm	55%	17 mW	2.8 mW	300
Hemmati (1987) Er:Tm:Ho:YLF	790 nm	33%	7 mW	56 mW	77
Huber et al. (1988) Cr:Tm:YSGG Cr:Tm:YSAG	647.1 nm	41%	@30 mW	320 mW	300
Kane et al. (1988) Tm:Ho:YAG	780 nm	20%	33 mW	90 mW	300

Table 3: A brief summary of some recent results of diode pumped holmium lasers.

Author/ Material	Pumping Wavelength (nm)	Slope Efficiency (%)	Threshold Power (mW)	Max. Output Power (mW)	Operating Temp. (k)	Remarks
Hemmati (1989) Tm:Ho:YLF (JLP)	792 nm	30% [1]	108 mW	26 mW	300 K	Output @77 K: 187 mW, S.E.=67% [1]
Esterowitz (1990) NRL, Tm:Ho:YLF	791 nm	42% [1]	3.1 mW @ 2.09 μm 10.5 mW @ 2.3 μm	6 mW	300 K	Both emissions @ λ=2.08 and λ=2.31 μm
McGuckin & Menzis (1992) (JPL)	792 nm	60% [1] 42% [1] Conversion efficiency	50 mW (extrapolated)	84 mW	275 K	Band width 6nm (≅16cm^{-1},) 200 μJ @ 150Hz
Bundi et al. (1992) Lockheed Sanders, Inc. Tm:Ho:YLF	792 nm	49,4% [2] 55% Conversion efficiency	≅0. 5 mW	2.12 Watt	77 K	Q-Switch @ 1 kHz Stable: 1-90 kHz
Stoneman & Esterowitz (1992) NRL	2.01 μm of Tm pumped @ 785 nm	42% [1]	120-180 mW	≅120 mW	300 K	Intracavity pumped configuration
NRL & Lightning Optical Corp. (1992-93)	785 nm Quasi-CW diode array	3-4% electrical to optical		15 Watt (av.)	270-300 K	25-30 Hz Rep. Rate High av. power diode pumped Ho laser
Kushawaha et al. Howard Univ. (USA) (1996) Tm:Ho:LuAG	Fiber-coupled CW-diode array @ 785 nm	35% [2]	200 mW	1.38 W	77 K	
Beach et al. LLNL (1996) Tm:YAG	End-pumped CW-diode array @ 805 nm	20% [2]	125 mW	26 W	Water cooled	Pumping at the absorption wing

(1) Slope efficiency calculated relative to absorbed power.
(2) Slope efficiency calculated relative to incident power.

4. Fluoride Laser Crystals

4.1 $YLiF_4$ (YLF)

Fluoride crystals are among the most important hosts for laser materials due to their special optical properties. Of the fluorides, we will concentrate on $LiYF_4$ (YLF) which is one of the most common rare earths doped laser materials with a variety of efficient mid-IR laser lines from the UV (Ce^{3+}:YLF) [41] to the mid IR. Generally, YLF has good optical properties with high transparency throughout the emission spectrum of the conventional sources used for pumping solid state lasers. YLF does not show UV damage, and it has lower non-radiative decay rates for transitions occurring between electronic levels participating in the pumping and lasing process. YLF also has a low two-photon absorption coefficient. Due to its low non-radiative rates, the material can be used for cascade emission [42] between intermediate levels as well as an up-converter, as will be discussed later.

YLF is also a good medium for mode locking at 1.06 and 1.313 μm due to its natural birefringence and low thermal lensing. Mode-locked pulses from YLF are shorter due to its broader linewidth, both for the 1.064 μm and 1.313 μm emission peaks. The crystallographic structure of $LiYF_4$ (or YLF) is the same as $CaWO_4$, which was developed as a potential laser material years ago [43]. However, when a trivalent rare earth ion substitutes for the Ca^{2+} ion, charge compensation is necessary. The process of charge compensation may results in different sites in the crystal, and is a source for the disordered structure in the crystal. YLF has the advantage that no charge compensation is necessary throughout the doping process, since the trivalent rare earth ion substitutes for the Y^{3+} ion. As a result, we expect to have a single undisturbed site. The crystal has tetragonal symmetry; the important optical and physical properties shown in Table 1 and Table 2 of the first chapter of this book. A schematic energy level diagram of those levels participating in the lasing process in Nd:YLF is presented in Figure 13.

4.1.1 Thermal and mechanical properties of YLF

The thermal and mechanical properties of αβHo:YLF were measured by Chicklis et al.[44] (Tables 1 and 2 of the first chapter of this book). The authors describe the estimated power loading at fracture. The method used by them is be described in Appendix A. The value of the power loading per unit length for YLF is 11 W/cm, while for YAG it is 60 W/cm.

Perhaps the most important factor affecting the laser performance of YLF is its refractive index. YLF exhibits a negative change of refractive index with temperature: dn/dT<0, where n is the refraction index and T is the crystal temperature. This minimizes the thermal lensing effects in the crystal and

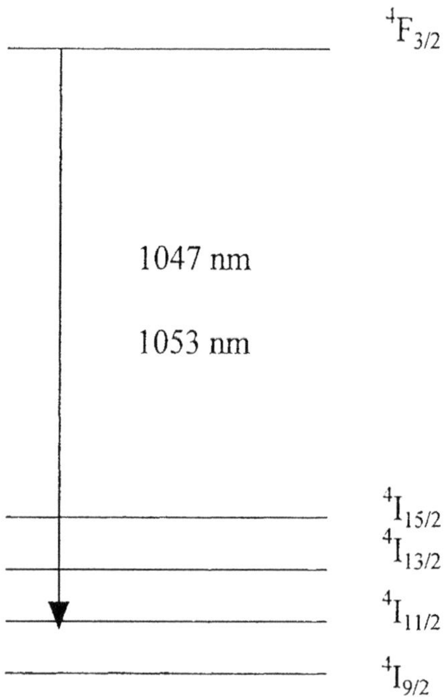

Figure 13. Schematic energy level diagram of electronic levels participating in the lasing process in Nd:YLF.

improves the fraction of the available power with the TEM_{00} mode (improving the beam quality). Let's assume we have an absorbing medium which is heated by radiation. If the temperature of the absorbing medium is increased by ΔT, the refractive index upon irradiation is:

$$n(\Delta T)=n(0)+(dn/dT)\cdot\Delta T \quad (1)$$

where n(0) is the refractive index at any point without pumping the absorbing medium. Let's assume we have a cylindrical lasing medium which is cooled through its surface. The temperature will have a maximum value at the axis and minimum value at the surface, and will drop gradually from the center to the peripheral region. If the condition dn/dT>0 is fulfilled, the axis region will be optically denser than the surface and the radiation along the axis will be focused. In the case of dn/dT<0, the periphery will be denser than the axis and rays propagating along the rod axis will be defocused.

A comparative study of thermal lensing measurements in YLF and YAG [45] show that at 700 W average input power, the extraordinary polarization (the polarized transition of Nd:YLF with emission at λ = 1047nm) has a focal length of 60±20 M while the ordinary polarization (σ polarization at λ=1053 nm) has a very weak thermal lensing effect with a lower limit for the focal length of about ±150 M. The measurements were taken by using a probe beam at 633 nm and an intracavity polarizer for selecting the 1047 nm or 1053 nm transitions, followed by imaging the beam coming from the surface of the laser crystal by a focusing lens and an aperture. Changes in the transmission of the probe beam passing through the aperture during the laser operation were measured and the appropriate focal lengths were obtained. The thermal lensing for both types of polarizations is much weaker than the thermal lensing in Nd:YAG (f_{th}=+8.3 M) [45]. The negative value of dn/dT cancels most of the stress-optics' contribution to the total changes in refraction index of the laser crystal. These changes are given by: $n(r)=n_0+\Delta n(r)T+\Delta n(r)\varepsilon$ where n(r) is the radial variation of the refractive index, n_0 is the refractive index at the center of the rod, $\Delta n(r)_T$ and $\Delta n(r)\varepsilon$ are the temperature and the stress dependent changes of the refractive index, respectively. Since the thermally induced focal length of a solid state laser material depends on the relative contribution of the above refractive index-changes, in YLF, $n(r) \cong n_0$, which implies that YLF will exhibit very weak thermal effects. A detailed investigation of the σ polarization with weak focusing effects was reported by Vanherzeele [46]. He measured the thermal lensing in two directions: parallel and perpendicular to the c-axis (the laser rod axis was along the crystallographic a-axis); Figure 14 illustrates the crystallographic directions of the laser rod in the experiment described in Ref. [46].

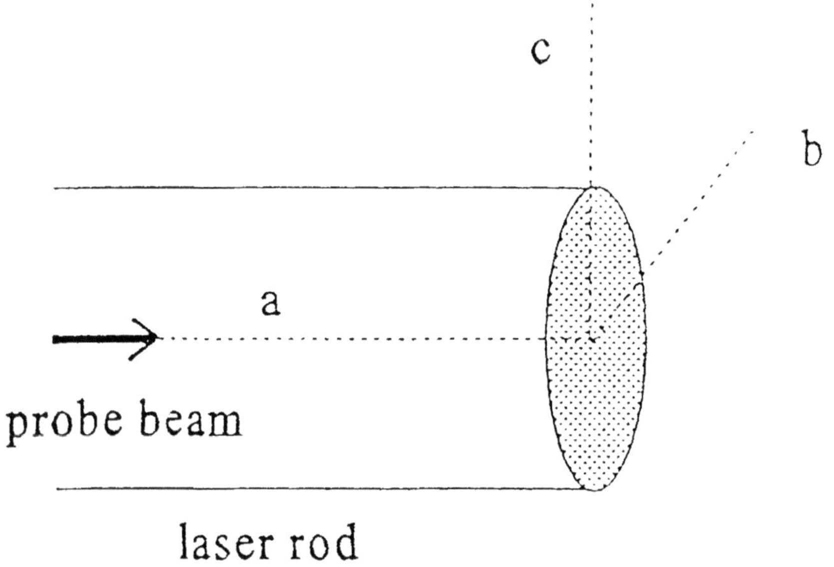

Figure 14. Crystallographic direction of the laser rod in thermal lensing experiment performed by Vanherzeele. (Taken from Ref [46].)

The method used in this experiment was similar to that reported in Ref. [45], except that the pinhole used was not circular but a narrow slit positioned parallel or perpendicular to the c-axis, namely, in the (a,b) plane and (a,c) plane. It was found that the thermal lensing is different not only for the σ and π polarizations but also for the (a,b) and (a,c) planes. For example, it was found that a σ polarization has a negative focal length, $f < 0$, for the (a,c) plane ($\sigma \| c$), and $f > 0$ for the (a,b) plane ($\sigma \perp c$). This effect will lead to astigmatism if not compensated for [43]. Temperature variations of the refractive index in YLF were investigated by Barnes and Gettemy [47]. The measurements were made in the visible spectral regime using an undoped YLF crystal prism. Results from the measured ordinary (n_0) and extra-ordinary (n_e) indices of refraction and their variation with temperature are presented in Table 4.

The temperature variation of the refractive indices of several oxide and fluoride crystals shows that fluorides generally have negative values of dn/dT, while oxides and other crystals (such as listed in Table 5) have a positive change of the refraction index with temperature and therefore will exhibit thermal lensing effects.

Table 4: Measured refraction indices and their variation with temperature for YLF [47]. Values of dn_0/dT and dn_e/dT were measured several times and averaged.

l, nm	n_0	dn_0/dT, $^0C^{-1}$	n_e	dn_e/dT, $^0C^{-1}$
435.8	1.46136	-0.54×10^{-6}	1.48389	-2.44×10^{-6}
546.1	1.45599	-0.67×10^{-6}	1.47826	-2.30×10^{-6}
578.0	1.45499	-0.91×10^{-6}	1.47705	-2.86×10^{-6}

Table 5: Temperature variation of refractive indices of several crystals (Taken from Ref. [47] and references therein).

Crystal	$dn/dT\times10^6$ ($^0C^{-1}$)	Crystal	$dn/dT\times10^6$ ($^0C^{-1}$)
Al_2O_3	13.1	BaF_2	-18.6
SiO_2	10	CaF_2	-16.0
CdTe	93	LiF	-17±0.2
Ge	396±40	KCl	36.5±0.2
Si	154±15	NaCl	-31
ZnSe	10.7±1		

4.1.2 Non Radiative Losses in YLF

Another important property of YLF is its non-radiative decay rates. YLF has lower non-radiative losses than oxides since the phonon energy of fluorides is lower than oxides. This has several effects on the performance of rare earth doped fluorides:

1. Rare-earth doped fluorides are generally more efficient luminescent systems.
2. There is less heat dissipated in the fluoride crystal as a result of lower non-radiative rates.
3. As a result of the low non-radiative losses in fluorides, rare earth doped fluorides exhibit a larger fluorescence lifetime of the emitting level than the rare earth doped oxides. For example, the $^4F_{3/2}$ lasing level of Nd^{3+} has a longer fluorescence lifetime in YLF than the same concentration in YAG (Table 6). This is an advantage for CW operation or Q-switch applications where Nd:YLF is superior to Nd:YAG in performance.

Table 6: Laser parameters for Nd:YLF and Nd:YAG [48].

Parameter/Host	**YLF**	**YAG**
Stimulated emission cross-section (σ), cm^2	1.2×10^{-19} (σ polarization); 1.8×10^{-19} (π polarization)	2.4×10^{-19}
Fluorescence lifetime of $^4F_{3/2}$ (τ), μsec	480	230
P_{th}(YLF)/P_{th}(YAG) P_{th}— power at threshold	0.85	
η(YLF)/η(YAG) η- slope efficiency	0.53	

Since YLF has lower thermal lensing than YAG it has a better beam quality even at high average power operating conditions. Therefore, we expect a higher fraction of the available laser power in the TEM_{00} mode in YLF than in YAG. The TEM_{00} mode volume for Nd:YLF is larger by a factor of four than the mode volume observed in YAG, whereas the TEM_{00} mode average power for YLF is higher by a factor of two than in YAG [48].

In this experiment the laser rods of both Nd:YLF and Nd:YAG were 6.3 mm (in diameter) x 76 mm (in length), doped with 1% Nd^{3+} (1.3×10^{20} cm^{-3} ion density), AR coated on both sides @1064 nm for YAG and 1050 nm for YLF. The pump cavities used for the experiments were in a gold or silver plated single elliptical configuration. The optical resonator was a confocal resonator for both YLF and YAG, pumped by a tungsten-halogen lamp. The curved - mirror resonator had an output coupling reflectivity of R=99%.

When pumped by a krypton laser, a pair of flat/flat mirrors defined the resonator which had an ouput coupling reflectivity of R=90%.

YLF has lower non-radiative rates within its electronic levels than the oxides, and it has a longer lifetime of the emitting levels. In the case of Nd doped fluorides like YLF or LaF_3, the longer lifetime of $^4F_{3/2}$ level of Nd:YLF is a source for excited state absorption (ESA). This phenomena is another channel for the depletion of the lasing level $^4F_{3/2}$. As a result of this depletion, yellow emission and upconversion lasing in the Nd doped LaF_3 has been observed. Seelert et al. [49] found that for pumping densities of $800W/cm^2$, the $^4F_{3/2}$ level is depopulated with a reduction of the lifetime by 30% , which also reduces the storage potential of this level.

4.1.3 Homium-Doped YLF

Crystals of αβHo:YLF with the following stoichiometric composition: $LiY_{1-x-y-z}Er_xTm_yHo_zF_4$, x=50%, y=5–6.7%, z=0.25%–0.5% (atomic percent), are grown by a top seeded solution technique in a high purity inert atmosphere; Ar or He are used as inert gases. When Ar was used, boules without bubbles and of uniform quality were obtained, while when helium was used as an inert atmosphere, isolated bubbles which were parallel to the c-axis appeared. These problems can be eliminated by replacing the helium with argon. Also, depending on the crystal growth conditions, scattering inclusions appear in YLF boules. These inclusions are formed as a result of chemical impurities such as Mg^{2+} and Ca^{2+} oxides, and hydroxides, as well as thermal instabilities in the melt. They function as fine bubbles of microscopic size which act as scattering centers. Chicklis et al. [33] note that there are two types of precipitation phenomena that should be distinguished: the first type are oxides that form rare earth oxyfluorides, while the second type are chemical complexes between impurities and oxygen. The rare earth fluorides used for the crystal growth should be of high purity without any metallic iron or carbon. The purity of the seeding and fluoride compounds are important in eliminating grain boundaries in the boule. The impurity level has an effect on the damage threshold of the laser crystal and on the degredation of the optical properties of YLF rods upon UV radiation.

The crystal $YLiF_4$ is grown from melt of LiF-YF_3, which melts incongruently at 819^{0}C. Yttrium in the LiF-YF_3 system can be substituted up to about 50% by other rare earth ions heavier than Sm^{3+}, without altering the physical properties of the crystal.

Devor et al. [50] described in details the preparation of YLF doped with Tm, Er, Ho in which the starting materials were R_2O_3 (R=Y, Er, Ho, Tm) of 99.999% purity which were converted to RF_3 by reaction with aqueous HF at room temperature. To complete the process of conversion, the residue was dried and treated with HF at 700-800^{0}C, where a conversion of 99.9% was achieved. The role of various oxides in the crystal growth and in the phase stability of fluorides was also investigated. It was found that levels as low as 10-15 ppm of oxygen would not alter the structure of fluorides. At higher concentrations, up to 0.2% mole, oxyfluorides are formed and this phase co-crystalyzes with the $LiYF_4$ phase at higher temperature. Upon cooling, a separate oxide phase could precipitate with the $LiYF_4$ crystal and this separate phase is incorporated into the YLF crystal.

Experiments conducted by Chicklis et al. [44] revealed that, even in the presence of oxide compounds, a clear scatter-free YLF can be obtained since the oxide phase is rejected from the YLF phase and is precipitated separately. Also, it was found that the relation between the trace amounts of optical scattering centers and the oxide content in the melt is not a simple relation, and that the formation of scattering centers depends also on the presence of other impurities such as OH^-, Mg^{2+}, and Ca^{2+}.

4.2 OTHER FLUORIDE CRYSTALS

BaY_2F_8 is another fluoride host for rare earth doped laser materials. BaY_2F_8 is a monoclinic crystal. The crystal axes are not orthogonal: its lattice constants are:
a=6.97, b=10.46, and c=4.26. The angle between the a and c axes is 99.7°. Axis b is normalized to the a,c plane. Figure 15 presents schematically the crystallographic and optical axes of BaY_2F_8 [51].

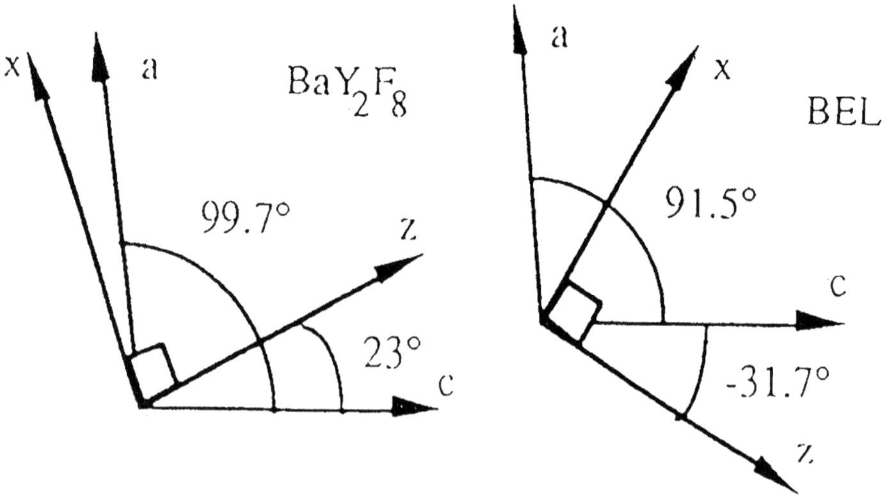

Figure 15. Crystallographic and optical axes in BaY_2F_8. (Taken from Ref. [51].)

Since the main crystal axes are not orthogonal, the optical axes of the refractive index tensor do not coincide with them. As a result, the real and imaginary parts of the refractive index, (which correspond to the refractive index of the material and to the absorption cross-section, respectively) may have different orientations in the (a,c) plane. This phenomena will lead to absorption and emission cross-sections in this material with different values, depending on the angle relative to C-axis. It was shown by Dinndorf et al. [51] that for $Nd:BaY_2F_8$ the absorption cross-section (σ_a) has a minimum value at 861 nm at an angle of $\theta_0=38^0\pm5$ relative to the c axis and a maximum for 861 nm at $25^0\pm4$. A similar phenomena was found for $Nd:La_2Be_2O_5$ (BEL). For BEL it was also found that there is an anisotropy of the absorption and emission cross-sections. This anisotropic behavior was found within the multiplets of the electronic transition $^4F_{3/2}$.

4.3 CASCADE UPCONVERSION EMISSION

In general the fluoride hosts have lower non-radiative rates than the oxide hosts. Due to the smaller non-radiative losses in fluorides, transitions between intermediate states become more efficient than in oxide hosts, for example. As in the case of $YLiF_4$ and other fluorides, BaY_2F_8 can be used for cascade emission to intermediate levels as well as for upconversion processes [52,53]. Although we will not analyze these processes here, we will outline some of the typical observations in fluorides hosts.

An example for cascade emission is the thulium YAG laser emission at 2.3 μm which was obtained by excitation at 781 nm (which is also the pump wavelength of the GaAlAs diode laser) [38]. This laser transition originates from the $^3H_4 \rightarrow {}^3H_5$ transition. The $^3H_4$ level of Tm^{3+} has a lifetime in the msec time regime. The lower lasing level $^3H_5$ has a lifetime of τ=1 msec and the lower Stark levels of this level are thermally unpopulated. The $^3H_4$ level can be sensitized by Cr^{3+} as well as via Yb^{3+} with anti-Stokes pumping of the $^3H_4$ level and lasing via $^3H_4 \rightarrow {}^3H_5$ and $^3H_4 \rightarrow {}^3F_4$ transitions. These laser transitions are inefficient and with high threshold energy. Direct excitation into the $^3H_4$ level of Tm^{3+} (see Figure 10) at 781 nm yielded 55 mJ output energy at 2.3 μm (the $^3H_4 \rightarrow {}^3H_5$ transition) with 15% overall efficiency at room temperature. Also, cascade emission at 3.5 μm ($^3H_4 \rightarrow {}^3F_4$) or $^3F_4 \rightarrow {}^3H_6$ (1.83 μm) were observed due to the long lifetime of the metastable levels which is a direct result of the low non-radiative processes in fluoride crystals.

Stimulated cascade emission lines in $Ho:BaY_2F_8$ were observed by Johnson and Guggenheim [54] for $^5F_5 \rightarrow {}^5I_5$ (2.4 μm), $^5I_6 \rightarrow {}^5I_7$ (2.6 μm), and $^5I_7 \rightarrow {}^5I_8$ (2.1 μm) at low temperature range, 22 K - 77 K. The schematic energy level diagram and the relevant laser transitions are presented in Figure 16.

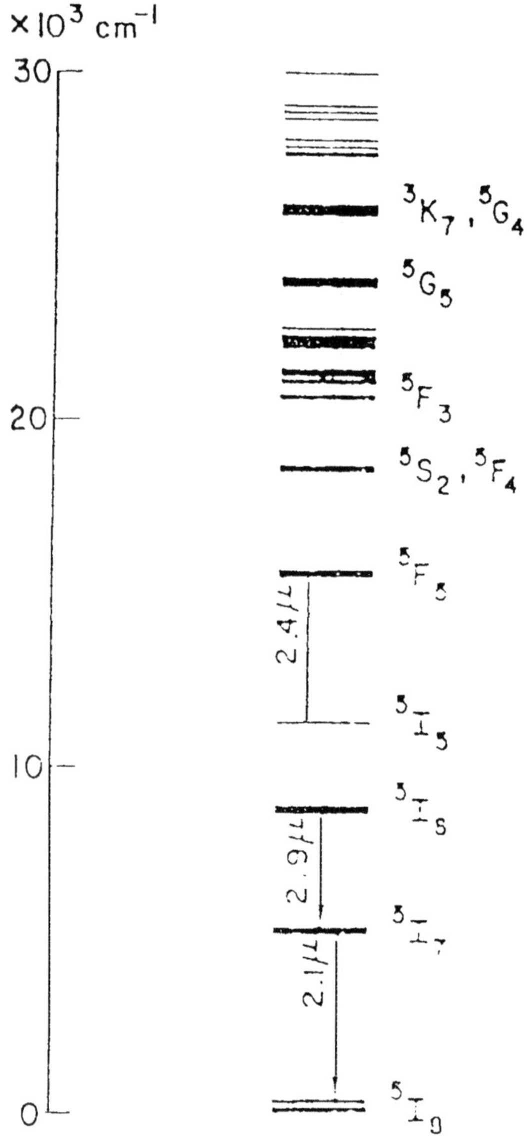

Figure 16. Energy level diagram and cascade transitions in Ho^{3+}:BaY_2F_8. (Taken from Ref. [54].)

Upconversion:

Upconversion is a multistep process involving low frequency photons which excite a metastable level that produces laser light with photons that are more energetic than the pump photons. It can originate in a singly-doped crystal or from cooperative energy transfer between two different rare earth ions. Consider the Yb-Er and Yb-Ho system, where visible emission both in Ho^{3+} and Er^{3+} via upconversion processes was observed in a BaY_2F_8 crystal codoped with Yb^{3+} and Er^{3+} or Yb^{3+} and Ho^{3+} [55,56]. The pumping wavelength in these experiments was at 960 nm (via the $^2F_{7/2} \rightarrow {}^2F_{5/2}$ transition). A schematic representation of the infrared pumped - visible upconversion processes in Yb-Er and Yb-Ho in BaY_2F_8 is presented in Fig. 17 below.

The Yb-Er system shows two laser lines: at λ =670.9 nm ($^4F_{9/2} \rightarrow {}^4I_{15/2}$ transition), and at λ = 550nm ($^4S_{3/2} \rightarrow {}^4I_{15/2}$). The red and the green emissions are three and two successive processes, respectively. An upconversion process via infrared pumping was found also in Ho:Yb:BaY_2F_8, while pumping was in the near IR. The 960 nm excitation ($^2F_{7/2} \rightarrow {}^2F_{5/2}$) yielded green emission of Ho^{3+} at 551.5 nm. In both cases, Yb-Er or Yb-Ho upconversion was observed at 77 K. The visible lasing in both Er and Ho was into the upper Stark component of the lowest electronic ground level of Er^{3+} ($^5I_{15/2}$) or Ho^{3+} ($^5I_8$). The upper Stark level in Er^{3+} is 410 cm^{-1} above ground level, while in Ho^{3+}, the highest Stark level of the $^5I_8$ ground state manifold is 385 cm^{-1} above the ground level. By lowering the temperature to 77 K, the population of this level will be decreased by a factor of ~300, and the system will become quasi-four-level.

An interesting phenomenon is a trio upconversion in the Er:YLF system [53], where the excitation is carried out via the $^4I_{15/2} \rightarrow {}^4I_{13/2}$ transition (1.55 μm), with upconverted emission at a wavelength of 550 nm ($^4S_{3/2} \rightarrow {}^4I_{15/2}$). This process is a cooperative energy transfer, where three ions are involved. The dependence of the upconverted 550 nm fluorescence on the incident intensity obeys a cubic power law. The risetime of the $^5S_{3/2}$ level is long, approaching that of the $^5I_{13/2}$ level. The CW upconversion laser emission can be explained by the excitation of one Er^{3+} ion to the $^2H_{11/2}$ level, followed by deexcitation of two neighboring $^4I_{13/2}$ ions.

Green emission at λ=549.8 nm ($^4S_{3/2} \rightarrow {}^4I_{15/2}$) at 77 K was also observed in Er:$YAlO_3$ by either two step infrared excitation or phonon assisted cooperative energy transfer [56]. The authors used two dye lasers for the consecutive excitation. The first one at 792.1 nm excited the $^4I_{15/2} \rightarrow {}^4I_{9/2}$ transition. This pumping was followed by non-radiative relaxation to populate the $^4I_{11/2}$ and $^4I_{13/2}$ levels which have lifetimes in the msec range.

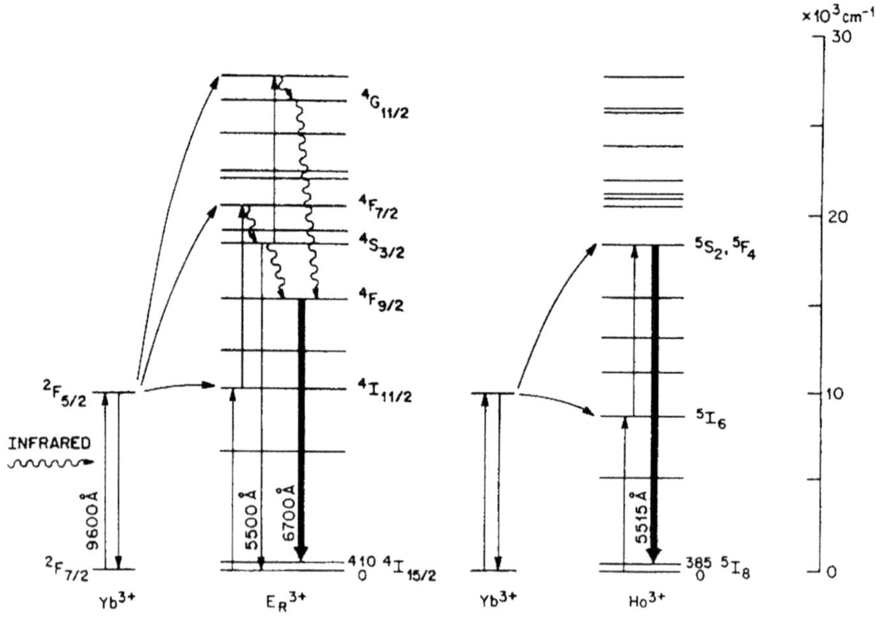

Figure 17. Infrared-to-visible upconversion in Yb^{3+} - Er^{3+} and Yb^{3+} - Ho^{3+} doped BaY_2F_8 (Taken from Ref. [55].)

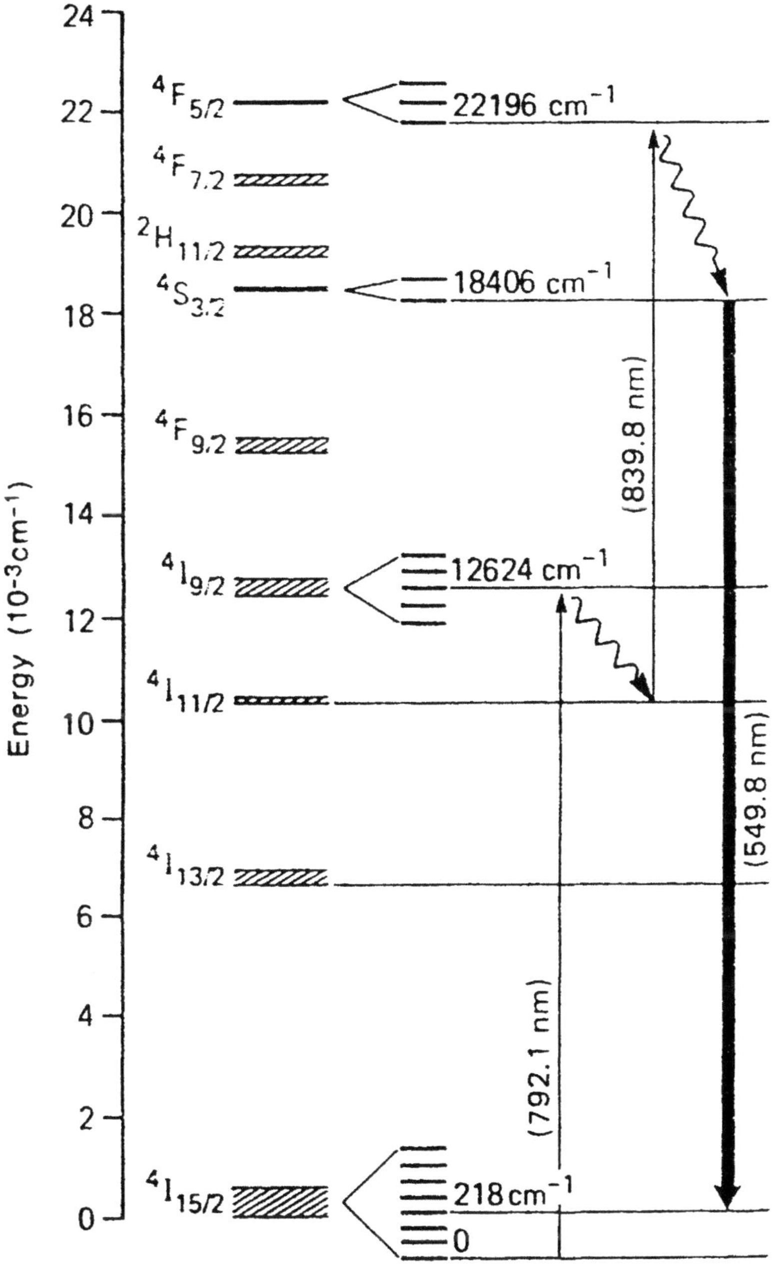

Figure 18. Energy level diagram of Er^{3+} with pumping and lasing pathways. (Taken from Ref. [56].)

A second dye laser at 839.8 nm excited the intermediate level $^4I_{11/2}$ into the $^4F_{5/2}$ level, with subsequent non-radiative decay to $^4S_{3/2}$ and a green emission $^4S_{3/2} \rightarrow {^4I_{15/2}}$. The pumping and the lasing pathways are displayed in Figure 18. In addition to the two-step scheme, the $^4S_{3/2}$ level is populated by cooperative energy transfer that involves two Er^{3+} ions in the $^4I_{11/2}$ state. This can be demonstrated by blocking the second dye laser beam and observing the green fluorescence originating from the $^4S_{3/2}$ level.

Other green and blue transitions resulting from an upconversion process were reported by Hebert et al. [58]. The laser crystal used for this upconversion experiment was 3 mm in length with a doping level of 1% at. of Er^{3+} and output coupling transmission of T=1%. They reported upconversion laser action at λ=469.7 nm and λ=560.6 nm in Er:YLF. The 560.6 nm ($^2H_{9/2} \rightarrow {^4I_{13/2}}$) laser emission was observed by excitation into $^4I_{9/2}$ (797 nm) or $^4I_{11/2}$ (968.9 nm). The transition was terminated on a Stark level which is 162 cm^{-1} above the lowest component of $^4I_{13/2}$; this level has a lifetime of 15 msec. Laser action at λ=469.7 nm is a result of multistep processes, where multiexcitations and energy transfer are involved.

Applications to upconversion:

To conclude this section we shall focus on some recent applications of upconversion. Upconversion is used in fiber laser technology to produce blue/green laser emission. The system is inexpensive, efficient, and a reliable source of light for a variety of applications in medicine, photography, and semiconductor wafer inspection [59]. Upconversion lasers are especially attractive when optical glass fibers are doped with rare earth ions. Relatively high optical intensities and efficient photon conversions can be extracted as a result of the high energy density of the pumping source and the long length of the gain medium. Upconversion is a multistep process which requires a long lived intermediate metastable level. This type of level can be found in low-phonon hosts such as fluoride glasses. The 544 nm upconversion laser emission in Er^{3+} can be obtained in Er doped fluoride glasses by pumping consecutively into the $^4I_{12/2}$ and $^4H_{11/2}$ levels by two 970 nm photons, followed by 544 nm emission ($^4S_{3/2} \rightarrow {^4I_{15/2}}$ transition). The lifetimes of the $^4I_{11/2}$ and $^4S_{3/2}$ levels in fluoride glasses are 9 msec and 0.45 msec, respectively. The appropriate lifetimes of Er doped silicate glass is 7 μsec for $^4I_{11/2}$ and 1 μsec for $^4S_{3/2}$. Therefore, in Er-doped silicate glass, any $^4I_{11/2}$ excited state will decay nonradiatively to $^4I_{13/2}$ to produce the 1.55 μm laser emission, rather than the upconversion process. Blue/green lasers can be obtained with holmium, erbium, thulium and praseodymium-doped fluoride glasses. Since the absorption bandwidth in glasses is broader than in crystals, the pump wavelength can vary by several nm over the absorption bandwidth without actually affecting the laser performance. Also, the broad emission bandwidth allows tunability of the laser output. Sometimes, the upconverted population by decays back to the first excited state by an efficient cross-relaxation process. This state can be used again for upconversion. This phenomena is called a loop mechanism and was demonstrated in Tm^{3+}:$Gd_3Ga_5O_{12}$ (GGG) [59]. In this example, Tm^{3+} was excited first by 660 nm photons into the $^3F_2$ state. The long lived $^3F_4$ level (17.1 msec at 77 K) absorbs a second photon of 638 nm wavelength into the $^1G_4$ level. The above process led to the blue emission $^1G_4 \rightarrow {}^3H_6$ (λ=484 nm) as well as to three ions in the $^3F_4$ level. The $^3F_4$ state was ready to reabsorb photons and continue the process. These processes are described schematically in Figure 19.

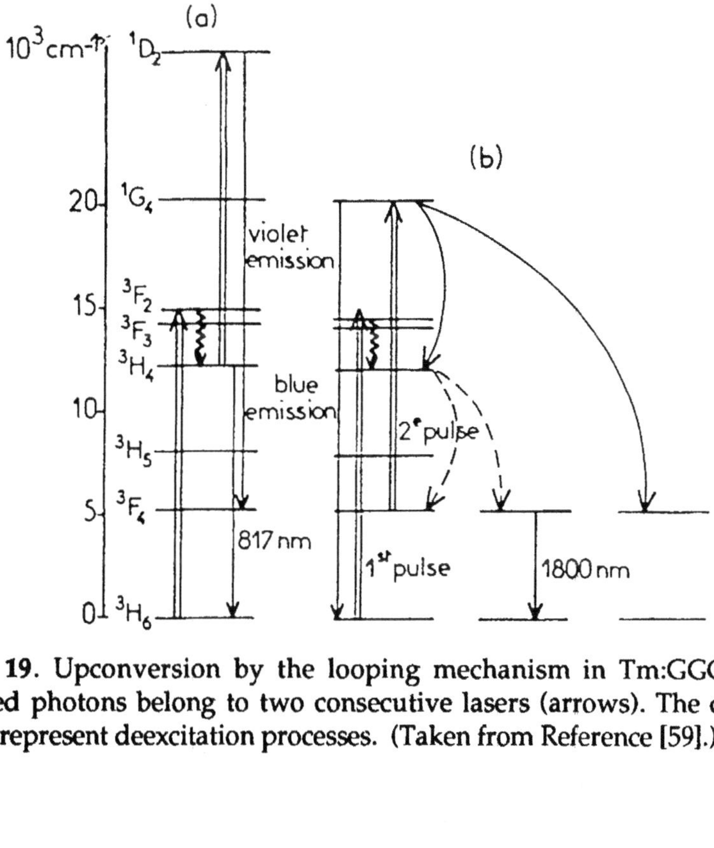

Figure 19. Upconversion by the looping mechanism in Tm:GGG. The absorbed photons belong to two consecutive lasers (arrows). The curved arrows represent deexcitation processes. (Taken from Reference [59].)

5. Conclusions:

The mechanical, physical and optical properties of garnets and various fluoride crystals were discussed and summarized in this chapter. The discussion encompassed the spectroscopic and lasing properties of neodymium and neodymium-chromium doped garnets. It was shown that the optical properties of Cr^{3+} depend on the chemical composition of the garnet. This can be used to enhance the energy transfer from chromium to neodymium to obtain a more efficient lasing system.

Garnets and fluorides can be doped with a variety of active ions, with emission ranging from 1 micron to 3 microns. One example of a 2 micron laser system was analyzed thoroughly, i.e. lasers based on thulium and holmium. The use of novel energy transfer schemes and new pumping methods to improve the laser efficiency and performance were discussed. Special emphasis was given to the fluoride crystals hosting the 2 micron lasing ions, which have unique optical and spectral characteristics. The low multiphonon decay rates in some fluorides (such as $YLiF_4$ or $BaYF_8$ permit multistep processes, with an extension of the emission wavelength to both the visible and mid-IR spectral regime. We expect that the future will bring many more candidates, both as hosts and ionic dopants, for new solid-state lasers.

Appendix A: Estimation of the Thermal Load for Fracture

The value of the thermal load at fracture of a uniformly heated laser rod cooled at the surface is an important parameter in estimating the average output power available from a given host. The radial temperature distribution is defined as:

$$T(r) = Q(r_0^2 - r^2)/4k \qquad \text{(A1)}$$

where T(r) is the temperature at a distance r from the rod axis, r_0 is the rod radius, k is the thermal conductivity, and Q is the heat per unit volume dissipated in the rod. A stress distribution gradient is accompanied by a temperature gradient as follows:

$$\sigma_r(r) = 2aE \cdot Q(r^2 - r_0^2)/[(1-\mu)16k] \qquad \text{(A2a)}$$

$$\sigma_\theta(r) = 2aE \cdot Q(3r^2 - r_0^2)/[(1-\mu)16k] \qquad \text{(A2b)}$$

$$\sigma_z(r) = 2aE \cdot Q(4r^2 - 2r_0^2)/[(1-\mu)16k] \qquad \text{(A2c)}$$

where the parameters appearing in Equations (A2a) – (A2c) are defined as follows:

$\sigma_r(r)$ – the radial stress at distance r
$\sigma_\theta(r)$ – the tagential stress at distance r
$\sigma_z(r)$ – the axial stress at distance r
μ – Poisson's ratio
E – Young's modulus
a – the thermal expansion coefficient

From these expressions, we observe that the radial component of the stress disappears at the rod surface while the tangential and axial components do not. Therefore, the rod is under tension which may cause it to crack.

References

[1] Denker, B.I., Osiko, V.V., Starikov, B.P., Timoshechkin, M.I., Shcherbakov, I.A., and Yablonskii, A. Ya. 1976: "Spectroscopic properties of neodymium-activated scandium garnets", *Sov. J. Quantum Electron.* **6** (3), pp. 334–336.

[2] Kokta, M.R. 1993: "Czochralski Growth of Oxide Laser Crystals", in: *Mat. Res. Soc. Symp.* Vol. 329, ed. by Bruce H.T. Chai and Steven A. Payne, (Fall Meeting November 29–December 3, 1993), pp. 33–43.

[3] Kokta, M.R. 1973: "Solubility Enhancement of Nd^{3+} in Scandium-Substituted Rare-Earth Aluminum Garnets", *J. Solid State Chem.* **8**, pp. 39-42.

[4] Brandle, C.D. and Vanderleeden, J.C. 1974: "Growth, Optical Properties, and CW Laser Action of Neodymium-Doped Gadolinium Scandium Aluminum Garnet", *IEEE J. Quantum Electron.* QE-**10** (2), pp. 67-71.

[5] Monchamp, R.R. 1971: "The Distribution Coefficient of Neodymium and Lutetium in Czochralski Grown $Y_3Al_5O_{12}$", *J. Cryst. Growth* **11**, pp. 310-312.

[6] Brandle, C.D. and Barns, R.L. 1973: "Crystal Stoichiometry and Growth of Rare-Earth Garnets Containing Scandium", J. *Cryst. Growth* **20**, pp. 1-5.

[7] Kiss, Z.J. and Duncan, R.C. 1964: "Cross-Pumped Cr^{3+}–Nd^{3+}:YAG Laser System, *Appl. Phys. Lett.* **5**, pp. 200-202.

[8] Balda, R., Fernandez, J., de Pablos, A., and Fernandez-Navarro, J.M. 1993: Cr^{3+}–Nd^{3+} Energy Transfer in Fluorophosphate Glass Investigated by Time-Resolved Laser Spectroscopy", *Phys. Rev. B* **48**, pp. 2941-2948.

[9] Smirnov, V.A. and Shcherbakov, I.A. 1988: "Rare-Earth Scandium Chromium Garnets as Active Media for Solid-State Lasers", IEEE *J. Quantum Electron.* **24**, pp. 949-959.

[10] Sugano, S., Tanabe, Y., and Kamimura, H. 1970: *Multiplets of Transition—Metal Ions in Crystals*, Academic Press, New York, NY.

[11] Tanabe, Y. and Sugano, S. 1954: "On the Absorption Spectra of Complex Ions", *J. Phys. Soc. Japan* **9**, pp. 753-766.

[12] Zharikov, E.V., Il'ichev, N.N., Laptev, V.V., Malyutin, A.A., Ostroumov, V.G., Pashinin, P.P., Pismenov, A.S., Smirnov, V.A., and Shcherbakov, I.A. 1983: "Spectral Luminescence, and Lasing Properties of Gadolinium Scandium Gallium Garnet Crystals Activated with Neodumium and Chromium Ions", *Sov. J. Quantum Electron.* **13**, pp. 82–85, and references therein.

[13] Zharikov, E.V., Zhitnyuk, V.A., Zverev, G.M., Kalitin, S.P., Kuratev, I.I., Laptev, V.V., Onishchenko, A.M., Osiko, V.V., Pashkov, V.A., Pimenov, A.S., Prokhorov, A.M., Smirnov, V.A., Stel'makh, M.F., Shestakov, V.A., and Shcherbakov, I.A., 1982: "Active media for high-efficiency neodymium lasers with nonselective pumping", *Sov. J. Quantum Electron.* **12**, pp. 1652–1653.

[14] Pruss, D., Huber, G., Beimowski, A., Laptev, V. V., Shcherbakov, I.A. and Zharikov, Y.V., 1982: "Efficient Cr^{3+} Sensitized Nd^{3+}:GdScGa-Garnet Laser at 1.06μm", **28**, pp. 355-358.

[15] Brenier, A., Boulon, G., Pedrini, C. and Madej, C. , 1992: "Effects of Ca^{2+} Zr^{4+} Ion Pairs on Spectroscopic Properties of Cr^{3+} Multisites in Cr^{3+}-Doped $Gd_3Ga_5O_{12}$ Garnets", *J. Appl. Phys.* **71** pp. 6062–6068.

[16] Grinberg, M., Brenier, A., Boulon, G., Pedrin, C. and Madej, C., 1993.: "Bandwidth and Time Evolution of the Cr^{3+} Fluorescence in (Ca,Zr) - Substituted $Gd_3Ga_5O_{12}$ ", *J. of Luminescence*, **55**, 303-314.

[17] Yun-Kui-Li, B. Zhang and Yu-xing Xia., 1993: ""Spectral Properties of a Novel Phonon Termated Laser Crystal YIGG:Cr^{3+} ",*Conference on Lasers and Electro-Optics*, Optical Society of America, p. 312, Paper CWJ37.

[18] Aravot, D., Lachman, E., Giler, S., Kaplan, I., Sagie, D., Kagan, J. and Kalisky, Y., 1987: "The Holmium Laser in Intravascular Coagulation", *7th Congr. Int. Soc. Laser Surgery Med.*, Munich, West Germany, June 22–26, 1987, p. 303.

[19] Henderson, S.W., Hale, C.P., Magee, J.R., Kavayd, M.J. and Huffaker, A.V., 1991: "Eye-Safe Coherent Laser Radar Systems at 2.1 μm Using Tm,Ho:YAG Lasers", *Opt. Letter.* **16**, pp. 773–775.

[20] Moody, S.E., 1987: "Evaluation of Laser Technologies for an Aircraft Wind Shear Detection," *Laser Radar II*, R.J. Becherer and R.C. Herney, Editors, *Proc. Soc. Photo-Opt. Instrum. Eng.* **783**, pp. 124–131.

[21] Barnes, N.P. , 1993: "Solid State Lasers for Monitoring the Health of Planet Earth", *Laser M2P* Lyon, December 8–10, 1993: Paper # VCP1.

[22] Johnson, L.F., Boyd, G.D. and Nassau, K., 1962: "Optical Maser Characteristics of Ho^{3+} in $CaWO_4$", *Proc. IRE (Corresp.)* **50**, pp. 87–88.

[23] Gandy, H.W., Ginther, R.J. and Weller, J.F. 1996., "Internal Q-switching of Ho^{3+}-Stimulated Emission in Iron Containing Glasses", *Appl. Phys. Letts.* **9**, pp. 277–279.

[24] Devor, D.P., Soffer, B.H. and Robinson, M., 1971: "Stimulated Emission from Ho^{3+} at 2 μm in HoF_3", *Appl. Phys. Letts.* **18**, p.p. 122-124.

[25] Johnson, L.F., Geusic, J.E., and Van Uitert, L.G., 1966: "Efficient High Power Coherent Emission from Ho^{3+} Ions in Yttrium Aluminum Garnet, Assisted by Energy Transfer", *Appl., Phys. Letts.* **8**, pp. 200–202.

[26] Chicklis, E.P., Naiman, C.S., Folweiler, R.C., Gabbe, D.R., Jenssen, H.P. and Linz, A., 1971: "High Efficiency Room Temperature 2.06 μm Laser Using Sensitized Ho^{3+}:YLF", *Appl. Phys. Letts.* **19**, pp. 119–121.

[27a] Antipenko, B.M., Vorykhalov, I.V., Sinitsyn, B.V. and Uvarova, T.V., 1980: "Laser Frequency Converter Based on a $BaYb_2F_8$:Ho^{3+} Crystal. Stimulated Emission at 2 μm", *Sov. J. Quantum Electron.* **10** , pp. 114–115.

[27b] Johnson, L.F. and Guggenheim, H.J., 1973: "Laser Emission at 3 μm from Dy^{3+} in BaY_2F_8", *App. Phys. Letts.* **23**, pp. 96–98.

[28] Tkachuk, A.M., Khilko, A.V. and Petrov, M.V., 1985: "Probabilities of Intracenter Spontaneous Radiative and Non-Radiative Intermultiplet Transitions in the Ho^{3+} Ion in an $LiYF_4$ Crystal", *Opt. Spectrosc. (USSR)* **58** , pp. 55–59.

[29] Reisfeld, R. and Jorgensen , C.K., 1977: *Lasers and Excited States of Rare Earths*, Springer-Verlag Berlin, Heidelberg, New York, Chapter 2.

[30] Koechner, W., 1992: *Solid State Laser Engineering*, Springer-Verlag, Berlin, Heidelberg, New-York, Third Edition, Chapter 2.

[31] Lotem, H., Kalisky, Y., Kagan, J. and Sagie, D., 1988: "A 2 μm Holmium Laser", *IEEE J. of Quantum Electron.* **24** (6), pp. 1193–1200.

[32] Beck, R. and Gurs, K., 1975: "Ho Laser with 50-W Output and 6.5% Slope Efficiency", *J. Appl. Phys.* **46** , pp. 5224–5225.

[33] Antipenko, B.M., Glebov, A.S., Kisileva, T.I. and Pis'mennyi, V.A. 1985: "2.12 μm Ho:YAG Laser", *Sov. Tech. Phys. Lett.* **11**, 284–285.

[34] Antipanko, B.M., Glebov, A.S., Kisileva, T.I. and Pis'mennyi, V.A., 1986: "A New Spectroscopic Scheme of an Active Medium for the 2 μm Band", *Opt. Spectrosc. (USSR)* **60**, pp. 95–97.

[35] Lotem, H., Kagan, J., Sagie, D. and Levine, L.A., 1985: "60 W 2.06 μm Ho^{3+}:YLF Laser", *J. Opt. Soc. Amer. A.* **2** , p. 107.

[36] Kalisky,Y., Kagan, J., Lotem, H. and Sagie, D., 1988: "Continuous Wave Operation of Multiply Doped Ho:YLF and Ho:YAG Laser", Opt. Commun. **65**, pp. 359–363.

[37] Kalisky, Y., Kagan, J., Sagie, D., Brenier, A., Pedrini, C. and Boulon, G., 1991: "Spectroscopic Properties, Energy Transfer, and Laser Operation of Pulsed Holmium Lasers", *J. Appl. Phys.* **70** , pp. 4095–4100.

[38] Esterowitz, L., 1989: "Diode Pumped Solid State Lasers at 2 and 3 micron for Biomedical Field", *French-Israeli Workshop on Solid State Lasers*, Boulon, G., Jørgensen, C. K. and Reisfeld, R., Editors, SPIE **1182**, pp. 49–58.

[39] Duczynski, E.W., Huber, G., Ostroumov, V.G. and Shcherbakov, I.A. 1986: "CW Double Cross-Pumping of The $^5I_7 - {}^5I_8$ Laser Transition in Ho^{3+}-Doped Garnets", *Appl. Phys. Lett.* **48**, pp. 1562–1563.

[40] Kalisky, Y., 1994: "Spectroscopic Properties, Energy Transfer Dynamics, and Laser Performance of Thulium Holmium Laser Systems", *Longer Wavelength Lasers and Applications*, Gabor Patonay, Editor, Proc. SPIE **2138**, pp. 5–16.

[41] Okada, F., Togawa, S., Ohta, K. and Koda, S., 1994: "Solid State Ultraviolet Tunable Laser: A Ce^{3+} Doped $LiYF_4$ Crystal", *J. Appl. Phys.* **75**, pp. 49-53.

[42] Esterowitz, L., 1994: Invited talk presented at *SPIE International Symposium, QE/ LASE'94 Meeting,* Los Angeles, CA, January 22–29.

[43] Nassau, K. and Broyer, A.M., 1962: "Calcium Tungstate: Czochralski Growth, Perfection and Substitution", *J. Appl. Phys.* **33**, pp. 3064-3073.

[44] Chicklis, E.P., Folweiler, R.C., Naiman, C.S., Gabe, D.R., Linz, A. and Jenssen, H.P., 1974: *Development of Multiply Sensitized Ho:YLF as Laser Material*, Report no. ECOM-73-0066-F.

[45] Murray, J.E., 1983: "Pulsed Gain and Thermal Lensing of Nd:$LiYF_4$", *IEEE J. Quantum Electron.* QE-**19**, pp. 488-491.

[46] Vanherzeele, H., 1988: "Thermal Lensing Measurement and Compensation in a Continuous-Wave Mode-Locked Nd:YLF Laser", *Opt. Letts.* **13**, pp. 369-374.

[47] Barnes, N.P. and Gettemy, D.J., 1980: "Temperature-Variation of the Refractive Indices of Yttrium Lithium Fluoride", *J. Opt. Soc. Amer.* **70**, p. 1244.

[48] Pollak, T.M., Wing, W.F., Grasso, R.J., Chicklis, E.P. and Jenssen, H.P., 1982: "CW Laser Operation of Nd:YLF", *IEEE J. Quantum Electron.* QE-**18**, pp. 159-163.

[49] Seelert, W., Kortz, H.P., and Yen, W.M., 1992: "Excited State Absorption and $^4F_{3/2}$ Lifetime Shortening in Diode Pumped Nd:YLF Q-Switched Lasers", *Advanced Solid State Lasers*, Optical Society of America, Washington, D.C., p. 209.

[50] Devor, D.P., Pastor, R.C., Robinson, M., Akutagawa, W.M. and Arita, K., 1974: *Reactive Atmosphere Growth of 2.06 μm Laser Crystals*, Report no. ECOM-74-0034-F.

[51] Dinndorf, K.M., Knowles, D.S., Gojer, M., Taylor, C.J. and Jenssen, H.P., 1992: "Principal Axes for Transitions in Monoclinic Crystals", *Advanced Solid State Lasers*, Optical Society of America, Washington, D.C. P. 270 and references therein.

[52] McFarlane, R.A., 1992: "Infrared-Pumped Visible Laser", *Advanced Solid State Lasers*, Optical Society of America, Washington, D.C., p. 275.

[53] Xie, P. and Rand, S.C., 1992: "Visible Cooperative Upconversion Laser in Er:LiYF4", *Opt. Letters*, **17**, pp. 1198-1200.

[54] Johnson, L.F. and Guggenheim, H.J., 1974: "Electronic- and Phonon-Terminated Laser Emission from Ho^{3+} in BaY_2F_8", *IEEE J. Quantum Electron.* QE-**10**, pp. 442-449.

[55] Johnson, L.F. and Guggenheim, H.J., 1971: "Infrared-Pumped Visible Laser", *Appl. Phys. Letts.***19**, pp. 44-47.

[56] Silversmith, A.J., Lenth, W. and Macfarlane, R.M., 1987: "Green Infrared-Pumped Erbium Upconversion Laser", *Appl. Phys. Lett.* **51** , pp. 1977–1979.

[57] Hebert, T., Macfarlane, R.M., Wannemacher, R. and Lenth, W., 1990: *Conference on Lasers and Electro-Optics*, Vol. **X**, 1990: OSA Technical Digest Series, Optical Society of America, Washington, D.C., p. 318.

[58] Fiehler, D., 1993: "Upconversion Processes Creates Compact Blue/Green Lasers", *Laser Focus World*, Nov. 1993, pp. 95–102.

[59] Brenier, A., Garapon, C., Madej, C. Pedrini, C. and Boulon, G., 1994: "Excited-State absorption in Tm^{3+} -Doped $LiNbO_3$:MgO and $Gd_3Ga_5O_{12}$ and Looping Mechanism," *J. of Lumin.*, **62**, pp. 147-156.

INORGANIC SCINTILLATORS

A. Lempicki [1], A. J. Wojtowicz [1,2], and C. Brecher [1]

[1] Chemistry Department, Boston University,
590 Commonwealth Avenue, Boston, MA 02215, USA

[2] Institute of Physics, N. Copernicus University,
Grudziadzka 5, 87-100 Torun, Poland

Abstract

Inorganic scintillators may well be the least well understood of all wide-gap luminescent materials. Despite a long and venerable history, the phenomenon of scintillation has not been the beneficiary of any unifying comprehensive theoretical support, nor have scintillator materials shown any consistent systematics. Nevertheless, such materials have found wide use in such diverse technological areas as medical radiology (PET), high-energy physics, nondestructive testing, and transportation security, and are becoming even more ubiquitous with each passing year.

In this chapter, we shall attempt to provide a coherent picture of the present state of understanding of the scintillation process and the crystalline materials that manifest it. The basic steps in the process will be described, along with an approach for evaluating them. The performance of a number of widely used materials will be quantified using this model, and the general state-of-the-art will be concisely summarized. Finally, we shall identify a number of vital issues that remain to be solved, and, hopefully, alert the broad and talented luminescence community to the challenges and opportunities lying on their doorstep.

To be published in:
Wide-Gap Luminescent Materials: Theory and Applications
Editor: S. R. Rotman
(Kluwer Academic Publishers, Inc., Norwell MA, USA).

INORGANIC SCINTILLATORS

A. Lempicki, A. J. Wojtowicz, and C. Brecher

Contents

1. Introduction

This chapter is devoted to an unusual phenomenon called *scintillation*, and to the particular class of wide-bandgap materials in which it occurs. Widely used for the detection and characterization of ionizing radiation, this effect is, in material terms, probably the most demanding of all luminescence phenomena. Although many wide-bandgap materials are capable of displaying intense luminescence, most of them, even with high quantum efficiency, perform very poorly as scintillators. There are many reasons for this, some simple and easy to understand, but others hidden in the subtle interplay of various physical properties of the particular system in question. Indeed, even the very good performance of some scintillator materials, although welcome and useful for radiation detection, is only poorly understood. While this chapter will touch upon many of the currently used materials, we do not intend to present a detailed survey of all scintillators used today. Instead we shall focus on the underlying processes that govern the entire phenomenon, identifying specific basic questions and problems within the framework of a more general picture, and emphasizing the link between physics of the scintillation effect and the material properties of the scintillators themselves.

Moreover, we shall devote the following discussion to scintillation as excited by high energy (γ- and X-ray) photons. We will not consider other forms of ionizing radiation such as electrons, α-particles, protons or neutrons, since most of the differences occur in the initial stoppage of the high energy particle. The greatest of these exceptions is presented by neutron radiation, which requires a special design of the scintillator to allow for a nuclear reaction between incoming neutron flux and the nuclei of the detector material. This intermediate step is needed in order to produce charged particles or photons capable of ionizing the material and producing detectable scintillation light. Although there are other cases in which light yields and pulse shapes of scintillations induced by charged massive particles are different from those caused by high-energy photons, most of the processes following the initial stoppage are virtually identical, and will find equal relevance in the forthcoming discussion.

Scintillation may be considered as essentially a photon conversion process. As in other luminescent processes it begins with absorption of a higher energy photon, raising the absorber to an excited state, whose energy is then dissipated by a sequential process that includes an efficient radiative relaxation step. The major distinction is the sheer magnitude of the energy of the exciting photon, and consequently of the primary electronic excitation, or *PEE*, state. The decay of this state is usually rather complex since it may follow many different routes but, in a good material, eventually culminates in literally thousands of relatively low energy photons and a flash of visible scintillation light. Therefore, although the term *scintillation* is usually associated with ionizing radiation, the phenomenon itself in many aspects closely resembles the other, more familiar, forms of luminescence.

The most important similarity between ordinary luminescence and scintillation is in the role played by so-called luminescence centers. In a simple case the center is created by an appropriately chosen dopant ion or *activator*. Sometimes a more complex center, produced by association of different entities (*e. g.*, donor-acceptor pairs or ions and defects), may also play a prominent role. The interrelations between host materials and dopant ions have been at the core of luminescence science, one of the oldest and still vital areas of research, and a large body of knowledge has been built up over the years [1]. Both empirical and theoretical work on various luminescence centers, including intrinsic luminescence (free and self-trapped excitons) have also been extensively reviewed [2], and will consequently not be addressed in any detail in this chapter. Here, we will focus instead on other, less discussed and less familiar, aspects of scintillation process.

The decay of the *PEE* state generates a shower of secondary particles, consisting mostly of fast electrons and X-rays, which in turn create a large number of electronic lattice excitations such as free electron-hole pairs and/or free or self-trapped excitons. Since few of these entities are likely to be close enough to the dopant ions serving as luminescence centers to excite them directly, a good scintillator must possess some efficient means of host-to-activator energy transfer in order to make productive use of the energy deposited in it by the ionizing particle. This is the single most important step in the scintillation process, and determines to a large extent the potential performance of all wide-gap materials. Although every class of such material exhibits some degree of host-to-ion energy transfer, they differ widely in efficiency, strongly influencing the speed and light yield of any individual scintillator material.

Several very useful treatises have been written on the subject of scintillators [3,4], which traditionally deal with the following aspects:

- Type of radiation (electromagnetic, charged or neutral particles);
- Physics of the *PEE* creation and decay (photoeffect, Compton scattering, energy loss by a charged particle, etc.);
- Characterization and classification of scintillator materials;
- Design aspects of detectors (*e. g.*, matching of scintillator and light sensing device, complex detectors); and
- Collection and interpretation of experimental data.

As mentioned above we will concentrate primarily on the physics of visible and near-UV light production through excitation by high-energy electromagnetic radiation (X- and γ-rays), with special emphasis on the mechanisms of energy transfer from the host to the luminescent ions. To comply with the goals of this book we will consider only inorganic crystalline solids, which by virtue of their high density provide the highest stopping power and play a central role in various applications of scintillators. Moreover, we will not include any discussion of amorphous or glassy materials which, as a class, are generally deficient in both stopping power and light output. While work on such materials is continuing, particularly in the area of high density glasses [5], we feel that the level of basic understanding is even poorer than in the case of crystalline materials (see Sec. 4.6).

This limitation of scope is dictated also by both historical developments and practical considerations. As in most fields of scientific endeavor, progress in scintillators has come in successive waves. The first was stimulated by the initial discovery of the phenomenon around the turn of the century. The second great wave was induced by the large scale applications of phosphors in medicine, fluorescent lighting, and television in the middle part of this century. Now, the last few years have brought about yet a third wave of scintillator research, dictated primarily by exigencies of high energy accelerators and new medical applications of γ-rays. In response to these needs, new scintillator materials have been discovered and developed, with high (and sometimes unrealistic) expectations of success. Indeed, the scintillation performance of these materials has revealed many unsuspected problems, which have also stimulated the revival of the field. This is the territory that we wish to cover in detail.

The history of scintillators begins in 1895 with Roentgen's discovery of the barium platinocyanide fluorescence excited by X-rays. In the following year Becquerel discovered that certain luminescent crystals containing uranium emit light spontaneously [6], and Pupin made first use of $CaWO_4$ in X-ray screens [7]. It was only shortly thereafter that the works of Maria Sklodowska- and Pierre Curie established the existence of radioactivity as a general phenomenon, stimulating an avalanche of discoveries that changed the face of modern physics. Crookes, in 1900, used ZnS to detect and count radioactivity, a method that was later (1909) used by Rutherford in his famous experiments on scattering of α-particles [8]. Undoubtedly Rutherford's research established scintillation as the principal detection method for high energy particles. While in the early days scintillations were counted visually, the combination of the photomultiplier tube (PMT) and new scintillator materials discovered after World War II established the scintillation counter as we know it today. It was Hofstadter's discovery of the thallium-activated alkali halides [9] that triggered a series of advanced research programs using modern solid state physics and spectroscopy, beginning, at long last, to develop a comprehensive understanding of the physical mechanisms behind the scintillation effect. The NaI:Tl scintillator, of unparalleled efficiency, has survived the test of time, remaining to this day the most widely used material. It is interesting how often a major first discovery, bursting unexpectedly on the scientific or technological scene, remains unbeatable despite years of subsequent painstaking work. (Other familiar examples of this pattern include silver halides in photography, the He-Ne and Nd-YAG lasers, and the cathode ray tube.)

The next great strides in scintillator development took place in the early 1970's, with the discoveries of CaF_2:Eu [10], BaF_2 [11], and BGO ($Bi_4Ge_3O_{12}$, bismuth germanate) [12]. BGO in particular has had an extraordinary history, originally proposed in the late 1970's for nuclear medicine applications (in positron cameras) [13], and culminating in 1990 when an unprecedented quantity (almost 12,000 crystals of 24 cm in length each, for a total volume of about 1.2 m^3) was used by the L3 collaboration to build an electromagnetic calorimeter for the LEP collider at CERN, Geneva [14].

During the last decade the need for improved radiation detectors began to grow rapidly. High energy physics urgently sought new extremely radiation-resistant scintillator materials capable of recording very small cross-section events produced by high repetition rate collisions in a new generation of highly luminous accelerators (LHC at CERN, SSC, SLAC, FNAL and Cornell in the USA) [15]. Medicine, where the evolution of diagnostic radiology from simple X-ray photography into CAT (Computerized Axial Tomography), SPECT (Single Photon Emission Computed Tomography) and PET (Positron Emission Tomography) stimulated the early development of new phosphors and scintillators, now began to demand even denser, brighter and faster materials. Other applications began to express their own unique needs:

Environmental concerns require highly sensitive scintillator systems to monitor soil or water contamination by radioactive or chemical spills. Oil well logging depends heavily on the radioactive probing of rock formations [16]. Nondestructive testing of massive metal objects (power conduits, sewer lines, airframes) are turning to higher energies, from X-rays to gammas. Perhaps the most challenging, in this age of rampant terrorism, is the need to inspect airport cargo for explosives and narcotics, requiring high energy discrimination to distinguish the characteristic signatures of the contraband from an unpredictable background [17]. Scintillators are now entering many aspects of everyday life and NaI:Tl is no longer the answer to all prayers.

In order to respond to these wide-ranging needs, there is no longer a single "best" scintillator adaptable to all. Each application has its own unique requirements, placing stringent demands on some properties while remaining tolerant of others. In medical applications, for example, light yield as measured in the number of scintillation photons emitted per MeV is of paramount concern, so as to minimize the radiation dose to which the patient is subjected. At such low radiation levels, the ability of the material to withstand damage from the absorbed radiation is only of peripheral interest. In the new generation of high energy physics experiments, however, these priorities are exactly reversed: the immense energies of the particles being measured makes the light yield per MeV far less important, but the high fluxes can render a radiation-sensitive crystal opaque in minutes. Additionally, the large quantities involved place a premium on low cost per unit volume. Both applications require high speed of response, and both share the same concern about the density of the material, since all the cross-sections governing the absorption of the incoming radiation by the scintillator material increase strongly with density [4]. Higher density also enables higher spatial resolution (particularly important in medical applications) and/or reduction of the overall size of the detector (important for high energy physics applications). Although the priorities may differ, there are a number of properties that are of interest to virtually every user:

1. *High light output* (measured in photons/MeV): By general consent, there are two benchmarks for this quantity: NaI:Tl, with a value of 38,000, and BGO, at about 8,200. High energy physics can be satisfied with a light output less than that of BGO, but for medical applications this represents the lower limit.

2. *Short decay time:* Most applications require decays in the nanosecond range.

3. *Emission wavelength:* As we shall see there exists a fundamental relationship between decay time and emission wavelength, so that items *2* and *3* are not independent. Additionally the wavelength is partially dictated by the properties of the light detector, PMT, photodiode or avalanche photodiode. Emission wavelengths below the cutoff of glass transmission ($\approx$300-320 nm) substantially increase the cost of the PMT.

4. *Density:* In general densities above 6 g/cm^3 are highly desirable.

5. *Chemical stability:* The scintillator should not be hygroscopic or otherwise chemically unstable. This is one of the primary deficiencies of NaI and CsI.

6. *Radiation hardness:* Radiation damage appears as absorption bands induced by ionizing radiation. Extrinsic damage is caused by the coloring of defects or

uncontrolled impurities that are already present in the lattice. Intrinsic damage occurs predominantly as a result of nonradiative decay of self-trapped excitons, which creates new defects that were not originally present. Extrinsic damage can in some cases be minimized by removing deleterious impurities, selective doping, or appropriate control of growth conditions; intrinsic damage, while often reversible after it occurs, is inherent to the given material.

7. *Cost.* The raw materials should not be exorbitantly expensive, and crystals should be fabricable through a growth technology capable of mass production. In nuclear medicine, a PET camera may utilize over a thousand crystals, imposing a benchmark cost of about $\$10/cm^3$ (at present the cost of BGO is about $\$5/cm^3$). For precision calorimetry in high energy physics tens of thousands of crystals are needed; with a total volume measured in tens of cubic meters, the acceptable cost is necessarily reduced, to about $\$2.5/cm^3$ [14].

These seven requirements are a condensation of the properties that current research and development efforts seek to satisfy. Reflecting these considerations, the most important scintillator discoveries during the last decade are undoubtedly YAP ($YAlO_3$:Ce) [18]; the fast component in BaF_2 [19] and other crossluminescence scintillators [20]; GSO (Gd_2SiO_5:Ce) [21]; CeF_3 [22,23]; LSO (Lu_2SiO_5:Ce) [24]; $LuPO_4$:Ce [25]; and LuAP ($LuAlO_3$:Ce) [26]. Although none of these materials satisfies all seven requirements, it is interesting to note that they include the heaviest, fastest, most radiation resistant and most chemically stable scintillators known today.

2. Theoretical Considerations

In this section we shall address the factors that govern the performance of scintillators. The performance limits may be either fundamental, imposed by well-established physical laws, or practical, imposed by considerations such as the wavelengths covered by photomultipliers. In particular we wish to establish which physical considerations limit the *efficiency* and *speed* of the scintillator. As we shall see, there are certain tradeoffs between efficiency and speed which do not allow both to be maximized independently, making it necessary to set priorities depending upon the ultimate application of the scintillator. Moreover, in terms of their usefulness for any particular task, scintillators of equivalent speed and efficiency must further be ranked according to their stopping power, as determined by the density and effective *Z* number of the material. Theoretical limits of efficiency are sometimes difficult to establish because of the complexity of events that occur between the absorption of a γ photon and the emission of scintillation light. Nevertheless, we shall attempt to show what room remains between what is theoretically possible in terms of scintillator performance and what has been achieved to date in the "real world". Some of the deficiencies can be addressed merely by improvement in material engineering, while others, unfortunately, are imposed by fundamental physical principles.

2.1. Scintillator Efficiency (η)

For phenomenological analysis, it is a convenient and generally accepted practice to divide the scintillation process into three essentially independent steps, as illustrated in Figure 1:

- Absorption of the γ- or X-ray photon energy by the material and conversion into electron-hole (e-h) pairs; the efficiency of this step will be expressed as the quantity β, the fraction of e-h pairs actually produced, relative to the total that could be produced if there were no other losses.
- Energy transfer from electronic lattice excitations to luminescence centers; its efficiency will be expressed as S, the fraction of the total e-h population that manages to get to luminescence centers.
- Emission of scintillation photons, with a quantum efficiency Q, representing the fraction of excited centers that actually emit scintillation photons.

The overall *process efficiency* is therefore the product of these three factors; following Lempicki, Wojtowicz and Berman [27], we define this as the quantity η:

$$\eta = \beta \cdot S \cdot Q , \qquad (\text{where } 0 \le \eta, \beta, S, Q \le 1) . \tag{1}$$

In the first step, γ- or X-ray photons penetrating the material lose energy by a variety of processes, such as pair production, Compton scattering and the photoelectric effect [4]. In pair production a gamma photon with an initial energy above 1.02 MeV simply disappears, replaced by an electron and positron with the same total energy. In Compton scattering the initial energy is divided between a scattered photon and a recoil electron. In the photoelectric effect, which dominates at relatively low energies, a γ- or X-ray photon is directly absorbed in the lattice, generating a very fast electron and a hole in a deep core level of an ion, with the two carrying all the energy of the original photon. In all cases, unless the energy deposited by means of the initial step can be transformed efficiently into electronic excitations (e-h pairs and/or excitons) the other steps in the process become irrelevant.

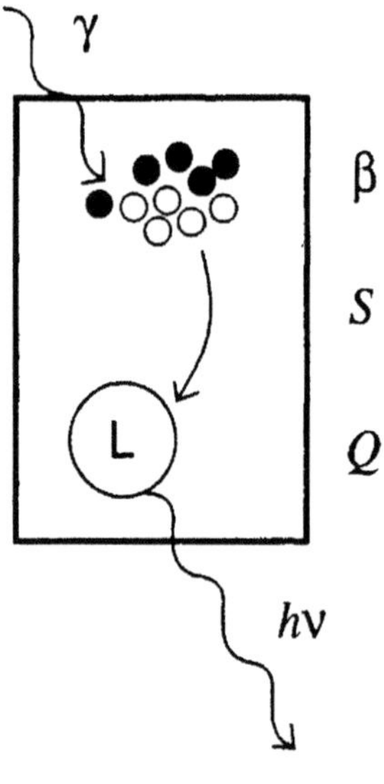

Figure 1. The basic steps of the scintillation process. The energy E_γ of the incoming particle is converted into electronic excitation with efficiency β, of which a fraction S is transferred to the luminescence centers. The quantum efficiency of the emission step is denoted by Q.

Let us define E_γ (in MeV) as the energy of the incoming particle and ξ (in eV) as the *average energy* needed to create one electron-hole pair. Then the *quantum efficiency* of the scintillation process, η_q, will be given by

$$\eta_q = \frac{E_\gamma \cdot 10^6}{\xi} \cdot S \cdot Q , \tag{2}$$

where $E_\gamma 10^6/\xi$ denotes the total number of e-h pairs. Since $10^6 \cdot E_\gamma >> \xi$, the quantum efficiency η_q is usually quite large (on the order of thousands) and depends on the energy of the gamma photon. Another quantity, the light yield $L = \eta_q/E_\gamma$, in photons per MeV, is often used to characterize the performance of a given scintillator material independently of the gamma energy.

The negatively charged electron will interact electromagnetically with both the electronic and ionic (lattice) subsystems of the material. There are two alternative approaches to describe this interaction, and hence the quantity β, some of which date back to the 1960's and 1970's [28-32]. We shall discuss these in more detail in the next section.

It should be emphasized that β is a direct expression of the efficiency with which the energy initially deposited into the lattice by the γ photon is converted into electron-hole pairs. Since, in the absence of all other losses, there is a minimum energy ξ_{min} necessary to create such a pair, β is nothing more than the ratio of this minimum value to the average energy actually expended in such pair generation:

$$\beta = \xi_{min}/\xi \ . \tag{3}$$

Similarly, if we take $n_p = E_\gamma \cdot 10^6/\xi_{min}$ as the maximum possible number of photons that can be produced by a given material in the absence of all other losses, the overall process efficiency η is simply the ratio of the number of photons actually produced (per absorbed gamma) to the maximum number possible; in other words,

$$\eta = \eta_q / n_p \ . \tag{4}$$

Still another parameter of future interest is the energy efficiency, η_e. This efficiency is the fraction of the total gamma energy E_γ that is converted into scintillation photons, and can be expressed as:

$$\eta_e = \frac{\eta_q \cdot h\nu_{em}}{E_\gamma \cdot 10^6} = \frac{h\nu_{em}}{\xi} \cdot S \cdot Q = \frac{h\nu_{em}}{\xi_{min}} \cdot \eta \ , \tag{5}$$

where $h\nu_{em}$ is the energy of a scintillation photon (in eV).

To summarize, the quantities β, S and Q express the respective efficiencies of the three basic steps in the scintillation process: conversion, transfer, and emission. Thus if it is possible to independently calculate or measure these quantities, one should be able not only to evaluate the behavior of a given scintillator, but to elucidate the essential details of the scintillation process and identify both the performance limitations and their causes.

2.2. Excitation and Conversion (β)

The primary goal of this section is to describe techniques for calculating the conversion efficiency β. Two major approaches have been developed for modeling the interaction between the crystal lattice and the fast electrons generated by the energy from the stopped gamma. One, the *polaron* model, considers only single-particle excitations of the electronic subsystem (electronic transitions from valence and core levels to continuum) and coupling to only one type of ionic (lattice) subsystem excitations (optical phonons) [31]. The second approach, in contrast, assumes that the major contribution of the electronic subsystem comes from many-particle valence electron excitations (*plasmons*), with the ionic subsystem playing only the role of the background for the collective electronic excitations and not considered explicitly [32,33].

2.2.1. Polaron model

The task of the theoretical effort is to develop expressions for ξ_{min} and ξ, from which, by means of Eq. 3, β can be obtained. The conversion of the energy deposited in the material by a γ- or X-ray photon is closely related to the problem of electron and hole multiplication, which was addressed for the first time by Shockley [28] and later adapted and developed by van Roosbroeck [29] and Klein [30]. Shockley considered electron-hole pair production in semiconductors assuming that emission of phonons by fast electrons (or holes) represents dissipation of the initial gamma energy (losses) while ionizations contribute to the final pool of electron-hole pairs. An electron or hole coupled to a polarization (optical phonon) cloud can be considered to propagate as a single quasiparticle called a *polaron*, which term we will apply to all approaches that assume that optical phonon coupling is responsible for most of the undesirable losses suffered by fast particles in the cascade. While such coupling may not be important for very fast particles, most of ionizations and losses occur in the final stages of the cascade, when kinetic energies are not very high.

We begin by examining the phenomenological process of electron-hole pair creation. When a gamma photon is absorbed in a lattice, thousands of such pairs are generated with a broad spectrum of energies. All will contain at least the ionization energy E_i, along with an unpredictable amount of kinetic energy. If this amount is itself greater than E_i, it will quickly go toward generation of an additional pair until, very rapidly, a population of $\underline{e}$-$\underline{h}$ pairs has been created with a distribution of kinetic energies all of which are less than E_i. This residual kinetic energy represents an unavoidable loss which is independent of any that might be introduced by interaction with the crystal lattice. Consequently, the minimum energy needed to generate a single $\underline{e}$-$\underline{h}$ pair can be expressed as

$$\xi_{min} = E_i + \langle E^e_{kin} \rangle + \langle E^h_{kin} \rangle \tag{6}$$

where $\langle E^e_{kin} \rangle$ and $\langle E^h_{kin} \rangle$ represent the average kinetic energies of those electrons and holes (mostly but not exclusively from the last stages of cascade) which are unable to ionize the material any further (*i. e.*, both kinetic energy terms are less than E_i).

Now let us consider the losses introduced by interaction with the crystal lattice. These additional losses, symbolized by ξ_ℓ, are characteristics of the particular material, and are measured in multiples of the longitudinal optical phonon energy:

$$\xi_\ell = N \cdot \hbar\omega_{opt} \tag{7}$$

where $\hbar\omega_{opt}$ is the energy of the longitudinal optical vibration and N is the number of phonons produced by each $\underline{e}$-$\underline{h}$ pair. In his model, Shockley also introduced the concept of mean free path between events, λ_{opt} being the mean free path of electron (or hole) between consecutive emissions of optical phonons, and λ_i the mean free path between ionizations. The ratio $N = \lambda_i/\lambda_{opt}$ represents simply the average number of phonons emitted at each ionization by an electron (or hole) with an energy above the ionization threshold E_i.

Van Roosbroeck [29] introduced the following parameters: K, the ratio of the average optical phonon loss to the ionization energy; L_f, the ratio of the average kinetic energy loss to the ionization energy; and Y, the fraction of the energy that actually resulted in producing e-h pairs:

$$K = \frac{N \cdot \hbar\omega_{opt}}{E_i}; \; L_f = \frac{<E^e_{kin}> + <E^h_{kin}>}{2E_i}; \text{ and } \; Y = \frac{E_i}{\xi}. \tag{8}$$

From these, we get:

$$\xi = \frac{E_i}{Y} = \xi_{min} + \xi_{\kappa} = (1+K+2L_f) \cdot E_i \tag{9}$$

Using a technique called "crazy carpentry", van Roosbroeck derived an approximate analytic expression for ξ in terms of the ionization energy and the optical phonon loss ratio,

$$\xi = E_i \cdot \frac{3 + K}{2 - K \cdot [1 - \exp(-1/K)]} \tag{10}$$

which, in the absence of optical phonon losses (K=0), reduces to $\xi_{min} = 1.5 \cdot E_i$. Thus in order to calculate ξ, we need only the ionization energy E_i and the loss ratio K.

Concerned with the differences in the relative masses of electrons and holes in semiconductors, Rodnyi *et al.* [33] developed an expression relating the ionization threshold energy E_i to the bandgap energy E_g:

$$E_i = E_g \cdot \frac{9}{7 - m_e/m_h}, \tag{11}$$

where m_e and m_h are the effective masses of electrons and holes, respectively. This gives, as limiting values, $E_i = 1.5 \cdot E_g$ for insulators ($m_h = m_e$) and $E_i = 1.286 \cdot E_g$ for semiconductors ($m_h >> m_e$). This, however, assumes equal momentum distribution after an ionization event. A better assumption (according to Landsberg [34]) is based on equality of group velocities rather than wave vectors, giving

$$E_i = E_g \cdot \frac{2\left(m_e/m_h\right) + 1}{m_e/m_h + 1} \tag{12}$$

This gives, as limiting values, $E_i = 1.5 \cdot E_g$ (for $m_h = m_e$) and $E_i = E_g$ (for $m_h >> m_e$), with corresponding ξ_{min} values of $2.25 \cdot E_g$ and $2.4 \cdot E_g$, respectively.

The question of calculating E_i and ξ_{min} has been addressed by other authors as well, each under somewhat different assumptions and resulting in slightly different values for ξ_{min}: Shockley [28], $3 \cdot E_g$; van Roosbroeck [29], $2.25 \cdot E_g$; Klein [30], $2.5 \cdot E_g$ - $2.83 \cdot E_g$; Rodnyi *et al.* [33], $2.25 \cdot E_g$ - $2.4 \cdot E_g$; and Robbins [31], $2.3 \cdot E_g$. We should note that while for small K there are some differences between these authors, all the models converge for large K (high phonon losses), and in practical terms are equivalent, yielding essentially the same results [35].

The problem of calculating the energy loss ratio K was addressed by Robbins [31], whose model can be considered as an extension of the "crazy carpentry" approach of van Roosbroeck. Robbins introduces the following additional assumptions:

i. Only phonon generation (not absorption) need be considered;

ii. the fast charged particles couple to longitudinal optical (LO) phonon modes;

iii. phonon dispersion can be neglected (single-phonon energy, $\hbar\omega_{opt} = \hbar\omega_{LO}$);

iv. the ionization rate for particles above threshold is constant;

v. free electron mass can be used (particles far from band extrema); and

vi. the proper value for the ionization threshold E_i is $1.5 \cdot E_g$ (*i. e.*, $m_h = m_e$);

He justifies the use of similar treatment for electrons and holes by pointing to the fact that the decay of immobile deep core holes produces X-ray photons or Auger electrons, which in turn create the next generation of hot electron-hole pairs. Thus the relaxation of the hot hole can be viewed as proceeding through the same intermediate states as the fast electron, so that it is not unreasonable (in the first approximation) to treat electrons and holes symmetrically. In addition Robbins assumes that optical phonon generation rate is proportional to the product of phonon frequency and the polaron coupling constant α:

$$\alpha = \frac{e^2}{\hbar} \cdot \left(\frac{1}{\varepsilon_\infty} - \frac{1}{\varepsilon_s} \right) \left(\frac{m}{2\hbar\omega_{LO}} \right)^{1/2} \tag{16}$$

where ε_∞, and ε_s are the high frequency and static dielectric constants. Since $K = N\hbar\omega_{LO}/E_i$, one can expect that K will be proportional to the quantity R, which plays the role of an index of relative efficiency for different materials:

$$R = \left(\frac{1}{\varepsilon_\infty} - \frac{1}{\varepsilon_s} \right) \frac{\left(\hbar\omega_{LO}\right)^{3/2}}{1.5 \cdot E_g} . \tag{17}$$

The constant of proportionality between K and R, assumed to be the same for all materials, was found empirically for ZnS and CdS:

$$K = \left(0.244 \times 10^4\right) \cdot R \tag{18}$$

defining the relation between K and the material constants ε_∞, ε_s, $\hbar\omega_{LO}$ and E_g:

$$K = 0.244 \times 10^4 \cdot \left(\frac{1}{\varepsilon_\infty} - \frac{1}{\varepsilon_s} \right) \frac{\left(\hbar\omega_{LO}\right)^{3/2}}{1.5 \cdot E_g} . \tag{19}$$

The procedure for obtaining ξ and ξ_{min} given in this section completes the calculation of β as defined in Eq. 3. In the next section we shall present a complementary view of energy losses in dielectrics which has been advocated by some authors [33] as directly applicable to scintillators.

2.2.2. Plasmon model

The concept of a plasmon as a quantum of collective oscillations of conduction electron gas (plasma) originated in energy loss experiments on metals, many of which clearly show multiple equidistant loss peaks. In these experiments the beam of incident electrons is transmitted or reflected from a thin metallic film. It is expected that incident fast electrons should couple strongly to longitudinal modes of plasma oscillations, showing loss peaks at energies equal to integral multiples of the plasmon energy. High densities of conduction electrons in metals correspond quite well to relatively high plasmon energies, as confirmed by optical measurements (ultraviolet transmission limits at plasmon energies). In semiconductors with lower densities of conduction electrons, (and variable by appropriate doping), the plasmon energies are clearly lower, corresponding to infrared rather than vacuum ultraviolet energies [36,37]. Nevertheless the high-energy electron loss peaks observed in some insulating materials have been interpreted in terms of the same mechanism assuming that valence (not conduction) electrons contribute to plasma oscillations in such materials. If $\varepsilon(\omega)$ represents the dielectric function, the energy loss by a photon, which usually corresponds to single-particle transitions, is proportional to its imaginary part $\underline{Im}\{\varepsilon(\omega)\}$; however, the energy loss by a fast particle (electron) is approximately proportional to the imaginary part of the *reciprocal* of ε, $\underline{Im}\{1/\varepsilon(\omega)\}$. Measured electron energy loss spectra agree well with energy loss functions calculated from the dielectric function, both of which show broad bands at energies (10 to 20 eV) corresponding to plasma frequency and not to single-particle transitions of any kind. This supports the conclusion that most of the loss suffered by fast particles in dielectric materials at the low energy end of the loss spectrum is due to collective excitations of valence electrons. Consequently, we take ξ, the average energy to create one $\underline{e}$-$\underline{h}$ pair, as approximately equal to one quantum of plasmon energy:

$$\xi = \hbar\omega_p = \hbar e\cdot\left(\frac{n_v}{m\cdot\varepsilon_0}\right)^{1/2} , \tag{21}$$

where e is the electron charge (in coulombs), n_v is the valence electron concentration (in m^{-3}), m is the electron mass (in kg) and ε_0 is the free space permittivity (equal to $10^7/4\pi c^2$ in SI units). While for some materials there is a correlation between $(n_v)^{1/2}$ and cathodoluminescence efficiency, Kingsley and Ludwig [38] point out that for many cases there is a substantial lack of agreement. The simple approach, in which one plasmon generates one electron-hole pair, can be improved by introducing an additional ionization by hot carriers generated by decay of plasmons. This becomes very important for materials with a high ratio $\hbar\omega_p/E_g$, for which the assumption of $\xi = \hbar\omega_p$ clearly overestimates the energy needed to create one electron-hole pair and, therefore, underestimates the efficiency of the scintillator material. Assuming a simple band structure model for direct-gap materials, Rothwarf [32] finds three different values for ξ, depending on the magnitude of $\hbar\omega_p$:

$$\xi \fallingdotseq \hbar\omega_p, \qquad \text{for} \quad 1 < (\hbar\omega_p/E_g) < 4; \tag{22}$$

$$\xi = \hbar\omega_p/3, \qquad \text{for} \quad 4 < (\hbar\omega_p/E_g) < 12; \tag{23}$$

and

$$\xi = \hbar\omega_p/9, \qquad \text{for} \quad 12 < (\hbar\omega_p/E_g) < 36. \tag{24}$$

It is interesting to note that in the last case as many as nine electron-hole pairs are created by the ionization process induced by products of plasmon decay. Clearly the ionization mechanism and the losses it generates become very important in the last stages of the cascade when $\hbar\omega_p >> E_g$, which the plasmon model fails to take into account. As noted by Vasil'ev [39], CeF_3 shows a significant contribution to the energy loss function *below* the plasmon energy $\hbar\omega_p$ that comes from one-electron transitions from the fluorine *2p* (valence band) and cerium *5p* (core) levels. This may be the case in other dielectric materials as well. On the other hand, Robbins points out that in the case of $\hbar\omega_p \approx E_g$ the distinction between single-particle and plasmon excitations becomes blurred [31]. In fact, when calculated plasmon energies are replaced by measured ones, the correlation between plasmon energy and cathodoluminescence efficiency often disappears [36]. This may be the reason the plasmon model has never advanced to the point of predictive capabilities. While plasmon excitations unquestionably contribute to energy loss by fast charged particles, the overall mechanism is more complicated, involving ionization processes at various stages including the last ones. The final verdict must be that the polaron model works better, providing a more comprehensive and useful description of the conversion process in scintillator materials.

2.3. Transfer processes S

The second parameter in Eq. 1 deals with energy transfer from the lattice to the luminescent center. Robbins, who introduced it formally into the equation, never considered cases where $S < 1$. Still, in the context of his theory, transfer implies a transport of energy over distance. The electrons and holes generated by ionizing excitation may retain their separate identities and mobility, or become trapped, either individually or in pairs. They can also produce excitons of various kinds: mobile, trapped or self-trapped, each of which may decay radiatively or nonradiatively. If the scintillator is to be efficient, however, their energy must eventually be emitted in the form of photons. The luminescent centers should provide the most probable recombination path. The process of transfer thus offers multiple opportunities for premature recombination of carriers, before the luminescent center is reached, reducing the efficiency of scintillation. Unfortunately there is no direct way of either calculating or measuring the transfer efficiency S; how it can be estimated will be shown in Sec. 2.6.

We wish here to present the physics of transfer in a greatly abbreviated form. This enormously complex problem will be dealt with by considering three generic types of transfer and giving some indication of their relative importance.

1) *Radiative transfer* occurs whenever there is an overlap between the absorption of the luminescent center that emits the scintillation light and another emission originating elsewhere in the lattice. The most common example of this is self-absorption, where both absorbing and emitting centers are chemically identical. The phenomenon consists of the emission of a real photon and its reabsorption before leaving the medium, for which reason it is also known as radiation imprisonment or trapping. The effect is most noticeable in materials characterized by resonance radiation and has been extensively studied in gases (primarily Hg vapor [40]). It has received some attention in dyes, where there is always an overlap of emission and absorption due to a finite Stokes shift [41], but has been treated to a much lesser degree in solids. We know of no concrete example where this contributes to the efficiency of scintillation; on the contrary, if present it tends to be rather harmful. We shall encounter this problem in stoichiometric Ce compounds described in Sec. 3.3.

The two characteristic signatures of radiation trapping are a geometric dependence of both decay time and emission profile. If there are successive emissions and reabsorption of photons, both the decay and spectrum will depend on the distance traversed by the luminescence light before it emerges. If the crystal has an elongated shape, the decay observed along the greater dimension will be slower than in the transverse direction. For the same reason the emission profile will be distorted on the short wavelength side due to self-absorption. In some applications where long scintillator crystals are used (for instance high-energy physics calorimeters), it is obviously very undesirable to have a spread of decay times depending on the length of path traversed by light. This problem plays a role in CeF_3 when considered for high energy applications [42].

To date, Stokes-separated absorption/emission such as occurs in solids has received no treatment comparable to that given resonance radiation. The one exception is the work by Sakai *et al.* [43], which however does not lead to a closed expression for the decay time. This is clearly an important area for further theoretical attention.

2) *Excitonic transfer* can occur whenever excitons are formed. If band excitons become trapped by a local or Coulomb potential of an activator, their annihilation at that point will obviously result in direct excitation of the luminophore. More often, however, self-trapped excitons (STE) will be formed somewhere else in the crystal; if these are capable of radiative decay and the necessary conditions of emission/absorption overlap are met, a Forster-Dexter type of transfer to an activator can also occur [44]. It is well known that the dipole-dipole rate of this transfer is inversely proportional to the sixth power of the distance between emitter (STE) and absorber, so that if the STE is not formed close to the activator ion itself, it will have to diffuse to the immediate vicinity for any efficient transfer of this type to occur. Consequently Forster-Dexter transfer will be efficient only for high activator concentrations or extremely rapid diffusion of the STE. This does appear to take place in BaF_2:Ce [45], as well as in the lanthanide-doped vanadates, tungstates and niobates [46]. It is interesting that in the latter materials, which contain molecular anions (for instance VO_4^{3-}), the STE is identified with a localized excitation of the complex group.

Excitonic transfer can also occur without trapping. A possible mechanism for this is fast "coherent" transfer, via highly delocalized excitons of the type considered by Davidov [47] and Nettle and Lempicki [48]. The reader is directed to the cited references for detailed treatment.

3) *Sequential trapping of free charge carriers* at activator ions, without the formation of excitons, is yet another mechanism of transfer. This is a rather common mechanism in many luminescent materials [49], and a comprehensive discussion of its role in rare-earth-doped scintillators and phosphors has recently been given by Wojtowicz [50]. Since we believe that this mechanism is responsible for the high efficiency of many modern scintillators, we shall describe it here in some detail, using the Tl-doped alkali halides [51] as an illustrative vehicle.

NaI:Tl is the archetypal scintillator, discovered in 1948 [9] and still in widespread use today. Its companions KI:Tl, KCl:Tl, CsI:Tl and others have played a very important role in our understanding of scintillator phenomena. It is sobering to realize that a series of papers analyzing scintillation in these materials, published between 1960 and 1975 [51-57], spell out in great detail all the nuances of transfer processes which we can apply today to more modern materials. No serious investigator in the scintillator field should be excused from studying this literature.

Pure NaI [58] is not a good scintillator, emitting only poorly at room temperature. At 78 K it shows an emission band around 300 nm. This is similar to most other alkali halides, whose emission spectra are typical examples of self-trapped exciton ($V_k + \underline{e}$) emissions [59]. All are strongly temperature quenched (the decay time becomes progressively shorter), reducing the efficiency of both radiative and excitonic transfer in most of these materials. A special case is that of CsI (see Sec. 3.1), which retains a significant level of luminescence even at room temperature. While temperature quenching keeps the quantum efficiency Q rather low, it also gives rise to a relatively fast (although nonexponential) decay of about 16 ns, making the material useful where speed is the primary concern [60]. Upon doping with Tl^+, however, the situation is changed radically. The Tl^+ ion, substituting at an alkali metal site, is a well known luminescent center, which because of its $5d^{10}6s^2$ configuration gives rise to a $^1S_0$ ground state and excited states $^1P_1$, $^3P_0$, $^3P_1$, $^3P_2$. The luminescence takes place from the lowest lying $^3P_0$ triplet and is therefore spin forbidden. This of course slows the decay of the Tl-doped scintillators and makes them less than desirable for many applications. The s^2 configuration is also often encountered among efficient luminophores (Bi, Pb, Sn) but suffers from the same problem as outlined above.

Nevertheless Tl-doped alkali halides are very efficient γ- and X-ray scintillators. Since concentration quenching limits the doping level to about the 1000 ppm range, it is obvious that nearly all of the energy of the gamma photon must be deposited in the crystal in the form of $\underline{e}$-$\underline{h}$ pairs, rather than exciting the Tl directly. Since radiative transfer is effectively quenched, and with the stability of the Tl^0 and Tl^{2+} charge states (which drastically reduces the lifetime of free electrons and holes) decreasing even further the already low excitonic contribution, both of these mechanisms can be dismissed as major factors. This leaves only the third mechanism, the migration of charge carriers, to be responsible for the emission band of NaI:Tl at 420 nm. Consequently, we should look more closely at such carrier-mediated transfer.

Immediately we come upon the problem that in alkali halides the mobility of carriers, especially holes, is very small due to self-trapping, *i. e.*, the formation of V_k centers (I_2^- molecules) which can undergo only temperature-dependent diffusive motion. Additionally, since charge states Tl^0 and Tl^{2+} are known to exist and have lifetimes longer than the radiative lifetime of Tl^+ [61], either holes or electrons can be trapped by the Tl^+ ions, which then have to wait for the second carrier to arrive. The various possible sequences are summarized below, with the symbols $\underline{e}$ and $\underline{h}$ designating thermalized electrons and holes in conduction and valence bands respectively:

Trapping Mechanisms:

1. Electrons trapped at Tl^+ $\qquad \underline{e} + Tl^+ \rightarrow Tl^0$
2. Holes trapped at Tl^+ $\qquad \underline{h} + Tl^+ \rightarrow Tl^{2+}$
3. Hole self-trapping $\qquad \underline{h} \rightarrow V_k$
4. V_k dissociation $\qquad V_k + Tl^+ \rightarrow Tl^{2+}$

These trapped carriers can recombine in a variety of ways. If the recombination is radiative then a $h\nu$ photon is emitted. This will of course depend upon the temperature; at least five bands (different frequencies) have been identified of which two are due to the Tl^+ ion, the rest to different processes. Exact emission wavelengths and temperature dependences vary somewhat among KI, KCl, NaI, etc., but the main issues of interest to us are the different possibilities listed below. For the sake of concreteness we have taken the case of KCl:Tl, which has been studied in greater detail because the recombination processes are slower [52]. The recombination mechanisms that may lead to photon emission are listed below [53,62]. The emission wavelengths, in nm, are indicated in parentheses.

Radiative Recombination Mechanisms:

5. Trapped electrons with free holes $Tl^0 + \underline{h} \rightarrow Tl^+ + h\nu$ (335, 420)
6. Trapped holes with free electrons $Tl^{2+} + \underline{e} \rightarrow Tl^+ + h\nu$ (335, 420)
7. V_k with free electrons, direct $V_k + \underline{e} \rightarrow h\nu$ (300, 370)
8. V_k with trapped electrons $Tl^0 + V_k \rightarrow Tl^+ + h\nu$ (335, 420)
9. Tl^+-bound exciton $Tl^0 + V_k \rightarrow Tl^+ + \{\underline{e} + V_k\} \rightarrow Tl^+ + h\nu$ (410)

It should be noted that process *9* is observed at low temperatures, when energy transfer from the bound exciton to the Tl^+ ion is suppressed [62]. Being a charge transfer (CT) process involving the tunneling of an electron, it can only occur at short distances, with the V_k situated next to Tl^0, and is a rare example of radiative decay by this means. All the processes *1* through *9* are manifestations of lattice-to-ion transfer mechanisms mediated by thermalized carriers, and illustrate the logic and temporal sequence implied in Robbins' treatment. However, for the sake of completeness we should also include the processes by which collision with a secondary electron $\underline{e}_s$ raises the activator (Tl^+) either to one of its higher energy levels (equivalent to below-bandgap optical excitation) [63], or to its next ionization state (as in an optical excitation with energy greater than the bandgap):

Direct Mechanisms:

10. Direct excitation $Tl^+ + \underline{e}_s \rightarrow (Tl^+)^* + h\nu$ (335, 420),
11. Direct ionization $Tl^+ + \underline{e}_s \rightarrow Tl^{2+}$; $Tl^{2+} + \underline{e} \rightarrow (Tl^+)^*$.

Each of the processes can be associated with an effective volume V_c centered at a Tl^+ ion. The meaning of V_c is that essentially all of the energy deposited in that volume can find its way to the ion. For hole capture (process 2), $V_c \approx 25$ unit cells in both NaI:Tl and KCl:Tl [53]. The concept of an effective volume will play an important role when we discuss Ce-based scintillators in Sec. 3.3 and 3.4.

We cannot discuss here the details of the vast literature on thallium-doped alkali halides. Our purpose is to show the complexity of the carrier-mediated transfer and to indicate which of the processes *1* through *11* are the major contributors to the scintillation light. As in all sequential processes, the overall rate is determined by the slow est step, which in the halide case is likely to be the motion of self-trapped holes (diffusing V_k centers resulting in processes *8* and *9*), or the detrapping of electrons (reverse of process *1*). Hole motion would be difficult to detect if not for the fortunate happenstance of a well defined 300 nm absorption band due to the Tl^{2+} ion. The intensity and time dependence of this absorption provide all the information needed to define the hole capture kinetics. The contributions of the various processes manifest themselves at different times, and can also be readily affected by temperature [56].

Early time: This contribution is largely due to the creation of an excited state $(Tl^+)^*$, as by process *11* or sequential prompt processes *2* and *6*. In both NaI:Tl and KI:Tl a Tl^+ ion captures a hole located in any of 25 neighboring unit cells, before that hole has had a chance to self-trap. This process is temperature independent and determines the initial decay rate of the pulse, with a time constant characteristic of the Tl^+ ion (≈200 ns). The contribution of this component to the total light output is not very large, on the order of 10%. The hole motion is presumably a random walk from cell to cell, but rapid enough to allow it to reach the ion before becoming self trapped. As we shall see, the process plays an important role in some Ce activated scintillators.

Intermediate time: Here the major process is a thermally-activated diffusion of V_k, allowing sequential processes *1* and *8*. This is the regime where individual halides differ because of the wide disparity in diffusion times, ranging from 10^{-9} sec for NaI, 10^{-7} sec for KI to 10^{-2} sec for KCl [61]. Consequently, the diffusion-controlled long component is considerably less important in NaI:Tl [51].

Long time: Here the dominant mechanism involves the prompt capture of an electron through process 1 and hole trapping through process 2, followed by thermal detrapping of electrons from Tl^0 and their subsequent capture by Tl^{2+} to form $(Tl^+)^*$. This is the origin of the very long component in halide scintillators, particularly evident in KI: Tl where it contains 75% of the light output and lasts over 200 ms.

Thus we can see that, despite their phenomenal scintillation efficiency, the thallium-doped halides suffer from an inherent and fatal flaw in terms of PET and high-energy physics applications: the slowness of the transfer and emission steps. The ramifications of this fact on the quest for improved materials will shortly become evident.

2.4. Luminescence yield Q

The luminescence process is described by the quantum yield Q, which is generally a well known quantity, particularly for the weak electron-lattice coupling characteristic of rare earth ions [44]. By painstaking effort the quantum efficiency can be measured directly, but it is equally valid and far easier to obtain as the ratio of the overall decay time τ_d to the radiative lifetime τ_r: $Q = \tau_d/\tau_r$. The theoretical understanding of Q, or, equivalently, the nonradiative rates, is well grounded, especially for narrow band emitters such as the rare earth ions. For broad band emitters such as BGO, the theory is not in quite as good shape [64], but empirical data are abundant. The interested reader is directed to the massive luminescence literature, some of which we have already cited.

In concluding our discussion of the factors that control the efficiency of scintillation, it is appropriate to recall that our treatment stands on the assumption that each step in the process can be considered independently of the others. While the validity of this assumption cannot be proven, its violation would imply the far less likely existence of a cooperative mechanism linking two or more of the steps, for which there is no experimental evidence. We feel that it provides a useful and effective tool for examining the overall phenomenon.

2.5. Temporal Response

As we have already seen, the temporal characteristics of the scintillation process are often at least as important as its efficiency. Since the primary excitation by the gamma photon is virtually instantaneous, the overall speed of a scintillator is governed by the rate constants of the transfer and emission processes, denoted R_{tfr} and R_r (in sec^{-1}), respectively. Since these two processes act in sequence, the overall rate constant of a scintillator, or inverse of its decay time, will be determined by the slower of the two. This is a property of a difference of exponentials, which describe the rise and decay time of the luminescence in a system of centers fed by a transfer [44].

There is no fundamental limitation on the transfer rate. It can be instantaneous when the luminescence center is excited directly (as presumably in stoichiometric materials), with no intermediate steps like excitons or e-h pairs. On the other hand transfer can still be efficient even when quite slow, as is undoubtedly the case in thallium-activated scintillators [54].

Unlike the case of the transfer rate, there is a fundamental physical law that limits the radiative rate R_r (inverse of the Einstein coefficient A). For electric dipole emission the radiative lifetime $\tau_r = 1/R_r$, in ns, is given by [44]:

$$\tau_r = 1.5\times10^{-5}\cdot\frac{9\,\lambda^2}{f\cdot n\cdot\left(n^2+2\right)^2} \tag{25}$$

where f is the oscillator strength of the transition (summed over the final states and averaged over the initial states), λ is the wavelength of the transition in nanometers and n is the refractive index. The quadratic dependence of τ_r on the wavelength, illustrated in Figure 2, is responsible for the substantial and often not fully appreciated speed advantage of blue or near-UV emitting scintillator materials.

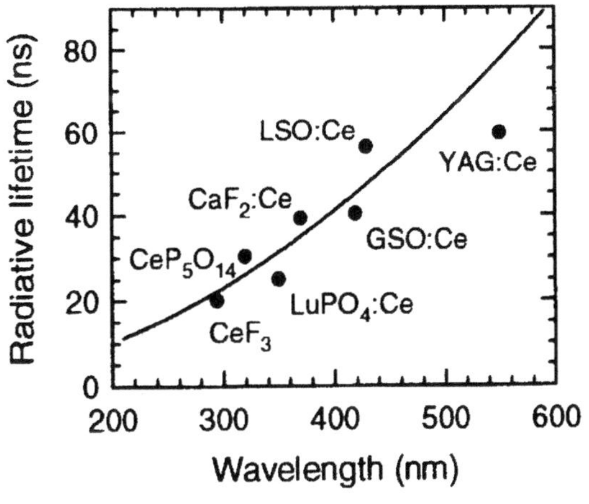

Figure 2. Radiative lifetime of Ce^{3+} d-f emission in various materials, as function of wavelength. In all these cases, the oscillator strength f was taken as 0.013, and the refractive index n as 1.7.

At this point we should note that speed can actually be increased arbitrarily, beyond that given by τ_r, by introducing nonradiative processes, but only at the expense of lowering the overall efficiency. For $Q < 1$, the emission rate becomes R_r/Q. It is likely that decay constants of less than a nanosecond, as observed in some organic scintillators, are due to partially quenched processes [65]. High oscillator strengths, of course, are clearly preferable.

For the fastest possible scintillators, therefore, two important conditions must be met: the optical transition must be fully allowed by all relevant selection rules, with a high oscillator strength and consequently short radiative decay time τ_r; and the transfer time has to be significantly shorter than τ_r, to allow the radiative decay to govern the speed. Consequently, such a scintillator cannot utilize slow transfer processes, regardless of how efficient they may be in sweeping long-lived electron-hole pairs. The condition of $R_{tfr} >> R_r$ imposes a severe restriction, since, as the example of Tl-doped scintillators demonstrates, relatively slow processes can often be quite efficient in collecting the energy deposited into the lattice and converting it to detectable light [54]. Conversely, an extremely fast process, such as direct excitation through inelastic electron-center collisions (see Sec. 3), may still be highly inefficient. Therefore we may have to reach a compromise between high speed and high efficiency, depending on the requirements of the specific application.

But how can we be assured of reaching the most favorable compromise, the best combination of speed and efficiency? Again the alkali halides offer us a logical approach: First, select an activator ion whose optical transition, unlike that of Tl^+, is fast and efficient; then distribute it in a lattice with a demonstrably high efficiency of transfer; and finally, increase the activator concentration to the stoichiometric limit, to minimize the distance over which the energy deposited in the lattice has to travel before it reaches an emitting site, thereby maximizing both the speed and light output. While this approach cannot be used with Tl^+, whose emission is severely quenched even at concentrations as low as 0.5%, it is perfectly feasible with Ce^{3+}, whose optically allowed emission is remarkably resistant to such quenching. As will be seen in Sec. 3, however, things are not always as simple as they seem.

2.6. Experimental verification of theory

Having discussed the physical meaning of the three phenomenological components (β, S, Q) of the overall process efficiency, we can now put them to use to test various theoretical predictions, as well as to compare the performance of various scintillators among themselves. With the evident incompleteness of the theoretical resources for calculating β, Robbins' theory gives us our only reasonable tool, providing a consistent way of approaching the subject, and leading to a definite methodology for calculating the fundamental process of conversion. More to the point, it has proven particularly successful in evaluating many phosphor and scintillator materials. Robbins was able to demonstrate a convincing correlation between measured and calculated efficiencies in all of the cases he considered. Certain oxides, such as Zn_2SiO_4, $CaWO_4$ or YVO_4, which have MO_4 tetrahedra characterized by strong molecular bonding and, consequently, relatively high-energy optical phonon modes, show significantly lower efficiencies, while alkali halides exemplified by CsI and NaI, which have much lower phonon energies, are consistently well known for their exceptionally good scintillator properties.

It was pointed out by Blasse [66] that Robbins' polaron model of the conversion process may prove to be useful in the case of new scintillator materials. In a series of papers Lempicki, Wojtowicz, and coworkers [27,67,68] pointed out that if a theoretical procedure for finding all three partial efficiencies were developed, it would greatly facilitate the evaluation of any scintillator material and its potential for improvement, providing an invaluable tool for future technological development. While a rigorous technique does not yet seem feasible, they have described a useful semi-empirical scheme, based on the Robbins model, and applied it to many known and some newly discovered scintillators. As already indicated, this approach assumes the separability and independence of the three steps in the overall scintillation phenomenon. Of the three, the luminescence efficiency is easiest to deal with since it can readily be measured and sometimes even calculated. Having calculated a conversion efficiency β from the Robbins theory, one is left with the transfer efficiency S as the only unknown parameter for characterizing the scintillation process.

With no direct way to measure S, we must instead infer it from the light output L, expressed as the number of generated photons N_p per MeV of γ energy. This is an experimental value (not to be confused with the theoretical quantity n_p of Eq. 4), and is the one commonly quoted in evaluating scintillators. If the number of e-h pairs actually generated is $n_{e\text{-}h}$, then the number of photons N_p that they produce is given by

$$N_p = n_{e-h} \cdot S \cdot Q = \frac{E_\gamma}{2.25 \cdot E_g} \cdot \beta \cdot S \cdot Q \tag{26}$$

and the light output by

$$L = \frac{N_p}{E_\gamma} = \frac{10^6}{2.25 \cdot E_g} \cdot \beta \cdot S \cdot Q \tag{27}$$

from which S can be calculated if E_g (in eV), β and Q are known and ξ_{min} (following van Roosbroeck and Rodnyi) is taken as $2.25 \cdot E_g$.

The procedure of Lempicki *et al.* [27] for describing the scintillation process is as follows: First, the conversion efficiency β is calculated using the Robbins model as described above. Then Q is measured or found in the literature. In the final step the transfer efficiency S is calculated using a measured value of the light yield and Eq. 27. The overall process efficiency η is the product of the previous three. In Table 1 we present results obtained using this procedure for some scintillator materials of interest, including the well-known Tl-doped halides, BGO and CaF_2:Eu, as well as certain more recently developed materials based on the Ce^{3+} ion in both dilute and concentrated cases. Most of numerical values were taken from Ref. 27.

Having found the *process* efficiency η, we can use Eq. 5 to calculate another parameter of interest: the *energy* efficiency η_e. Since the energy efficiency of emission can be no greater than the ratio of the energy of the emitted photon to that of the band-gap, a minimum value for ξ of $2.25 \cdot E_g$ (van Roosbroeck) limits the maximum energy efficiency possible for a scintillator to 1/2.25, or 44%, of that ratio. Thus the ratio $h\nu_{em}/2.25 \cdot E_g = \eta_e\,(max)$ is the maximum energy efficiency that can be expected from any given scintillator in the best circumstances imaginable. From the values listed in Table 1 for the various materials (ranging between 0.15 and 0.27), it is clear that a value of 20-25% is the best that one can reasonably expect from a good scintillator. In many cases the efficiencies are much lower because of low luminescence yield (BGO), poor conversion ($LuPO_4$:Ce), or poor transfer (CeF_3) [69]. Of these three possibilities transfer is the dominant concern because we know least about it.

Table 1.
Parameters characterizing the scintillation process in various materials, based on the polaron model.

	E_g	ξ_{min}	ξ	β	S	Q	η	$h\nu_{em}$	$\eta_e(max)$	η_e	L_c	L_m	τ_d	M
CsI:Tl	6.4	14.4	15.0	0.96	0.97	1.0	0.94	2.3	0.16	0.15	66,700	65,000	600	≤400
CsI:pure	6.4	14.4	15.0	0.96	1.0	0.03	0.029	3.9	0.27	0.008	66,700	2,000	16	125
NaI:Tl	5.9	13.3	15.2	0.87	0.58	1.0	0.50	3.0	0.23	0.11	65,800	38,000	230	≤165
BGO	5.0	11.3	16.5	0.68	1.04	0.13	0.09	2.6	0.23	0.021	60,600	8,200	300	27
CaF_2:Eu	12.2	27.5	44.3	0.62	1.06	1.0	0.66	2.9	0.11	0.070	22,600	24,000	900	26
CeF_3	10.4	23.4	39.3	0.60	0.13	1.0	0.08	4.2	0.18	0.014	25,400	3,200	20	160
CePP	8.7	19.6	66.6	0.29	0.27	1.0	0.08	3.9	0.20	0.016	15,000	4,000	30	133
LSO:Ce	6.4	14.4	38.2	0.38	1.0*	1.0*	0.38	2.9	0.20	0.076	26,200	18,500	46	402
$LuPO_4$:Ce	8.26	18.6	45.1	0.41	1.0*	1.0*	0.41	3.5	0.19	0.077	22,200	17,000	24	708
LuAP:Ce	8.0	18.0	37	0.49	1.0*	1.0*	0.49	3.4	0.19	0.093	27,000	21,200	17	1247

Parameter abbreviations: *E_g, bandgap energy in eV; ξ_{min}, minimum energy needed to create one e-h pair; ξ, average energy spent to create one e-h pair; β, conversion efficiency; S, transfer efficiency; Q, quantum efficiency of the luminescence center; η, overall efficiency of scintillation process; $h\nu_{em}$, scintillation photon energy; $\eta_e(max)$, maximum possible energy efficiency even if process efficiency were unity; η_e, calculated energy efficiency; L_c, calculated light output on assumption of S=Q=1; L_m, measured light output in photons/MeV; τ_d, decay time in nanoseconds; M, figure of merit in photons/MeV-ns, based on measured light output and decay time. The asterisks in columns S and Q for LSO, $LuPO_4$, and LuAP indicate assumed values; see text.*

Material abbreviations: *BGO, $Bi_4Ge_3O_{12}$; CePP, CeP_5O_{14}; LSO, Lu_2SiO_5; LuAP, $LuAlO_3$.*

Lempicki *et al.* have introduced yet another parameter, a figure of merit M, which is clearly very useful in those applications where timing is important [27]:

$$M = \frac{L}{\tau_d} \tag{28}$$

where L is the light yield (in photons per MeV) and τ_d is a decay time. For an exponentially decaying scintillation pulse and for shaping times longer than τ_d, M will be equal to the magnitude of the scintillator decay curve extrapolated back to zero. In the case of complex decays, the effective τ_d is a properly defined average value for all components. Since $\tau_d = \tau_r \cdot Q$, where Q is quantum efficiency and τ_r is a radiative lifetime of the luminescence center, we can use Eqs. 3 and 4 to obtain:

$$M = \frac{10^6}{\xi} \cdot S \cdot \frac{1}{\tau_r} = \frac{10^6}{\xi_{min}} \cdot \beta \cdot S \cdot \frac{1}{\tau_r}. \tag{29}$$

The figure of merit, as given in Table 1, has particular relevance to LSO, $LuPO_4$ and LuAP. Because of the large uncertainty in the light yields reported for these three materials, we found it necessary to modify the procedure. Rather than calculating S, we assume that both S and Q are equal to unity; then, with E_g and β from the table, we estimate the maximum possible light output (in photons per MeV) using Eq. 27, arriving at values of 26,200 for LSO, 22,200 for $LuPO_4$ and 27,000 for LuAP. As we shall discuss more fully in Sec 3.4, these estimates are above the reported light yields, indicating that these materials are not performing up to their theoretical potential, and suggesting that they can still be improved to reach their maximum possible yields. The figures of merit represent values based on reported light output and decay times. Although the conversion efficiency β of LSO is lower than that of $LuPO_4$, the bandgap is also lower, giving a higher light output. However, the decay time of Ce emission in LSO (46-48 ns [24]) is longer than in $LuPO_4$ (22 ns [25]) and results in a higher figure of merit for the latter. In terms of figure of merit LuAP appears to be the clear leader among the Ce-based materials.

The situation is different in the cases of CeF_3 and CeP_5O_{14}, where $Q \approx 1$, but the measured light output reveals a very poor utilization of the conversion (see Sec. 3.3). From our definitions of β, S, and Q, we formally ascribe the inefficiency in these materials to poor transfer. Thus, although the concept of transfer may need further clarification in this case, the result points to a serious deficiency in the stoichiometric limit of these two Ce compounds. Further consideration of transfer in these materials are found in Sec. 3.3 and Refs. 69 and 70.

To summarize, we have defined seven important phenomenological parameters to describe the scintillation process: the process efficiency η, the quantum efficiency η_q, the energy efficiency η_e, the figure of merit M, and, of course, the efficiencies β, S and Q of the three constituent steps. It is clear from Eqs. 1 - 5 that the last three, plus the quantity ξ_{min}, are all one needs to obtain a full description of any scintillator material (aside from density, chemical stability, radiation damage and cost). While some of these parameters can either be measured directly or inferred from such quantities as light yield L or decay time τ_d, it would obviously be particularly useful if we could relate them to, or even calculate them from, known material constants. Unfortunately, there are at present no theories or models that can enable us to do so. Only Q, the luminescence quantum efficiency, can be measured and, in many cases, also calculated. Other parameters are more difficult to deal with although, despite the remaining uncertainties, theoretical progress on the conversion problem (ξ_{min} and β) is clearly being made. The transfer step, although critical for performance of many scintillators, remains the least understood and most challenging aspect of scintillator performance. In the following sections we shall make use of the procedures described here and demonstrate their value to guide further research.

In summary, this approach provides a framework for discussing the scintillation properties of any material, and can be both a starting point and a very useful tool in developing new materials. In particular, it decisively identifies weak links in the process: a transfer step in the case of NaI:Tl, CeF_3, and CePP; a poor conversion in the case of CePP, LSO, $LuPO_4$ and LuAP; and a poor quantum efficiency of the luminescence step in the case of BGO. Unfortunately, it does not provide an explanation for some interesting questions, such as why transfer is very good in BGO but not in CeF_3 (both are concentrated or stoichiometric materials in which the luminescent center is not a dopant but a constituent of the lattice). It does, however, help us to understand others, such as the relative inferiority of NaI:Tl in comparison to CsI:Tl or poor conversion in materials containing molecular groups, and we expect it to become even more useful with further development.

3. Material Issues

The current revival of scintillator research has produced a torrent of publications on new materials. In particular, the interested reader is referred to book publications listed in Ref. 71. In this section we shall review only a limited number of materials, chosen to illustrate the principles and phenomena discussed in Sec. 2. Our purpose is not to present a status report on scintillator materials in general, but rather on the degree of understanding of the principles involved in the phenomenon. Among the things we shall examine is the extent to which Robbins' semi-empirical theory of conversion predicts the observed results. Since we never directly measure the number of electron-hole pairs produced by a stopped gamma, but rather the light yield, we can obtain the former only if we have good reason to believe that the transfer S and the luminescence efficiency Q are unity. We can afford to exclude the problem of Q since in most materials of interest, Q is known and equal or close to unity (BGO being a singular exception). Therefore the comparison with experimental results leaves an unavoidable ambiguity: In case of conflict, we have no way to know whether the calculation of the conversion efficiency is at fault, or the transfer efficiency S is less than unity. This, unfortunately, is the present state of affairs, caused by the approximate nature of Robbins' theory and a lack of any quantitative theory to describe the transfer process.

Compounding the problem further, we have to remember that the product $\beta \cdot S$ obviously depends upon the perfection of the crystal. The performance of a material invariably improves with the time expended in its development, and hence at any point, β, S, or both can be adversely affected by imperfections in the crystal. A second source of uncertainty derives from the measurement of light output itself. This subject is treated at some length in Ref. 72 and we will not dwell upon it here; suffice it to say that the absolute measurement of photons (per unit of gamma energy) emitted in one burst is fraught with difficulties. Usually such a measurement is not even attempted, with only a comparison with a "standard" such as BGO being reported. This "standard", however, is not very satisfactory because absolute measurements of the light output of BGO, as found by various investigators, range from 5,500 to 10,000 photons/MeV [73,74]. Consequently, we must reluctantly admit that the science of scintillators is not as precise as we would like to think, and that measurements are not uniformly reproducible from laboratory to laboratory. Nevertheless new materials with higher performance are continually being developed, and what follows in this section is a qualitative attempt to interpret this progress.

We shall find it useful to divide the scintillator materials into two classes:

- *Activated materials*, in which the luminescence center is a well defined impurity ion, introduced into a host lattice. The concentration of the impurity is usually of the order of 0.1%. A prime example of this kind of material is NaI:Tl; others include the more modern materials utilizing Ce^{3+} as the activator; and
- *Stoichiometric materials*, in which the concept of a clearly identifiable luminescent center is no longer applicable. Luminescence may be the property of the lattice itself or of a constituent of the lattice, which however cannot be regarded in isolation, *i. e.* as immersed in a host medium. Examples are CsI, $Bi_4Ge_3O_{12}$ (BGO), CeF_3, $PbWO_4$ and BaF_2.

Obviously the border between activated and stoichiometric materials is not very rigid and may even be crossed deliberately. An example is provided by such materials as $Ce_xLa_{1-x}F_3$, which can be prepared with active ion (Ce) concentration anywhere between 0 and 100%. Interesting and not yet completely clarified physics accompanies the variation of x, some of which will be discussed in the following sections.

3.1. Intrinsic Scintillators

In this section we shall describe a small number of scintillators whose light emission properties are uniquely determined by their lattice alone, without the inclusion of any dopants. Among these are two that have played a unique role in the saga of scintillator development and, in spite of a long and venerable history, are still widely used today. These are $Bi_4Ge_3O_{12}$ (BGO) and CsI. Other examples of this class include the tungstates $ZnWO_4$ and $CdWO_4$, and certain more recently developed lead compounds such as $PbWO_4$ and $PbSO_4$, the former now the prime candidate for use in the LHC calorimeter being built at CERN. Selected properties of these materials are listed in Table 2. Crossluminescence scintillators, such BaF_2, are considered a separate category, and will be discussed separately.

As a class these stoichiometric compounds are not distinguished by an exceptionally high light output, for which reason their primary application is in the field of high energy physics where radiation levels are very high and efficiency is not at a premium. Despite its shortcomings in light output and speed, however, BGO is still reluctantly accepted as the mainstay of commercial medical PET cameras as well. It

Table 2. Selected properties of stoichiometric scintillators

	$Bi_4Ge_3O_{12}$	$ZnWO_4$	$CdWO_4$	$PbSO_4$	$PbWO_4$	CsI
Density (g/cm^3)	7.13	7.87	7.99	6.4	8.23	4.51
Emission (nm)	480	480	480	340	460-520	315
Decay (ns)*	300	5,000	5,000	1.8-425	1.5-93	16
Light yield (photons/MeV)	8,500	10,000	14,000	4,900	140	2,000

* *When decays are strongly nonexponential a range is given.*

is probably the recognition of the deficiencies of this type of material that has been the stimulus for the recent spurt of activity in the quest for better materials. No appreciation of the present state-of-the-scintillator-art can be achieved without some knowledge of these systems.

Other than their stoichiometric nature, these compounds have little in common to unify the physics of their light emission processes. Bi^{3+} and Pb^{2+} are well known activators with s^2 electronic configuration and $^1S_0$ ground state. In the excited state the *sp* configuration gives rise to $^3P_0$, $^3P_1$, and $^3P_2$ states resulting in spin-forbidden transitions whose long decay times are decreased only by nonradiative recombination. At low concentrations as activators, the s^2 ions have played an important role in luminescent materials, specifically phosphors [2,75]. In the case of Bi and Pb their emissions are characteristic of the individual ion, and are maintained even at concentrations up to the stoichiometric limit. In the case of the tungstates, however, the dominant emission originates in the WO_4^{2-} group [76]. Finally, the emission of CsI is a true lattice property with no direct linkage to either of the constituent ions.

BGO: First discovered by Weber and Monchamp [12] in 1973, this material has assumed a very important role as scintillator because of its convenient emission wavelength, high density and a moderately fast decay. Probably the most spectacular use of BGO was in the CERN calorimeter which utilized almost 12,000 crystals, each 24 cm long and approximately 7 cm^2 in cross-section [14]. BGO is also extensively used in medical radiology, where a PET camera may use hundreds, even thousands of crystals [77]. In both applications the limiting factor in BGO proved to be its modest speed, resulting in count pileup and limited time resolution, and (in medical applications) requiring an undesirably high dose of radiation. While the light output is more than enough for high energy physics it is definitely inadequate for modern PET.

A voluminous body of literature exists on the luminescence properties and mechanisms in BGO, which, although quite interesting in its own right, is of limited relevance to scintillation. The book by Blasse and Grabmaier [2] gives a good summary. For our purposes BGO, being a quintessential stoichiometric material, provides an interesting case for testing the Robbins theory [31]. The calculated number of electron-hole pairs per MeV, $n_{e\text{-}h}$, was found to be 60,200. In this particular material it is reasonable to assume that the transfer efficiency S must be close to unity, since whatever is considered to be the luminescent center in BGO is present in every lattice cell, obviating the need for energy transfer over a distance. Given a room temperature luminescence efficiency Q of about 0.13 [78], we obtain for the number of photons $n_p = n_{e\text{-}h} \cdot S \cdot Q$ a value of 7,800 photons per MeV. This is in remarkably good agreement with the generally accepted measured value of 8,200 photons [72]. Thus, at least in the case of BGO, the Robbins theory produces a very satisfactory result.

CsI: Cesium iodide, discovered in the early 50's [79] occupies a unique position among scintillators because when doped with Tl it has a light output of 65,000 photons/MeV, the highest of all known materials [74]. Because of the light that they shed on the mechanisms of scintillation, the Tl-doped halides have already been discussed in some detail in Sec. 2.3. Unique among the halides however, CsI is a remarkable scintillator even undoped, with a rapid decay and respectable light yield on the order of 2,000 photons/MeV, as indicated in Table 2 [80], enabling it to play a role in high energy physics [81]. The emission of CsI is characterized by two bands at 315 and 450 nm. The shorter wavelength band is of intrinsic origin, being due to a self trapped exciton; its decay is 16 ns, shortened by temperature quenching. The longer wavelength band is due to I^- vacancies and its intensity is heavily dependent on conditions of crystal growth and post heat treatment [80]. It is certainly to the shorter wavelength (near UV) emission that CsI owes its success.

$PbSO_4$: Scintillator properties were discovered by Kirk in a natural crystal of anglesite [82]. Further work by Moses *et al.* [83] has shown that the material has a remarkable efficiency at low temperature (68,500 photons/MeV at -145 C) and highly nonexponential decays at all temperatures. Efforts to perfect crystal growth of this material are continuing [84]. This material would provide a particularly good host lattice if only one could find a suitable dopant to which transfer of energy from the lattice would be sufficiently rapid and efficient at room temperature to compete successfully with the rate of nonradiative processes.

$PbWO_4$: In his guidelines for the search for new scintillators, Blasse [85] states that "It is not realistic to propose this compound as a possible luminescent material at room temperature". While the statement is perfectly true, it illustrates the divergence of criteria as to what constitutes a good luminescent material and a good scintillator. $PbWO_4$ has now been chosen in preference to the more efficient CeF_3 for the construction of the calorimeter for the Compact Muon Solenoid (CMS) of the LHC at CERN. The very low light output of this material is of little consequence for the high radiation environment [86,87]. When it comes to very large scale applications, requiring thousands of precisely fabricated crystals, criteria of cost and manufacturing capability become dominant. The tungstate's reputed (but unconfirmed) ability to resist radiation-induced damage surely contributed to this decision.

3.2. Crossluminescence Scintillators

A unique class of undoped scintillators consists of those materials that exhibit the phenomenon of crossluminescence. The most prominent of these materials is BaF_2, whose scintillation was first observed in the early 1970's [11]. More than a decade later, Laval, Moszynski, *et al.* discovered that this scintillation contains a very fast decay component which makes possible a time resolution on the order of 100 ps [19]. This emission was in the UV, from 180 to 240 nm, and had a decay time of about 600 ps, some three orders of magnitude faster than the conventional luminescence.

While its light output was modest (≈2000 photons/MeV) and its density (4.88 g/cm^3) not particularly high for an inorganic crystal, BaF_2 immediately attracted a great deal of attention as a competitor to plastic scintillators, which although somewhat faster, have densities about a factor of four lower. The nearest competitor in the inorganic world was CsI, which, however, is hygroscopic while the fluoride is chemically and mechanically a very robust crystal.

The existence of this unusually fast decay component proved to be of fundamental importance since it represented a new kind of luminescent transition. The mechanism involves the excitation of a core level of Ba^{2+}, rather than the more usual cases of an impurity level situated in the band gap or a transition originating in the valence band [88,89]. The phenomenon takes place only when the excitation exceeds 18 eV, a value much larger than the band-to-band ionization threshold of about 10 eV.

The energy level scheme of BaF_2 and the origin of the luminescence are shown in Figure 3. At an excitation energy of about 10 eV, an electron-hole pair can be generated in a normal interband ionization process. The resultant exciton becomes self-trapped and recombines radiatively, generating a long wavelength component of emission peaking at around 300 nm. When the excitation energy reaches 18 eV two short wavelength bands appear, at 195 and 220 nm. This shorter wavelength emission, characterized by the 0.6 ns decay time, arises when an electron is raised from the Ba^{2+} $5p$ core level to the conduction band, momentarily raising Ba to the *3+* charge state. The hole generated in the core level then recombines radiatively with an electron from the valence band, giving rise to the fast component and leaving a secondary self-trapped exciton to recombine more slowly. Since the excited electron "crosses over" the valence band, the emission is often referred to as crossover luminescence or simply crossluminescence. The process can be considered a form of charge transfer transition, wherein a valence electron tunnels from the F^- anion to the hyperionized Ba cation. Since parity selection rules do not apply to such interionic transitions, the oscillator strength can be quite high and the decay correspondingly rapid. For the appearance of such crossluminescence the energy difference $E_{(C-V)}$ between the valence band and the core level must be smaller than the band gap E_g (≈10 eV in the case of BaF_2), so as to prevent either the absorption of the photon

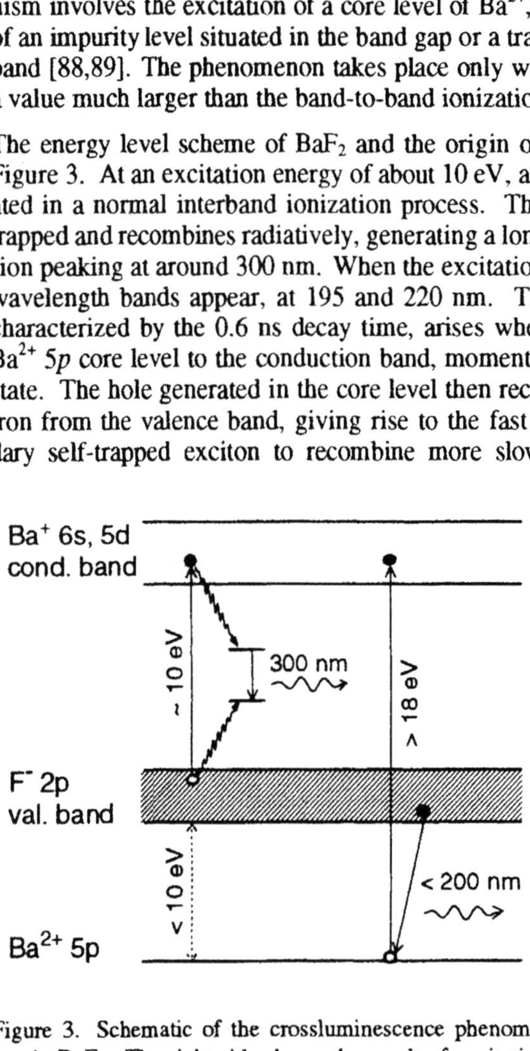

Figure 3. Schematic of the crossluminescence phenomenon in BaF_2. The right side shows the result of excitation greater than about 18 eV, with a valence electron dropping into the vacated core hole. The left shows formation of a self-trapped exciton at lower excitation energies.

or a Auger process from valence to conduction band. This condition is rather restrictive and limits the number of materials in which crossluminescence can be observed.

Excellent reviews of the whole subject of crossluminescence and its relevance to scintillation have recently appeared [20,90,91], to which the reader is referred for detailed descriptions of the recent experimental and theoretical work. While cluster calculations have been helpful in accounting for band shapes and even transition moments, the overall low yield of crossluminescence scintillators is not fully understood. Since the decay time of the fast component of BaF_2 shows practically no change with temperature, the low light output cannot be ascribed to thermal quenching. A reasonable presumption is that the limiting factor is the concentration of holes in the outer core band. To quantify this would require a kind of conversion theory similar to that provided by Robbins for electron-hole pairs in the valence and conduction bands.

In Table 3 we reproduce a listing of compounds (originally given in Ref. 20) which were the subject of search for crossluminescence. The second column gives the

Table 3. Properties of Crossluminescent Materials [20]

	Density (g/cm^3)	$E_{(C-V)}$ (eV)	E_g (eV)	λ (nm)	Decay (ns)	Light yield (photons/MeV)
KF	2.5	7.5-10.5	10.7	156		
KCl	2.0	10-13	8.4			
KBr	2.75	10-13	7.4			
KI	3.1	9.5-14	6	STE		
RbF	3.6	0-7.5	10.3	203, 234	1.3	1700
RbCl	2.86	4-9	8.2	190		1
RbBr	3.35	6.7-9.5	7.4			
RbI	3.55	5-10	6.1	STE		
CsF	4.1	0-4.5	9.9	390	2.9	2000
CsCl	4.0	1-5	8.3	240, 270	0.9	900
CsBr	4.4	4-6	7.3	250	0.07	20
CsI	4.5	0-7	6.2	STE		
CaF_2	3.2	12.5-17.3	12.6	STE		
SrF_2	4.2	8.4-12.8	11.1	STE		
BaF_2	4.9	4.4-7.8	10.5	195, 220	0.8	1400
$KMgF_3$	3.2			140-190	1.3	1400
$KCaF_3$	3.0			140-190	<2	1400
KYF_4	3.6			170	1.9	1000
K_2YF_5	3.1			170	1.3	300
$KLuF_4$	5.2			170-200	1.2	200
KLu_2F_7	7.5			165	<2	200
K_2SiF_6	3.1			140-250		
$CsCaCl_3$	2.9			250, 305	1	1400
$CsSrCl_3$				260, 300	1	
$LiBaF_3$	5.2			190, 230	1	1400
$BaMgF_4$	4.5			190, 220		1000
BaY_2F_8	5.0			165-310	0.9	
K_2LiGaF_6				140-250		
K_2NaAlF_6				140-250		

energy differences between the core level and the bottom and top of the valence band. Whenever this value is known and the condition $E_{(C-V)} < E_g$ is fulfilled, crossluminescence occurs. The entry STE indicates that only a self-trapped exciton emission is observed. Although the highest light output is observed for CsF, this crystal is too hygroscopic for practical use. Of all the materials exhibiting crossluminescence only BaF_2 has played a practical role. For some time it was considered to be the prime candidate for the SSC detector until questions of radiation resistance [92] and supply (among others) tilted the balance in favor of a liquid argon detector. Eventually the issue became moot because of the cancellation of the SSC.

We should remark that the efficiency of crossluminescence is necessarily limited by the presence of a still unrecombined electron-hole pair in the conduction and valence bands after the primary photon has been emitted. If this recombination is at least partially radiative a relatively slow component of the decay is bound to appear, which, since it occurs at longer wavelengths, is difficult to filter out optically. Attempts to reduce the magnitude of the slow decay component by the incorporation of other impurities to promote the nonradiative recombination of the self-trapped exciton have met with limited success [93]. Van Eijk, one of the pioneers of crossluminescence, has expressed the opinion that materials significantly surpassing the performance of BaF_2 are not likely to be found [20].

3.3. Stoichiometric Cerium Compounds

In this section we first encounter Ce^{3+} as an important constituent of scintillators. One of its remarkable properties is that it can maintain a high luminescent efficiency both as a stoichiometric constituent and as a dopant. In fact most of the research currently being pursued deals with materials containing this ion.

As the first of the lanthanide series, containing one *f* electron ($4f^1$ configuration), Ce^{3+} has a rather simple electronic energy level structure. As shown in Figure 4, the ground state is split by spin-orbit interaction into $^2F_{5/2}$ and $^2F_{7/2}$ multiplets, separated typically by some 2300 cm^{-1} [94]. Crystal field splitting within the two multiplets is on the order of 100 cm^{-1} and plays no role in scintillation. Two other features, however, are of major significance: One is that the lowest excited state belongs to the 5*d* configuration, making the radiative transition to ground Laporte-allowed [95] and therefore extremely fast. Equally important is that the excited state typically lies some 40,000 cm^{-1} above the ground state [96]; this huge energy gap makes the probability of nonradiative deactivation by multiphonon mechanisms almost vanishingly small, so that in the absence of other loss processes (such as rapid energy migration to sinks),

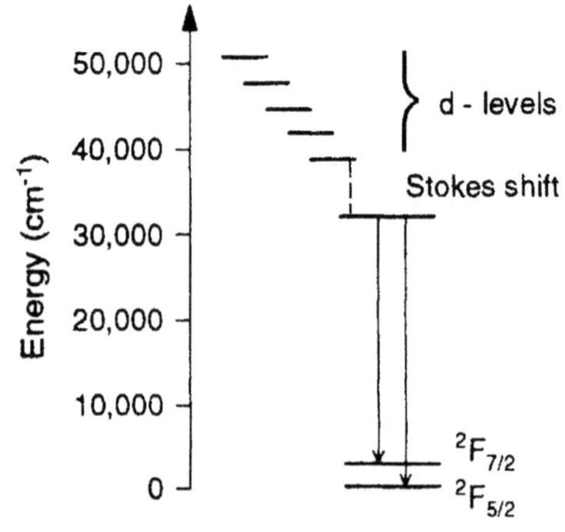

Figure 4. Energy level structure of the Ce^{3+} ion.

the quantum yield Q is often close to unity, even at concentrations up to the stoichiometric limit.

Thus Ce^{3+} offers an opportunity not available from the Tl-doped halides, whose emission is both concentration-quenched and relatively slow. With a luminescence quantum efficiency close to unity, and an atomic weight sufficient, at high enough concentration, to provide superb stopping power for the incident gamma photon, an increase of Ce^{3+} content to the stoichiometric limit should eliminate any possible losses attributable to transfer over a finite distance, resulting in a major step forward in both speed and efficiency. Consequently, we shall begin the cerium story by first examining two materials, each of which, at one time or another, was regarded as leading to a maximally efficient "ultimate" scintillator.

Cerium Pentaphosphate, CeP_5O_{14}, is one of the most remarkably efficient stoichiometric luminophores. Nd pentaphosphate was widely investigated as a minilaser material in the 1960's [97]. This Ce compound has been known as a fast cathodoluminescent phosphor [98] and was the subject of a rather detailed study as a scintillator [70,99]. Given its low density (3.5 g/cm^3) and the low yield from the flux technique by which it is grown, this material was never expected to become a viable scintillator in any practical application. It did, however, offer important advantages as a test bed for new ideas since dilution with lanthanum (to form $Ce_xLa_{1-x}P_5O_{14}$) provided a range of Ce concentration and a continuous transition from stoichiometric to dilute activated materials. As shown in Figure 5, for $x=0.05$ the emission is a textbook case of Ce luminescence, with the expected doublet structure. As the Ce content increases toward $x = 1$, the short wavelength peak gradually decreases in intensity due to red shift of the absorption. The decay times for the two concentrations, shown in Figure 6, while remaining strictly exponential, show distinctly different trends. While the decay time of the concentrated sample increases nearly linearly with temperature, that of the dilute sample remains constant. The convergence of the two decays at low temperature indicates the absence of any concencentration quenching. This behavior has been semiquantitatively explained in terms of self-absorption and radiative transfer [70]: If an excited Ce ion is equally likely to emit a photon into either band, and if the overlap between emission and absorption increases with temperature, then the shorter wavelength photon becomes more likely to be reabsorbed and re-emitted, effectively enhancing the longer wavelength component at the expense of the shorter. This lengthens the observed

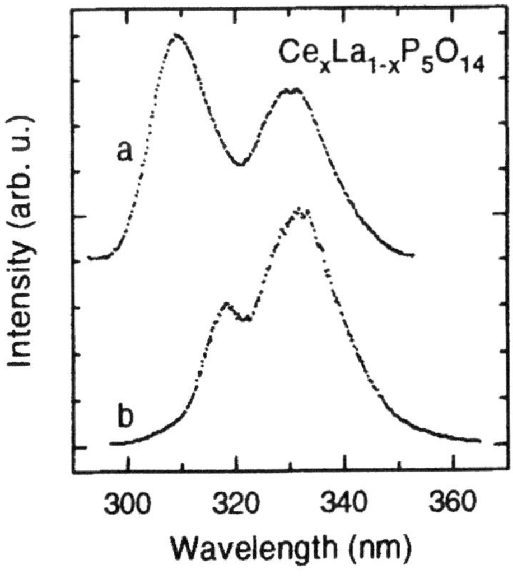

Figure 5. Emission spectrum of $Ce_xLa_{1-x}P_5O_{14}$ at two Ce^{3+} concentrations: Upper trace (a), $x=0.05$; lower trace (b), $x=1$.

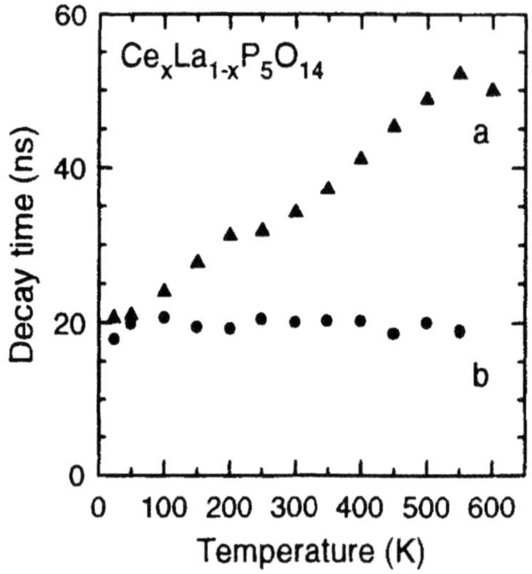

Figure 6. Decay times of emission of $Ce_xLa_{1-x}P_5O_{14}$ at two Ce^{3+} concentrations, as function of temperature: Triangles (a), $x=0.05$; circles (b), $x=1$.

decay time τ_o over that at high dilution τ_d, by a quantity proportional to the degree of overlap: $\tau_o = (1+\zeta)\cdot\tau_d$, where ζ is the probability of reabsorption.

Probably the most interesting and unexpected result of the work on cerium pentaphosphates is shown in Figure 7, which gives the light output (per MeV) as a function of concentration. The strictly linear dependence on x shows no indication of any change from a dilute activated to a stoichiometric material. The constancy of the light output per ion, even at the extrema of x, suggests that transfer is essentially negligible. This inference is strengthened by the very modest light output of 4,000 photons/MeV, in spite of the much larger number (see Table 1) of 14,800 e-h pairs calculated from conversion. Apparently cerium pentaphosphate is not capable of taking full advantage of the e-h pair generation, a cardinal sin for scintillators.

Unlike the alkali halides, the pentaphosphate provides no independent information on carrier or exciton self-trapping, preventing any detailed analysis along the lines developed in Sec. 2.3. If hole self-trapping by the polyphosphate chain (perhaps by breaking of bridge bonds) were rapid and efficient, then all excitons resulting from holes localized on the anionic sublattice would be lost to the Ce^{3+}, with relaxation of the lattice preventing any kind of transfer to the emitting center. Only if the hole were initially created in the immediate vicinity of the Ce^{3+} ion itself would the resultant Ce-bound exciton be able to transfer its energy to the *f-d* electronic structure of the ion and produce luminescence. Thus all excitation of the Ce ions would have to occur by some sort of direct process (including also collisions with secondary electrons, as in processes *10* and *11* of Sec. 2.3). This of course means that the

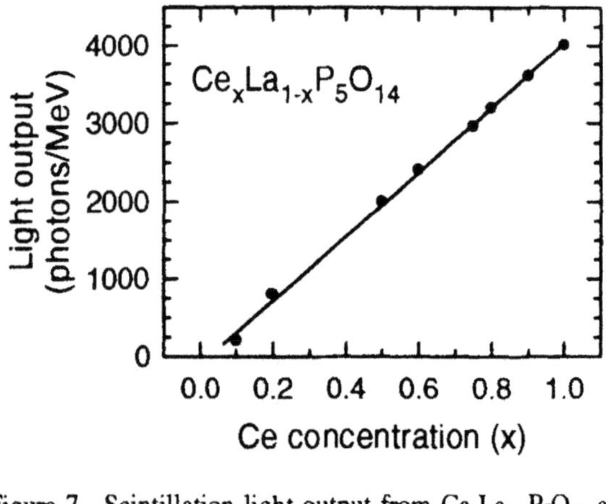

Figure 7. Scintillation light output from $Ce_xLa_{1-x}P_5O_{14}$ as function of cerium concentration.

interaction volume, which in the Tl-doped halides extends as far as 25 unit cells, has shrunk in the case of the pentaphosphate to essentially the volume of the Ce ion [99].

It is clear, therefore, that the simple reasoning that led to the consideration of stoichiometric compounds is far from adequate. The poor transfer is puzzling but undeniable, and in absence of information on the kinetics of electrons and holes in this lattice, we can do nothing more than speculate about the reasons.

Cerium Fluoride, CeF_3, is in many ways an extremely attractive crystal for scintillator applications. It has a tysonite structure, belonging to the D^4_{3d} space group, with a single rare earth site of C_2 symmetry. It has a density of 6.16 g/cm^3, superb chemical stability and a well-developed technology for crystal growth. It was first proposed as a scintillator by Anderson [22] and independently by Moses and Derenzo [23]. The great initial interest in this material was caused by the observation of an ultrafast decay component of about 2 ns, which suggested potential for both medical and high energy physics applications. It was soon shown, however, that the fast component was due to a nonradiative quenching process that actually decreased the light output [69,100,101]. The quench involved an energy transfer from Ce^{3+}, which played the role of a donor, to an unidentified acceptor. The light output, amounting to 50% of BGO, initially seemed capable of significant improvement. Effort in the United States dealt mostly with medical applications, while in Europe the material became a candidate for the LHC calorimeter [102].

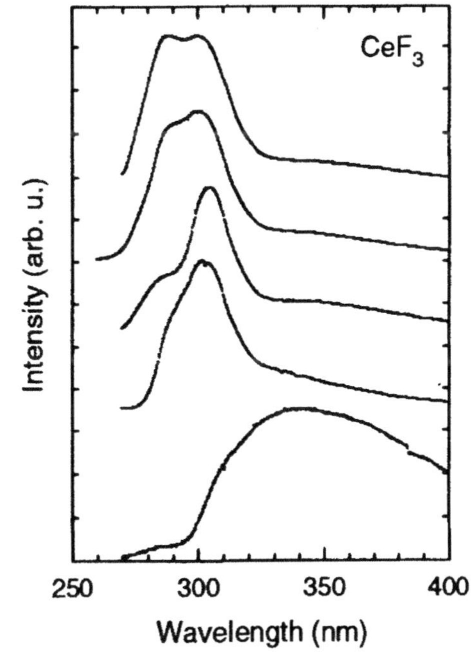

Figure 8. Emission spectra of CeF_3 crystals grown between 1990 (bottom) and 1993 (top). Note the progressive reduction of the long-wavelength emission centered at about 350 nm and the corresponding increase of the normal Ce doublet at 285 and 305 nm, related to a decrease in fluorine vacancies.

We shall not attempt to review the chronology of the research, but rather try to present a balanced view of its present status. Figures 8 and 9 display the Ce^{3+} emission from a number of different fluoride specimens. The latter figure shows the effect of cerium concentration in $Ce_xLa_{1-x}F_3$ under both optical and ionizing excitation. The two series, virtually indistinguishable from each other, were obtained at room temperature from "state-of-the-art" specimens grown in 1993 [103], and show the characteristic Ce doublet at small x. Yet, as illustrated in Figure 8, it was not always so simple. Here we see optically excited spectra of CeF_3 sample grown between 1990 and 1993 (progressing from bottom to top). Note that aside from the doublet whose components at 285 and 305 nm comprise the short wavelength lumi-

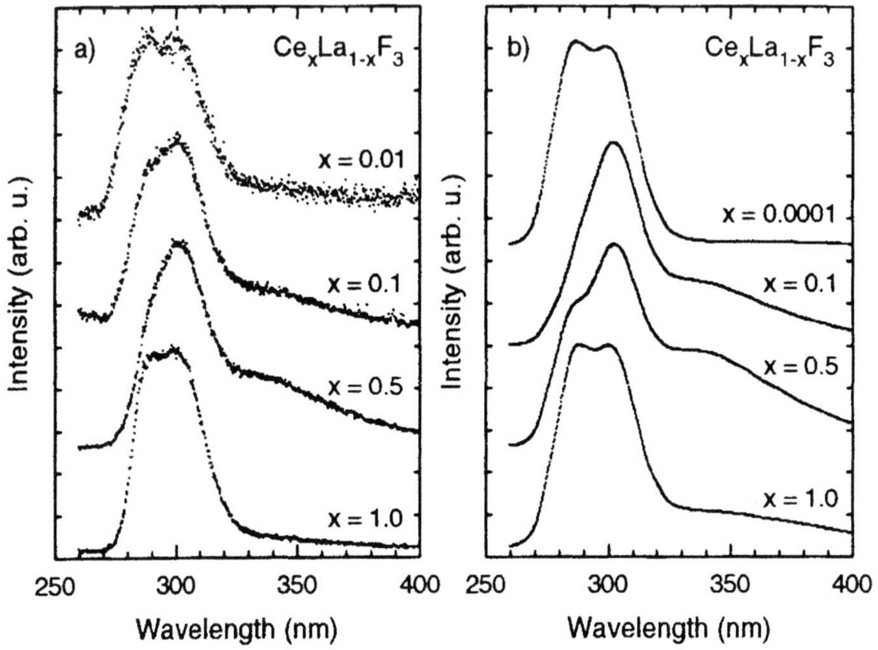

Figure 9. Emission spectra of $Ce_xLa_{1-x}F_3$ at various Ce concentrations. Left (a), excitation by gamma (0.55 MeV) irradiation; right (b), excitation by optical (250 nm) radiation.

nescence (SWL), there is an additional luminescence at longer wavelengths (LWL) extending to 400 nm and beyond, with a decay time at least twice that of the SWL. It was first proposed in 1991 that the LWL, which in the oldest sample (bottom curve in Figure 8) completely dominates the spectrum, was due to an emission from a "perturbed" Ce site [100]. Development of improved fabrication techniques [103] brought about the gradual emergence of the short wavelength component at 285 nm.

There has been much speculation about the origin of the "perturbed" emission [69] which, in view of the existence of only a single Ce crystallographic site, must be caused by a neighboring defect. While the nature of this defect has never been definitively established, its concentration has clearly been continually reduced over the years as the improved fabrication techniques lowered both the impurity content (mostly other rare earth ions and oxygen) and vacancy concentration of the crystals. A detailed analysis of the spectroscopy and interaction between normal and perturbed Ce, based on a series of $Ce_xLa_{1-x}F_3$ samples, has been reported in Ref. 69. The spectroscopic results can be summarized as follows:

1. There is no evidence for quenching by either concentration or thermal effects, at least up to 500 K.
2. Despite the absence of energy migration between Ce ions in normal sites, efficient nonradiative transfer does take place from normal to perturbed Ce, causing the SWL to show a nonexponential decay.

3. The concentration of perturbed Ce ions was found to be $\approx 3\times10^{20}$ cm^{-3} or $\approx 1.6\%$. of the total Ce content. This puts the concentration of perturbers (assuming distribution on anionic sites) at about 6×10^{19} cm^{-3} or 1000 ppm.
4. Chemical analysis indicates an oxygen content of 30-50 ppm and a level of cationic impurities of < 100 ppm, both of them too low to account for the effect.

Thus, for lack of better knowledge, we tentatively identify the perturbers as F^- vacancies, consistent with the known superionic conductivity of the lattice [104].

The mechanisms leading to the formation of excited cerium ions $(Ce^{3+})^*$ have been considered in Ref. 69 and can be related to processes 10 (direct excitation) and 11 (direct ionization) given in Sec. 3.1. In these cases the critical excitation volumes V_c are less than one unit cell. On this basis a calculation was made of the light output of $Ce_xLa_{1-x}F_3$ as a function of x (Figure 10). It is quite obvious that as x increases, the excitation volumes will begin to overlap. This is precisely the origin of the light output saturation. The dashed curve (a), which represents this calculation, agrees well with experimental results for small values of x. The steep rise indicates that for $x < 0.05$ there is a some degree of transfer from the lattice, a conclusion confirmed by the excitation spectra of Moses *et al.* [105]. As x increases, however, the light out put soon saturates, at a level roughly 10% of the predicted conversion. This must be attributed to a deficiency in at least one of the three steps in the overall process:

(i). *Inadequate Generation of Carriers:* This would imply that Robbins' calculation of conversion is not applicable to CeF_3. This is rather doubtful, as there is no reason to expect this one material to be exempted from the general laws applying to conversion. Whatever the limitations of the Robbins treatment, they should apply across the board, not just to a single compound. The real reason must be sought among processes that render part of the *e-h* population ineffective in exciting Ce to its emitting level.

(ii). *Deficiencies in Charge Carrier Capture:*. These could assume a number of different guises. One would be a process similar to that in NaI:Tl which reduces the intensity of the fast component. It was proposed in Ref. 69 that electrons can be trapped by Ce^{3+} ions to form Ce^{2+} (analogous to Tl^+ becoming Tl^0); this would require a well of stability for the Ce^{2+} ion, with a state located below the edge of the conduction band. Ce ions in the 2+ charge state have in fact been detected

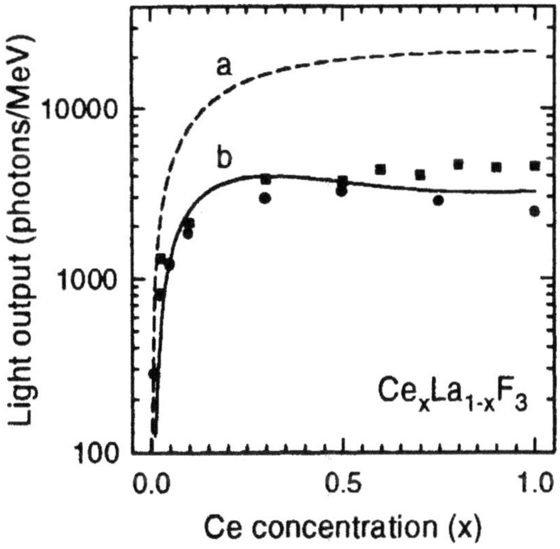

Figure 10. Scintillation light output of $Ce_xLa_{1-x}F_3$ as function of Ce content x. Dashed curve (a), calculated on basis of excitation volume; solid curve, fit to experimental points, calculated from charge carrier trapping model.

in a number of lattices [106]. Such trapping, similar to that in the pentaphosphate, has never been definitively proven and should lead to both long components of decay and thermoluminescence, neither of which are observed in CeF_3. Nevertheless the trapping model with two adjustable parameters does indeed correctly reproduce the experimental data as shown by the solid curve (b) in Figure 10.

A different explanation has been offered by Pedrini *et al.* [107], who ascribe the low scintillation efficiency of CeF_3 to an inability of the Ce ion to capture valence holes. This would mean that only direct excitation or ionization of the Ce ion can be useful in producing luminescence, as discussed in Ref. 69, with most of the energy deposited in the lattice being dissipated elsewhere. This hypothesis is supported by a simple single-configuration-coordinate model [108] which predicts a relatively high energy barrier for hole capture by the Ce ion in CeF_3. This, however, conflicts with the observation of very efficient emission from Ce^{3+} in LaF_3 even at extremely low concentrations [109,110], indicating a relatively high capture cross section for holes.

(iii). *Nonradiative Processes:* Quite obviously, if nonradiative recombination of $\underline{e}$-$\underline{h}$ pairs occurs *before* excitation of Ce then the conversion will not be fully utilized. No evidence has yet been found that such processes do occur, and it is far more likely that quenching of Ce ions takes place *after* excitation [42]. Thus we have an apparent paradox: How can quenching of the excited Ce^{3+} ion take place, when it has been well established that the luminescence quantum yield Q is essentially unity?

In fact, no conflict actually exists. The evidence for unit quantum efficiency comes exclusively from optical excitation into the multiplet containing the emitting level, and says nothing about any states that may lie at yet higher energies. Thus it is fully conceivable for a mechanism to exist that provides a nonradiative route from the higher-lying state to ground, effectively bypassing the emitting state without materially affecting its quantum efficiency. Let us examine how this can occur:

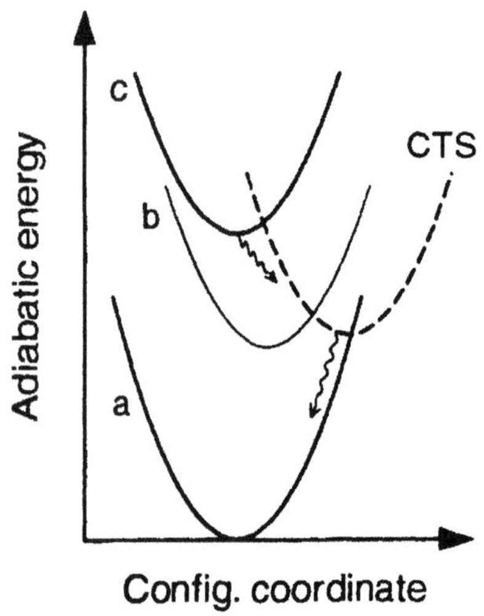

Figure 11. Mechanism for nonradiative de-excitation by means of a charge transfer process. Parabola (c) represents a high excited state, and (a) the ground state, of Ce^{3+}. Intersection with the CTS parabola allows excitation to bypass the emitting state (b).

Under ionizing excitation, when free carriers are present, Ce^{3+} can trap either electrons or holes (process *1* or *2*, Sec. 2). It has been shown qualitatively in Ref. 69 that the *order* of trapping may be critical, with process *1* leading to nonradiative recombination, while process *2* leads to emission. The concept of different results depending on the sequence of trapping is not new, and has been proposed to explain the origin of different emission lines in Tl-doped alkali halides [53]. The initial trapping of an electron by Ce^{3+} can also be viewed as effectively transforming it into a Ce^{2+} ion: $Ce^{3+} + \underline{e} + \underline{h} \rightarrow Ce^{2+} + \underline{h}$. This is rather similar to the formation of a charge transfer state (CTS), which, depending on its energy, is known often to decay nonradiatively [108]. In this case, however, the hole has been self-trapped *before* its capture by the Ce^{2+} ion, introducing an additional degree of stabilization to the bound hole-ion system, and thereby raising the probability of nonradiative decay. As illustrated in Figure 11, this mechanism involves the electronic structure and lattice relaxation of only a *single* Ce ion, with concentration effects depending only on the relative mobilities of electrons and holes.

Interactions between *pairs* of Ce ions can provide additional nonradiative pathways [108]. Such quenching can begin with the simultaneous occurrence of processes *1* and *2*, *i. e.* the trapping of a hole and an electron by *different* Ce^{3+} ions, in effect generating Ce^{2+} - Ce^{4+} pairs. If the two ions were located adjacent to each other, electron exchange may annihilate the pair, leaving two Ce^{3+} ions in their ground state. Such processes are known to occur between different ions, such as Ce^{4+} - Eu^{2+} [111] and Ce^{4+} - Yb^{2+} [112] and of course would be favored in stoichiometric materials. For $x < 1$ a large mobility of the charge states would have to be assumed.

Another way self quenching may occur is between adjacent Ce^{3+} ions that are initially unexcited. Should one of them become excited by secondary impact to a high hydrogen-like state or one of the higher *d* states, a resonance with a high-lying 2+ state of the adjacent ion would allow transfer of an electron to form a Ce^{4+} - Ce^{2+} pair, which in turn annihilates by reverse exchange as described above. This may well be the simplest mechanism of self-quenching because it does not require any spatial diffusion of the unequal charges or the stabilizing factor of a Ce^{2+} state located well below the conduction band edge. Such a mechanism has been proposed in Ref. 108.

Although there is no shortage of hypotheses, the inconclusive (and occasionally contradictory) state of the current evidence does not allow a definitive choice. It is clear that none of the simple models gives a fully satisfying explanation, and it would not be surprising if we found more than one mechanism to be operative.

The failure of the two stoichiometric materials CeF_3 and CeP_5O_{14} can be considered counterintuitive. However, the increased concentration of Ce in itself probably has little to do with this failure. We have seen that in both lattices the dilution of Ce by La does not lead to any improvement. In $Ce_xLa_{1-x}P_5O_{14}$ the light output *per ion* stays constant over the entire range of x, while in $Ce_xLa_{1-x}F_3$ it gradually decreases. This observation is particularly puzzling in view of the spectacular efficiency of other dilute Ce compounds, described in the next section.

3.3. Cerium-Activated Materials

This section deals with some of the brightest and fastest scintillator materials developed in the past decade. The first indication that scintillators with much higher light output might be found among activated materials lightly doped with Ce was the discovery of the gadolinium and lutetium oxysilicates Gd_2SiO_5:Ce (GSO) [21] and Lu_2SiO_5:Ce (LSO) [24]. This was followed by lutetium orthophosphate, $LuPO_4$:Ce [25], and the yttrium and lutetium orthoaluminates $YAlO_3$:Ce (YAP) [18] and $LuAlO_3$ (LuAP) [26]. In all these cases the Ce concentration is below 1 mole percent, but their light output is impressively high. Unfortunately the literature of some of these modern scintillators is largely limited to accounts of performance and is singularly devoid of the kind of detail that allowed for a deeper understanding of Tl-doped halides and the stoichiometric Ce compounds.

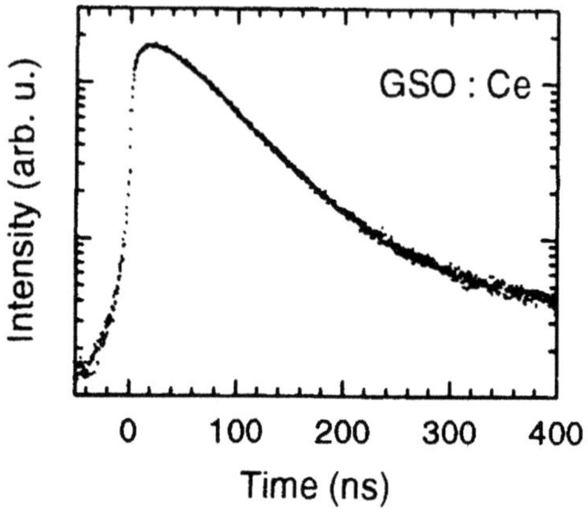

Figure 12. Total scintillation light pulse from Ce-doped gadolinium oxysilicate (GSO). The rounded top of the pulse reveals delayed Gd-to-Ce transfer.

GSO: This oxysilicate, first described in 1983 [21], is a precursor of the current Ce-doped scintillators. It has a respectable density of 6.71 g/cm^3, is chemically very stable and is available commercially [113]. Its structure is monoclinic, belonging to space group C^5_{2h}, with two distinct rare earth sites surrounded by 7 or 9 oxygens, respectively. The Ce emission has a composite band peaking at 425 nm; its lifetime was first reported as 60 ns, not as short as in some other Ce doped materials, emitting at higher energies. Figure 12 gives a typical scintillation pulse, excited by ionizing radiation, without any wavelength discrimination. More detailed properties are reported by Melcher *et al.* [114] and Suzuki *et al.* [115,116]. Suzuki resolved emissions from the two crystallographically distinct Ce sites (Ce*1* and Ce*2*); these emissions behave quite differently at low temperature (11 K), peaking respectively at 380 and 470 nm. The quenching rates of the two centers are vastly different, the emission of Ce*1* being dominant at room temperature. The reasons for the different spectroscopy of Ce*1* and Ce*2* have been discussed by Lammers and Blasse [117]. The presence of Gd in the lattice leads to severe complications since it is well known to have 4*f*-4*f* transitions from its $^8S_{7/2}$ ground state to $^6P_{7/2}$, $^6P_{5/2}$, and $^6P_{3/2}$ excited states, which occur in GSO at 313, 303 and 302 nm and a higher ^6I_J multiplet at 275 nm, partly overlapping a 4*f*-5*d* absorption band of Ce^{3+} at 284 nm. The decay times depend significantly on excitation: While optical excitation at 284 nm (directly into Ce*1*) produces a fast risetime and a single exponential decay of 24 ns; optical excitation into the Gd ^6I_J multiplet at 275 nm shows a risetime of a few tens of nanoseconds and a slow decay. The latter is even more evident in the gamma-excited decay (Figure 12), which clearly shows a delayed buildup and a slow component of several hundred nanoseconds. Whichever way the Gd ions are excited, two effects well known from the theory and practice of energy transfer in lanthanides [44] are immediately triggered. One is the "donor-donor" migration among Gd ions, which occurs because of the resonant nature of 4*f*-4*f* transitions. This has been studied by Suzuki *et al.* by partial substitution of Gd by Lu (which has no excited states in the Ce absorption region). Such partial substitution slows down the Gd-Gd transfer, resulting in a longer buildup and even slower long component. Because the radiative lifetime of Gd is so long, energy migration through the Gd sublattice can ultimately terminate on nonradiative recombination centers. A second effect, donor (Gd) to acceptor (Ce) energy transfer, will also occur because of the partial overlap of the Gd emission and the Ce absorption. This process is of course strongly dependent on Ce concentration, which controls the mean Gd-Ce distance.

It is quite evident that all of these complications can be avoided if Gd is completely replaced by a filled shell ion such as Y, La or Lu. Unfortunately lanthanum oxysilicate decomposes when melted and cannot be grown [118]. Y_2SiO_5:Ce has been investigated as a scintillator to some extent [115], but attention quickly turned to LSO, which we discuss next. Nevertheless GSO remains a viable scintillator, due primarily to the advanced state of crystal growth technology [119].

YSO: Y_2SiO_5:Ce is isostuctural with LSO (which follows). It appears to have been investigated only superficially [115] by the Schlumberger group who specialized in silicates and developed LSO. Having a density of only 4.54 g/cm^3 and a light output only one-third as great as LSO, it evidently attracted much less interest.

LSO: Lu_2SiO_5 is historically the first material in which the substitution of Lu for Y has produced a significant improvement in performance, therefore writing a new chapter in scintillator development. Lutetium offers a number of significant advantages: Its ionic radius (0.85 Å) is nearly the same as that of Y^{3+} (0.89 Å), making a structural change far less likely than if the substituent were La^{3+} (1.02 Å). Lutetium also brings about a substantial increase in density and consequently stopping power. The removal of gadolinium certainly eliminates any interference with the $4f$-$5d$ absorption of Ce. Clearly such a substitution should be pursued whenever possible.

In any case the discovery of the LSO scintillator [24,120] was instantly received as a major success because of its high density (7.4 g/cm^3), rapid decay (40 ns) and high light output (75% of NaI:Tl, which according to Ref. 72 would amount to 28,000 photons/MeV). More recent work has reported a smaller light output of 18,500 photons/MeV [121], still quite respectable. It is evident, however, that the light output of LSO crystals can vary enormously, perhaps because of the presence of an uncontrolled defect also responsible for a substantial thermoluminescence [122]. Indeed, at room temperature LSO that has been exposed to ordinary room light can glow for hours. It is standard practice to anneal samples at 150 C for one hour before taking measurements [123]. It has also been shown that LSO yields significantly less light per MeV when excited by lower energy radiation (X-rays up to 60 KeV) than at higher energies, making for a strongly nonlinear response [124]. A comparison of the luminescence properties of the various oxysilicates is given in Ref. 115.

The nature of the transfer process that makes LSO such a good scintillator is far from clear. The low level of Ce doping and the reported weakness of the host emission in the undoped material [125] indicate that excitonic transfer is unlikely, leaving charge trapping, described in Sec. 2.3, as the dominant mechanism. This is an area that lies at the heart of the modern scintillator problem and obviously requires more work. As we shall see below, only in the orthophosphates [112] and orthoaluminates [50] has an efficient charge trapping mechanism received some independent confirmation..

Yttrium orthophosphate, YPO_4:Ce, like YSO before it, has been pushed aside by its heavier Lu brother, again because of its lower density and light output. Emission and excitation spectra of the two compounds are similar and need not be discussed here. There is, however, one very significant difference: While in $LuPO_4$ the risetime of the scintillation pulse is unmeasurably fast, (which fact, as we shall see, has played a role in interpreting the transfer as being extremely fast and efficient), the corresponding yttrium compound shows a distinct risetime, as illustrated in Figure 13,

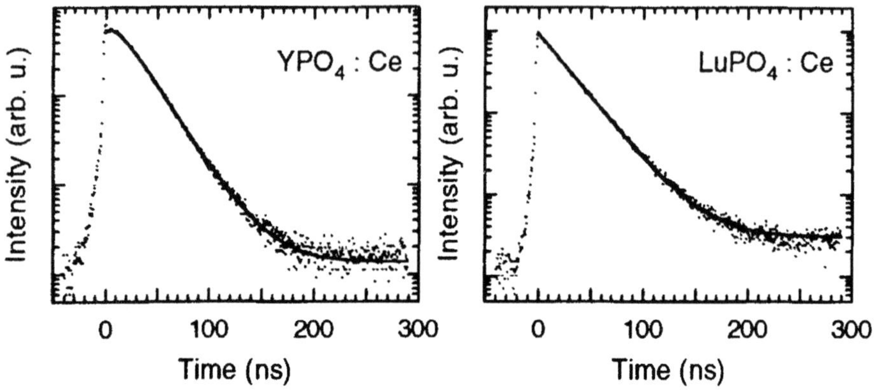

Figure 13. Scintillation light pulse of UV emission from Ce-doped yttrium orthophosphate (left) and lutetium orthophosphate (right). Note the rounded peak of the YPO_4 pulse, compared to that from $LuPO_4$, indicating a slow rise time.

indicating that the transfer process in YPO_4 is considerably slower. Why this should be is not understood, and in an environment driven more by science than by pragmatism would be an interesting object for study.

<u>*Lutetium orthophosphate*</u>, $LuPO_4$:Ce [25], is a member of another class of compound where substitution of Lu for Y has been carried out successfully. This material is particularly important because, like NaI:Tl, it has provided valuable insight into the mechanisms involved in efficient scintillators. Unfortunately, it is unlikely to find practical use because the only fabrication method currently available is growth from a lead pyrophosphate flux, which produces only small plate-like crystals at most 1 mm thick and 2 to 3 mm along the sides. The crystal structure is zircon, space group D_{4h}^{19}, with one eightfold-coordinated rare earth site of symmetry D_{2d}. The density of the material is 6.53 g/cm^3. Ce^{3+} can be incorporated in the lattice up to more than 1 mole percent but the light output saturates well below this concentration. Figure 14 shows a room

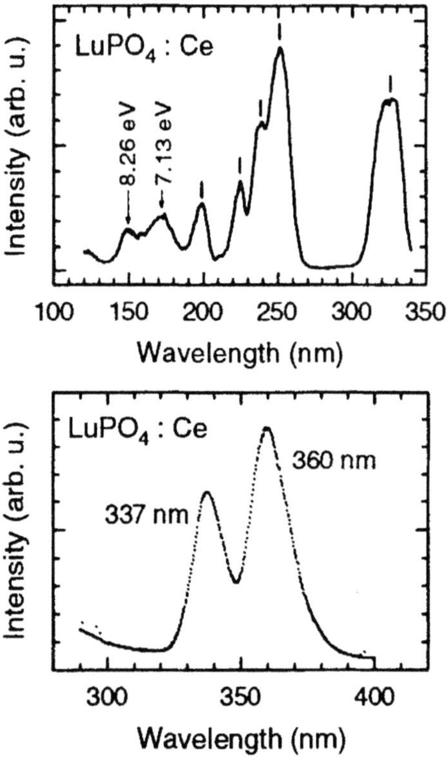

Figure 14. Emission (bottom) and excitation (top) spectra of $LuPO_4$:Ce. Five of the peaks in the latter correspond to $4f$-$5d$ transitions, one to photoionization of Ce^{3+}, and one to exciton formation by elevation of PO_4^{3-} electrons from the valence band.

temperature emission spectrum of Ce, whose shape remains the same regardless of whether the excitation is ionizing or optical. The excitation spectrum, in the same figure, reveals the five symmetry-split components of the *5d* state, plus two others (indicated by arrows) upon which we comment below. The pulse under ionizing excitation, with a decay time of 28 ns, has already been shown in Figure 13.

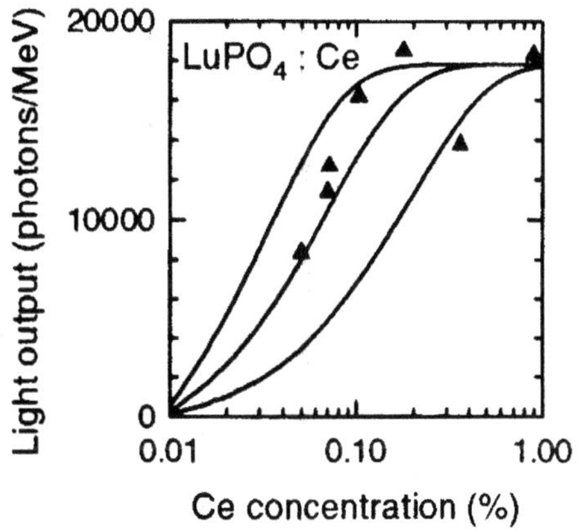

Figure 15. Light output of cerium-doped lutetium orthophosphate. Curves, calculated for excitation volumes of 500, 1500, and 3000 unit cells; triangles, experimental measurements.

A most revealing factor is the dependence of the light output on Ce concentration in $LuPO_4$ illustrated in Figure 15. Notice the extremely rapid increase of the light output, which reaches saturation at a Ce content well below 1%. The curves *a*, *b*, and *c* were calculated using a method described in Ref. 69, assuming excitation volumes of 500, 1500 and 3000 unit cells respectively. There is evidence that the excitation volume is even larger than these values. Note that at concentrations on the order of 0.1% (mean Ce-Ce distance 40 Å), this material shows a light output of 17,000 photons/MeV. Even a nominally pure crystal, with a measured Ce content of only 6 ppm, shows a light output of 4,000 photons/MeV [112]. This concentration translates into a mean Ce-Ce distance of 200 Å or one Ce ion per 40,000 unit cells! It is particularly significant that scintillation light pulses observed in $LuPO_4$, including the crystal just mentioned, show no measurable risetime (*i. e.*, < 1 ns) and therefore the energy transfer over these large distances must be extremely rapid. Presumably something similar occurs in LSO and YAP, both of which show remarkably high light output at low Ce concentration.

Again, the mechanism of transfer is far from clear. If the host emission were involved, it should at least partially overlap the Ce absorption, but no such overlap is found. On the contrary, the host emission from undoped $LuPO_4$ crystals, measured at low temperatures, occurs at 430 nm, completely out of range of the Ce absorption [112]. This is quite different from lanthanide-doped vanadates, tungstates and niobates [46,76] where lattice emissions overlapping the rare earth ion absorptions are clearly evident.

If nonradiative transfer is to be eliminated from consideration we must turn to the possibility of carrier-mediated transfer. Nakazawa and Shiga [126] were the first to point out that the *europium*-doped Y, La, Gd, and Lu phosphates have an excitation peak around 150 nm, which corresponds to a PO_4^{3-} absorption from the valence band. As seen in Figure 13, the excitation spectrum of $LuPO_4$:Ce also shows such a peak at the same energy, which presumably has the same origin. This means that the peak at 174 nm is due to the photoionization of Ce^{3+}, thereby locating the ground state of that ion just above the valence band, where it can readily capture holes generated by ionizing radiation [108]. Additional evidence that we may be dealing with sequential trapping of carriers came from a rather unexpected source. In order to understand better the role of closed shell cations (Y, La, Lu), Wojtowicz *et al.* [112] investigated a series of crystals of the general formula $Lu_xYb_{1-x}PO_4$:Ce, including $LuPO_4$ with varying Ce/Yb doping. Yb has three important characteristics relevant to our subject:

1. It is an inverse analog of Ce in the sense of having one hole in the 4*f* shell, rather than one electron.
2. The first excited state is at 10,000 cm^{-1}, followed by a huge gap extending to about 100,000 cm^{-1}. In that sense it should exhibit an optical "neutrality" as exhibited by the closed shell ions, in contrast to an ion like Gd.
3. It is significantly less expensive than Lu, an important practical consideration.

Given these three characteristics, Yb-based scintillators could be of great interest.

Since completion of a closed shell is energetically favored we can expect that Ce^{3+} should be a good hole trap (going to Ce^{4+}) while Yb^{3+} should be a good electron trap (going to Yb^{2+}). From measurements of emission and decay times of the Lu-Yb crystals at different temperatures, Wojtowicz *et al.* [112] concluded that fast trapping of holes by Ce competes effectively with hole self-trapping. Having bound a hole, Ce^{4+} becomes an attraction center for electrons, whereupon an excited Ce^{3+} is formed. Quite the opposite happens with Yb. Only after it has bound an electron does it start looking for holes, by which time the holes may have managed to self-trap. This entire process is greatly slowed down at low temperatures, as evidenced by the sluggish rise and decay times of Yb-doped $LuPO_4$. It should be noted that this is quite analogous to the thermally-activated search for holes in Tl-doped halides, responsible for the intermediate time components of the decay, discussed in Sec. 2.3

Thus on the basis of radically different temperature dependence of decay times of Ce and Yb in $LuPO_4$ Wojtowicz *et al.* [112] advanced the hypothesis that transfer in the phosphate is mediated by carrier transport and trapping rather than nonradiative transfer. This conclusion runs somewhat counter the conventional picture of transfer in molecular crystals but gains some support from the observation that in some materials hot holes can traverse hundreds of angstroms in picoseconds [127]. How this picture of transfer applies to other efficient Ce-doped scintillators is one aspect of the still unsolved and interesting problems in this class of materials.

The class of materials which we shall describe next are yttrium aluminates, which all have played a role in solid state laser development, primarily when doped with Nd [128]. Because the ionic radii of Ce^{3+} and Nd^{3+} are 1.034 Å and 0.995 Å, respectively, substitution of Ce for Nd was not expected to play a major role. However, the yttrium compounds are not distinguished by high enough density to play a role in many scintillator applications. Thus, as in the oxysilicate and phosphate, the focus has been on the substitution of Lu for Y. Nevertheless yttrium compounds are not to be dismissed, as evidenced by the next two entries.

<u>*YAG*</u>: Yttrium aluminum garnet, $Y_3Al_5O_{12}$, when doped with Nd, has a long history as the most important solid state laser material. Because the properties of the lattice and the technology of YAG growth are very well understood it was logical to substitute Ce for Nd and investigate the scintillation properties. YAG has a cubic

garnet structure (space group O_h^{10}, with Y at D_2 sites) and a rather low density of 4.55 g/cm^3. There is a vast literature on various aspects of YAG:Ce, including its phosphor properties (see in particular the bibliographies of Ref. 129), as well as its characteristic defect structure [130]. The luminescent properties of YAG:Ce have been the subject of some controversy. It has a host emission in the near UV (270 to 330 nm depending on temperature) [131]. On the basis of temperature dependence the dipole-dipole, host-to-Ce transfer has been ruled out by Robbins *et al.* [132] in favor of thermally activated holes from cation vacancies. This model was in turn challenged by Aizenberg and Rotman [133], who showed plausibly that dipole-dipole transfer is possible if one takes into account the increased overlap between host emission and Ce absorption at higher temperatures. The authors also claimed that the defect center is an electron trapped at an oxygen vacancy.

The scintillation properties of YAG have been carefully described by Moszynski *et al.* [134] who find a decay time constant of between 65 to 70 ns, and an appreciable light output of 20,300 photons/MeV. The Ce emission in YAG is at an exceptionally long wavelength of 550 nm, a fact theoretically justified by cluster calculations of Andriessen *et al.* [135]. Because of its rich and well studied defect structure [130], YAG retains some validity for the understanding of Lu substituted aluminates to be discussed shortly. As a practical scintillator, however, we believe it is of no more than marginal interest.

LuAG: Lutetium aluminum garnet, $Lu_3Al_5O_{12}$, is the isostructural analog of YAG. This Ce-doped material has added attraction in that the technology of garnet crystal growth is so well known. The density of this material is 8.34 g/cm^3. The emission spectra differ significantly depending on whether excitation is by means of optical (below bandgap) or ionizing radiation. Figure 16 shows emissions under a series of optical excitations. The band in the green at 500 nm is almost certainly due to Ce, its position being similar to that in YAG. For excitation shorter than 350 nm a double peak occurs in the 380 to 430 nm region with a separation of 2118 cm^{-1}, strongly suggestive of the Ce^{3+} spectrum seen in most other crystals. Conceivably we are dealing here with a Ce on an aluminum rather than a rare earth site. Attempts to measure decay times of this doublet were unsuccessful. Under gamma excitation (Figure 17)

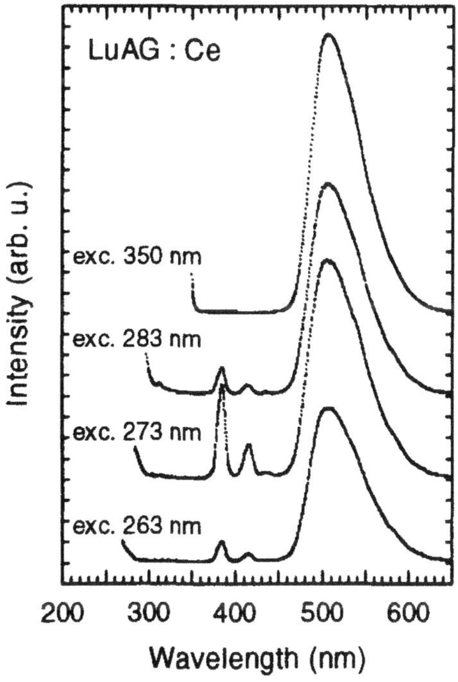

Figure 16. Emission spectrum of cerium-doped lutetium aluminum garnet, for various excitation wavelengths.

the green band is unchanged but a strong new emission appears between 260 and 420 nm. This is undoubtedly the defect emission also found in YAG. The sharp dip at 350 nm coincides with the Ce absorption. As we have seen in the case of YPO_4 the decay time of Ce^{3+} when excited optically is exponential (in this case with a time constant of 50-60 ns, similar to YAG) but under gamma excitation it is strongly nonexponential with evidence of a risetime, as shown in Figure 18. The decay of the blue (defect) emission shows a fast transient followed by slow decay of several hundred nanoseconds.

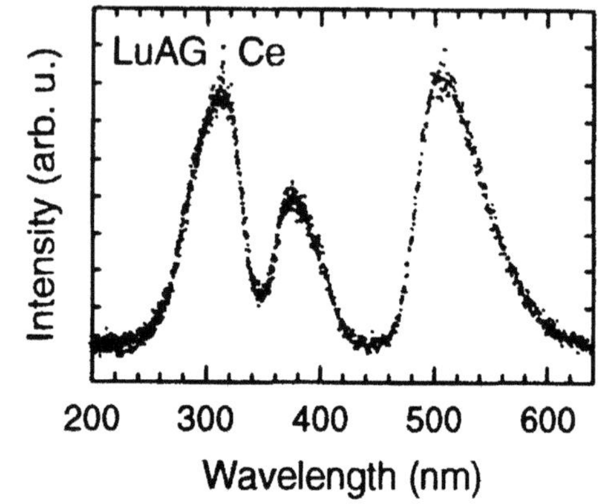

Figure 17. Emission spectrum of cerium-doped lutetium aluminum garnet, excited by ionizing radiation. Broad UV emission is split by Ce absorption.

These observations have several interesting aspects: First, the presence of Ce does not totally quench the defect emission in spite of a complete overlap between this emission and the Ce absorption (at 350 nm). This indicates either a large Ce-to-defect distance or a slow crossrelaxation within the defect emission band, raising unresolved questions regarding the energy transfer processes. All in all, the unfinished spectroscopy of LuAG could possibly shed some important light on transfer, but again pragmatism takes precedence over research.

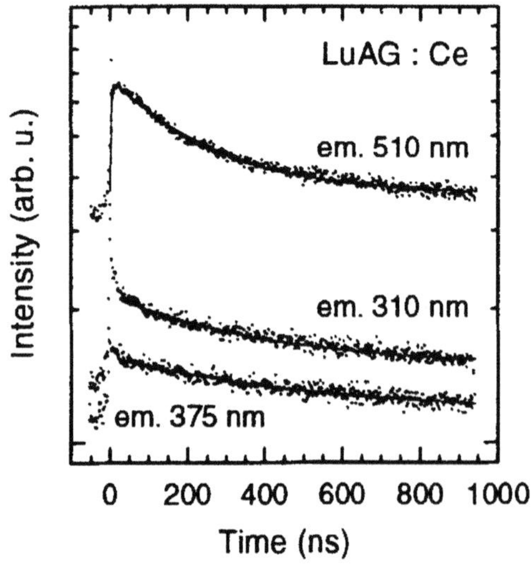

Figure 18. Decay traces of gamma-excited light pulses in Ce-doped lutetium aluminum garnet, at various emission wavelengths.

YAP: Yttrium aluminum perovskite $YAlO_3$ (*aka* YALO) also has a history as a neodymium host for lasers. This orthorhombic structure belongs to space group D_{2h}^{16}, and has a lanthanide site symmetry of C_s,. Doped with Ce, the material was first investigated by Weber [136], who found an emission at around 370 nm and a decay time of 16 ns. The crystals were grown by the Czochralski method in an argon atmosphere. Scintillator interest in YAP:Ce was stimulated in 1991 by Ziegler *et al.* [137], Baryshevsky *et al.* [18], and Korzhik *et al.* [138].

In the latter two publications the decay time is given as 28 ns, suggesting that the emission from the Weber crystals was somewhat quenched. Being chemically inert and having a light output of 40% of NaI:Tl (or 15,200 photons/MeV), YAP is an interesting scintillator material. Our own measurements indicate that the light output may be even higher, some 21,200 photons per MeV [26] for thick (>1 cm) crystals, and a phenomenal 33,000 photons per MeV [139] for thin (<0.5 mm) specimens. The dependence of light output on thickness is indicative of a parasitic absorption, discussed in more detail in connection with LuAP, below. Trower [140] has been pioneering the use of YAP for inspection systems designed to detect explosives and narcotics in cargo [141]. The main drawback of this material is its relatively low density of 5.55 g/cm^3, which excludes it from some applications.

LuAP: Lutetium aluminum perovskite, $LuAlO_3$, which is the isostructural analog of YAP, is the latest entry in the field of Ce-doped scintillators [26]. This material displays a remarkable confluence of desirable properties, including high density, high light output, and rapid decay (see Tables 1 and 4). The first crystals of LuAP doped with Nd and Ti were grown by Kaminskii *et al.* [142]. Ce-doped crystals were grown more recently in the United States [26,143].

Considering that the Ce-doped crystals have been under development only from autumn of 1994, LuAP:Ce carries great promise, particularly when several problems are understood and solved. Not the least of these is the growth of the crystal itself. Recent experience has shown that, defying melt stiochiometry, the garnet phase often nucleates in preference to the perovskite, so that many of the early aluminate specimens were far from single-phase material [144]. Indeed, Bondar *et al.* [145] suggest that the perovskite phase may not even be thermodynamically stable. The demonstrated ability to grow large single crystals casts serious doubt on this assertion, and it now appears that the so-called phase instability may be a surface, rather than a bulk, phenomenon. Clearly the technology of perovskite crystal growth needs considerable refinement.

Some of the features of the early specimens of LuAP serve as vivid examples of the problems often encountered in the development stage of a new material. Figures 19 and 20 show the emission and excitation spectra taken at room temperature. Note that the doublet structure of the Ce emission is completely obliterated, probably indicating a higher degree of electron-phonon coupling than in, say, CeF_3 or $LuPO_4$. The positions of the emission and excitation peaks are virtually identical to those of its yttrium analog [136]. The decay time shows no significant change between room temperature (17.2 ns) and 38 K (17.8 ns), indicating a high quantum efficiency of luminescence. Figure 20 shows a superposition of the excitation and absorption spectra, plotted on the same scale. Here we see an important difference. If the only absorbing species were Ce^{3+} ions (as would normally be expected), the two spectra should be essentially identical. Note, however, that while the exci-

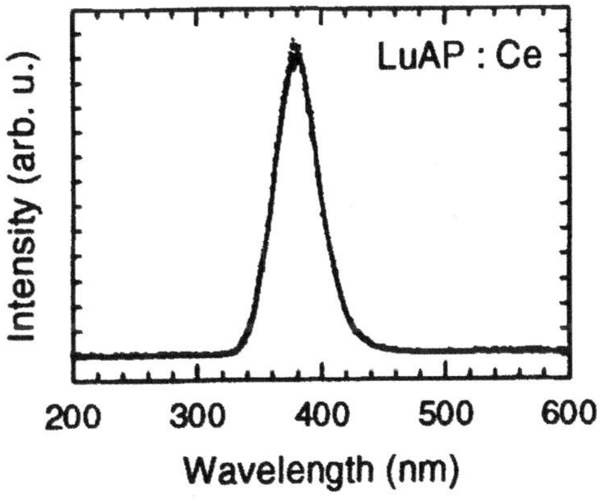

Figure 19. Emission spectrum of cerium-doped lutetium orthoaluminate, excited optically at 300 nm.

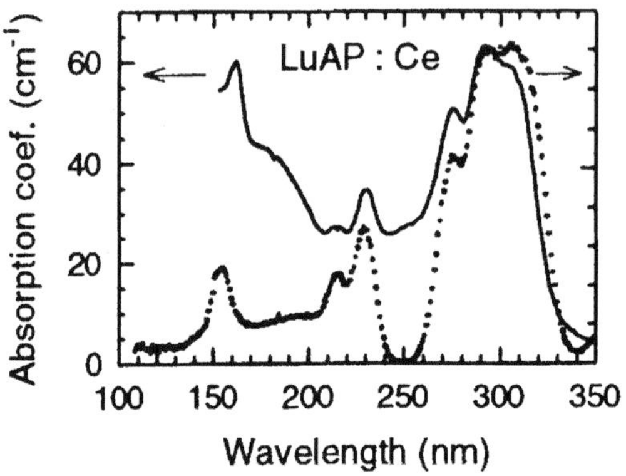

Figure 20. Excitation spectrum (dotted curve) and absorption spectrum (solid curve) of Ce-doped lutetium orthoaluminate (LuAP). Note the rising baseline of the latter toward the deeper UV, indicating an underlying parasitic absorption.

tation spectrum goes to zero at 250 nm (corresponding to the gap between the crystal field split 5*d* states), the absorption spectrum shows a pronounced rise in the underlying background, strongly suggesting that these crystals contain some species that absorb but do not contribute to emission. This phenomenon is not unique, and is clearly evident in the YAP measurements reported by Weber [136]. This crucial difference between absorption and excitation spectra has been the subject of a pair of perceptive papers by Gumanskaya *et al.* [146] and Baryshevsky *et al.* [147]. These show that the similarity between the excitation and absorption spectra can be greatly influenced by the reducing or oxidizing nature of the conditions under which the crystal is grown. Particular attention is drawn to the multiple valence states of Ce, with oxidizing conditions leading to the presence of Ce^{4+} and a charge transfer absorption in the UV (possibly caused not by Ce^{4+} itself but by a charge balancing defect center that accompanies it [147]). Reducing conditions, by contrast, could convert some of the Ce^{3+} to Ce^{2+}, lowering the output by simply decreasing the population of the emitting ion.

This parasitic UV absorption is found in both orthophosphate analogs, YAP and LuAP alike. Besides the disparity between excitation and emission spectra, this phenomenon also introduces geometric effects. For example the light output has been found to decrease with sample thickness, accompanied by a distortion of the characteristic Ce emission spectrum on the short wavelength side, depending on the length of the crystal path traversed by the measured light [139]. This is illustrated in Figure 21, which shows the emission of LuAP measured perpendicular or parallel to a plate-like crystal wafer. Such differences would normally be ascribed to self absorption due to overlap of the Ce resonance absorption with the Stokes-shifted emission from the same state. The decay traces (Figure 22), however, show no evidence of the requisite lengthening in the long direction, which, as discussed in Sec. 2.3, effectively eliminates self absorption and radiation trapping as the causal mechanism.

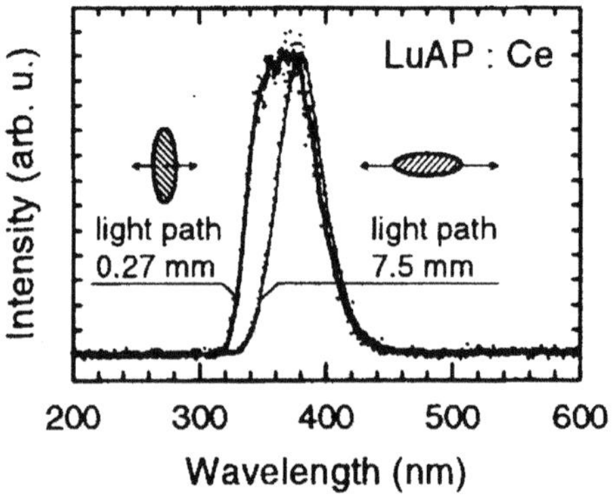

Figure 21. Spectra of gamma-excited emission from cerium-doped lutetium orthoaluminate (LuAP), measured along two different light paths.

Aside from thickness considerations, the light yield also shows a significant dependence on Ce content. Both relationships are illustrated in Figure 23, which shows the output as a function of thickness for three different Ce dopant levels (two in LuAP, one in YAP), as defined by chemical analysis. Note that the three Ce concentrations span a full order of magnitude, corresponding to about a four-fold increase in light output. Although the three roughly parallel curves are not separated by quite the same factors as the concentrations, it is clear that the Ce content is far from optimal in either material. Extrapolating the values obtained from the thin (almost absorption-free) specimens, we find that the potential output of either material could exceed 40,000 photons per MeV.

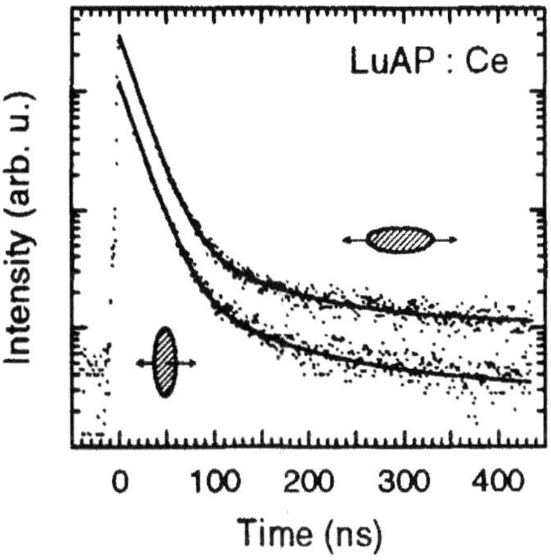

Figure 22. Decay traces of gamma-excited emission from cerium-doped lutetium orthoaluminate (LuAP), measured along two different light paths. See Figure 21.

There is clearly a strong similarity between YAP and LuAP, giving reason to believe that proper growth conditions, similar to those described for YAP in Refs. 146 and 147, have not yet been achieved. The story of LuAP as presented here is a good illustration of the difficulties and problems encountered in the early development stage of a new scintillator. It has been shown repeatedly that such development takes several years before all the various crystal growth and material parameters can be brought under control and the properties optimized for the intended use. None of the current mainstays of the field have been exempt from such "growing pains", and some, such as LSO, are enduring them yet.

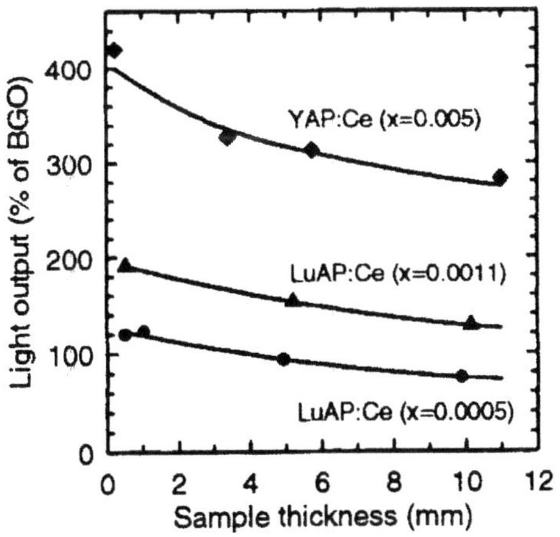

Figure 23. Measured light output of cerium-doped YAP and LuAP under γ-excitation, as function of sample thickness. Note also its dependence on the atom fraction (x) of Ce, increasing by about a factor of four for a ten-fold higher concentration.

If one surveys the current scintillator scene, the two materials LSO and LuAP come closest to an ideal scintillator for most applications. The ultimate choice between them cannot be made until the issues of thermoluminescence and light sensitivity (in LSO) and Ce concentration and parasitic absorption (in LuAP) have been resolved. However it is only fair to state that these materials share two problems, one hopefully transitory and one intrinsic. The first is economic: At the present time the starting material, Lu_2O_3, costs on the order of \$4/g for a purity of 10 ppm. This translates to about \$40/cm^3 for the material alone. Compared with \$10/cm^3 for grown and cut BGO crystals, this is evidently too high for many applications. However, lutetium is actually more abundant than mercury, and its cost is largely artificial, dictated almost entirely by the process of its separation from other rare earths and its (up to now) minimal usefulness. Paradoxically, it is the very lack of demand that has served to keep the price high, since the market did not justify investment to achieve economy of scale. As illustrated by past cases of other rare earths (such as europium), this cost should drop significantly when the market expands to the point where the law of supply and demand can fully function [148]. Moreover there is reason to believe that the purity of Lu_2O_3, especially with respect to Yb, the element closest in atomic weight, can be substantially reduced without harmful effect.

The second problem is caused by the ^{176}Lu isotope. Although most of the element is the stable ^{175}Lu, the heavier isotope, present at 2.59%, is naturally radioactive, decaying in a variety of β and γ modes with a half-life of 2.6×10^{10} years. This would contribute an extra background estimated at about 300 counts/sec per cm^3 of crystal. This is little more than a "nuisance factor" for most applications, and becomes irrelevant where, as in PET, coincidence counting can be used. However, the ^{176}Lu isotope also has a high absorption cross-section for thermal neutrons (about 2050 b or 1.26×10^{23} b/g). This could cause complications in neutron-activated systems, such as for detection of explosives and narcotics [149], but can probably be addressed by shielding and energy discrimination techniques.

In summary, it is clear that cerium-doped lutetium-based materials represent a major advance in the scintillator field. This is not to say that superior classes of materials will never be found, only to recognize that the requirement of high host density coupled with nuclear stability leaves rather limited choices and prospects.

4. Outstanding Problems

In this section we shall summarize a number of nagging questions that have arisen during the last few years and which may determine the future course of research and progress. Having exhausted the store of information presently available, we shall take the opportunity to speculate on some of the issues that remain to be resolved. We shall limit these remarks primarily to Ce-doped materials.

4.1. Role of Lutetium

As we have seen, lutetium-based compounds have produced a number of outstanding scintillators, of which LSO and LuAP are likely to reach a significant level of practical use. As is obvious from the foregoing discussion, the Lu^{3+} ion is particularly favored as a major constituent of the host lattice because of its high atomic number, imparting a high density and concomitant stopping power to the lattice. All other things (*e. g.*, optical quality, crystal structure) being equal, it is clear that superior stopping power gives a lutetium-based lattice a critical advantage over, say, its yttrium-based counterpart (see Figure 24). But are all other things equal, or does lutetium have a more subtle property that provides even greater benefit? This question is raised by the fact that Lu, by virtue of its full 4*f* shell, has a much higher-lying core level (6.9 eV in the free atom) than do the other spectroscopically inert rare

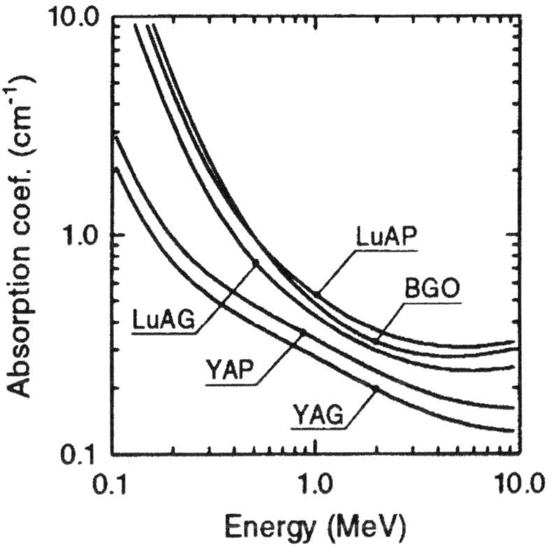

Figure 24. Absorption coefficient for gamma irradiation (*i. e.*, stopping power) for various scintillator materials.

earths (Y, 25.6 eV; La, 14.4 eV) [150]. While Robbins' theory of conversion does not consider the core levels, such an easily ionizable electronic state could conceivably have a significant influence on the conversion and transfer processes. Any attempt to answer this question is hindered by the fact that, with the one exception of YAP, yttrium-based crystals (*e. g.*, YSO and YPO_4) have received far less developmental effort than their Lu counterparts. This not surprising, since much of the research has been driven by pragmatic considerations, in which the higher density of Lu crystals plays an important role.

As outlined in Sec. 2, the theory of scintillation contains nothing specific to the electronic structure of the cations, except for their indirect influence on dielectric constants and vibrational frequencies. However, while the longitudinal optical phonons, as a source of loss, play an important role in Robbins' theory, their energies are open to question. For crystals containing more than two atoms per unit cell, the distinction between LO and TO phonons becomes blurred, and the Lyddane-Sachs-Teller relation for calculating the LO frequencies is not valid for these low symmetry systems. Obviously it is the frequencies that contribute most to ε_∞, rather than any internal vibrations of molecular groups, that should be used in Eq. 19, but it is not always clear which of the experimentally reported vibrational frequencies are the appropriate ones. Despite these uncertainties, however, we provide in Table 4 a comparison of three pairs of materials in which the Y-to-Lu substitution has been made.

Here we present a summary of the data that can be used to calculate the conversion and, under the assumption that $S = Q = 1$, the light output in photons/MeV as well. The primary data that we need are the dielectric constants ε_0 and ε_∞, $\hbar\omega_{LO}$ and E_g. Even for this limited number of compounds these quantities are not always known. Where they are known we have made proper references in Table 4. Some, however had to be measured or calculated as part of this exercise, and required judicious selection of techniques and occasional arbitrary judgments. For example, we took as the effective ω_{LO} the measured IR cut-off frequencies (see Figure 25); this is not rigorously correct because the desired LO vibrations do not couple directly to electromagnetic radiation. The band gap E_g was identified with the shortest wavelength peak in the excitation spectrum. We assume here that this energy corresponds to optical ionization, with the spectral resemblance of a peak because for still shorter wavelength the absorption rises so rapidly that no more light-producing excitation is possible. The optical dielectric constant ε_∞ was obtained from the

Table 4. Effect of substituting Lu for Y in isostructural scintillator materials

	Silicates		Phosphates		Aluminates		
	Y_2SiO_5 (YSO)	Lu_2SiO_5 (LSO)	YPO_4	$LuPO_4$	$YAlO_3$ (YAP)	$LuAlO_3$ (LuAP)	$Bi_4Ge_3O_{12}$ (BGO)
Space group	C^6_{2h}	C^6_{2h}	D^{19}_{4h}	D^{19}_{4h}	D^{16}_{2h}	D^{16}_{2h}	
Site Symmetry	C_1	C_1	D_{2d}	D_{2d}	C_s	C_s	
Coordination number	6&7	6&7	8	8	12	12	
Band Gap Energy (eV)	6.4 [a]	6.4 [a]	8.2 [g]	8.7 [g]	7.65 [g]	8.0 [g]	5
Density (g/cm^3)	4.54	7.4	4.31	6.53	5.55	8.34	6.71
Effective Z number	39	66.4	32.5	63.7	33.5	64.94	62.91
Calc. Absorption Length (cm) for 511 KeV Photons	2.63	1.22	2.77	1.5	2.13	1.05	1.04
Calc. Photofraction for 511 KeV Photons	0.038	0.34	0.038	0.28	0.043	0.3	0.41
Emission Wavelength (nm)	420 [a]	420 [a]	360 [b]	365 [b]	365 [c]	375 [c]	510
Decay Constant (ns)	37/82 [a]	46 [a]	≈24 [b]	24 [b]	24 [c]	17 [c]	400
Static Dielectric Constant	9.94 [d]	9.59 [d]	11.24 [d]	10.58 [d]	16.83 [d]	16.83 [d]	16.3
Mean Refractive Index	1.7 [a]	1.82 [a]	2.04 [e]	2.04 [e]	1.97 [g]	1.97 [g]	2.15
Highest LO Phonon (est.)	1050?	1050 [g]	1150 [g]	1158 [g]	800 [g]	800 [g]	362
Calc. Number of Electron-Hole Pairs per MeV	20,300	23,100	22,600	23,000	26,300	25,700	60,200
Reported Light Output (photons/MeV)	9,500 [a]	18,500 [f] 27,000 [a]	8,000 [b]	17,000 [b]	20,100 [c] 32,800 [g]	21,200 [c]	8,200
Figure of Merit (photons per MeV-ns) Observed	257	402 [f] 587 [a]	333	708	838 [c] 1367 [g]	1247	27
Figure of Merit (photons per MeV-ns) Max. Possible	580	502	941	1020	1079	1511	200

a. *Ref. 115*
b. *Ref. 112*
c. *Ref. 26*
d. *Ref. 152*
e. *Ref. 151*
f. *Ref. 121*
g. *Ref. 139*

literature (or directly measured in one or two cases), and averaged for anisotropic crystals. The static dielectric constant ε_0 represents a problem because its values are not always readily available for these compounds. Whenever necessary we have therefore used the computational procedure of Shannon [152]. In view of the other uncertainties, we made no attempt to take into account the exact oxygen molecular volume in these compounds, but used the polarizability of oxygen as 2.01 cm^3. Information gaps were filled as consistently as possible, with particular attention to the pairwise Y-Lu comparison.

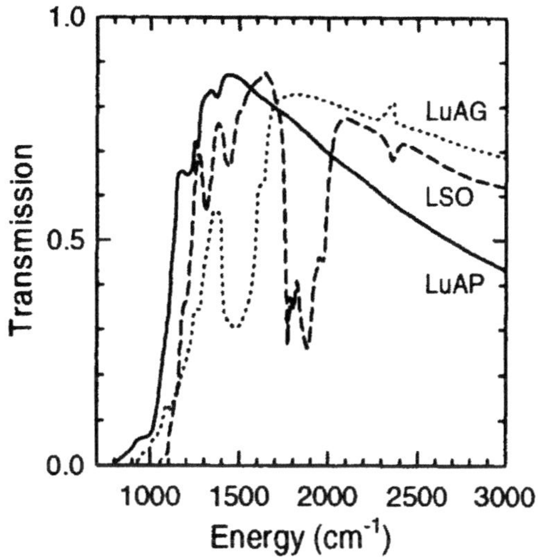

Figure 25. Infrared absorption spectra of various lutetium based crystals.

The main results of Table 4 are of course the calculated numbers of electron-hole pairs/MeV and the measured light output. Under the assumption that $S = Q = 1$, these two numbers should be equal, to the extent that Robbins' theory of conversion is correct. It is interesting that the predicted numbers of electron-hole pairs for this entire group of compounds lie in the range of 20,000 to 26,000 per MeV. None of them surpass, or even come close to matching, the alkali halide scintillators. Nevertheless these are respectable numbers which, translated into light output in photons/MeV, can satisfy many current requirements, with the added bonus of much higher figures of merit.

While the exercise summarized in Table 4 has not given a definitive answer, it teaches an important lesson. Although the calculated numbers of electron-hole pairs are essentially identical for each pair of yttrium/lutetium analogs, we have already seen at least one instance (thin YAP) where the measured light output substantially exceeds the predicted value. Thus the Robbins model, despite its usefulness, falls well short of a comprehensive theory. The uncertainty in assigning unambiguous values to $\hbar\omega_{LO}$ and the neglect of other than valence-conduction band ionizations are serious limitations, so that the values derived for the conversion can at best be considered little more than qualitative. Consequently the next step in the calculation of the light output, which requires the assumption that $S = 1$, may represent an overoptimistic view of material quality. We must therefore conclude that much remains to be done before a reliable quantitative theory of scintillator performance can be achieved. Furthermore it is essential to recognize that comparison between materials is and will be valid only if they represent the same level of material perfection (freedom from impurities, traps, nonradiative recombination centers, etc.). Indeed, while lutetium substitution clearly improves the absorption length and photofraction, the somewhat higher reported values of light output may be attributable merely to more extensive development.

4.2. Binary oxides

Oxide-based lattices are the major focus of attention for new and improved scintillator materials. There are two very practical reasons for such attention. One is that most oxides are particularly inert to ambient conditions of handling and use. Typically they are quite hard, chemically inert, insoluble in water, and particularly amenable to optical processing. In contrast, most halides tend to be relatively soft, water soluble, and vulnerable to attack by atmospheric moisture. While fluorides are less susceptible to these deficiencies, they are subject to a second deficiency: they are far more difficult than oxides to grow in the pure form, and are particularly sensitive to the presence of even minute amounts of oxidic contamination, which severely degrades their emission properties. Thus it is hardly surprising that oxide-based lattices have received the lion's share of attention from investigators in the field.

One of the perplexing issues is that favorable scintillation properties are not generally transferable among isostructural families of materials. Thus bismuth germanate (BGO) is accepted as the benchmark material, against whose performance all the others are currently measured; its analog, bismuth silicate, comes up short. Lutetium oxysilicate (LSO) is one of the promising recent discoveries, but, perversely, not the germanate. Lutetium orthophosphate shows remarkably favorable characteristics; the isostructural vanadate, far less so. Nor are patterns any more evident if we limit ourselves only to compounds with the same elements: Thus cerium-doped lutetium orthophosphate is a superb scintillator, while the related pentaphosphate (with a higher proportion of P_2O_5) is disappointingly poor. But the orthoaluminate and the garnet, differing from each other in Al_2O_3 content and lattice structure, are both excellent scintillators, regardless of whether the rare earth is yttrium or lutetium.

Comparison of oxide with halide scintillators does, however, reveal one persistent pattern. Excluding any activators, virtually all halide scintillators are simple *binary* compounds (*e. g.*, NaI, CsI, BaF_2, and CeF_3), while oxide-based scintillators, including such diverse materials as $Bi_4Ge_3O_{12}$, Lu_2SiO_5, $Y_3Al_5O_{12}$, and $LuAlO_3$, are invariably *ternary* in nature. Not a single binary oxide is in the group: BaO and zirconia, both with superb chemical and optical properties (and the latter readily grown commercially), are conspicuous by their absence. So too is yttrium oxide, a well-known CRT phosphor with high transfer efficiency. A better understanding of why some classes of materials do *not* make good scintillators would be of immeasurable help in understanding why others do.

4.3. Role of Lattice Defects

Also high on our list of unanswered questions is an identification of the lattice defects that interfere with the various steps in the scintillation process and an elucidation of their roles. Their potential for mischief spans the entire process, providing recombination centers to reduce the population of carriers, trap sites to inhibit their mobility, or absorption centers to reduce the emitted light. As we have already seen, the light output of a number of excellent scintillators (*e. g.*, the orthoaluminates) is clearly being restricted by the presence of parasitic absorptions that have been identified with charged point defects. It is important to know not merely how to control them (a chemical issue), but the manner of their interaction. Anything that casts light on these loss processes will be most welcome, and should necessarily lead to improved scintillator performance.

4.4. Theory

One of the particularly vexing difficulties with scintillator materials is their tenacious resistance to generalization (including, perhaps, to this one as well).

These materials have a uniqueness that almost jubilantly defies characterization. The necessary confluence of far too many properties into far too narrow a range has made the search for new and better scintillator materials a particularly frustrating and humbling experience.

The most interesting and important aspect is the transfer process, which shows the rather capricious character of being either almost totally absent or extremely fast and efficient over large distances. In Sec. 2.3 we have listed three categories of transfer, of which the carrier-mediated type turned out to be of greatest importance. A successful theoretical treatment, however, cannot be achieved without a full and complete picture of the material properties. When this is available, as in NaI:Tl, solid state physics can usually provide a credible and coherent description of the transfer processes. Predictive capabilities, however, are not likely to emerge; this deficiency is not limited to the scintillator field, but rather is endemic to all electronic materials (as witness high-temperature superconductors). There is simply no theoretical substitute for a continuous materials research effort.

A second area of great interest is the generalization of the Robbins theory of conversion. While we have seen that with minor modifications [31] it is applicable to both wide band gap and semiconductor type materials, the model still remains essentially empirical. A better approach is to be found in re-examining the rate of direct excitation of the luminescent centers by collision with the energetic primaries or secondaries. Any success in this area would be of immeasurable help in systematizing the search for new and better scintillators.

A third area for theoretical attention is the phenomenon of self-absorption. This effect is a recognized enemy of good scintillators for two reasons: It can result in both lower light output and longer decay time. This is particularly undesirable for high energy detectors which utilize long crystals, producing a spread of decay times as a function of the path traversed by the emitted photon [70]. Although the phenomenon is quite common in luminescent materials, the only fundamental mathematical treatment is limited to fully resonant radiation [40]. There appears to be no comprehensive theory of effects due to self-absorption for materials which show only *partial* overlap of absorption and emission (limited Stokes shift). Development of appropriate mathematical treatment would improve our ability to quantify scintillator performance, as well as other luminescent processes.

4.5. Radiation Damage

A fifth area, and one that we have only touched upon in this review, is radiation damage. While of fundamental importance for the entire field of scintillators, it is particularly relevant to high energy physics applications where radiation fluxes are very high. Radiation damage is in principle impossible to avoid when the material is traversed by particles whose energy far exceeds the ionization and bond energies. The distinction between radiation-soft and radiation-hard materials is therefore largely a matter of degree of damage and rate of repair. Moreover, the damage may be either "extrinsic", involving the coloration of point defects already present in the crystal (*e. g.*, by trapping of carriers at vacancies), or "intrinsic", due to the decay of self-trapped excitons, which can develop even in otherwise perfect crystals.

Obviously crystal perfection and purity are important, so as to minimize the chance of extrinsic damage. It has also been shown that shunt processes which may lower the light output through nonradiative recombination may contribute to radiation hardness [57]. The subject is just too vast to be disposed of by a few words, and the interested reader is directed to Refs. 92 and 153 for more detailed treatment.

4.6. Glasses

Although we have omitted glasses from our discussion of scintillator materials, we would be remiss if we did not at least recognize both the problems and opportunities that they present. In principle, glasses have a major technical advantage in that, once the composition and process has been defined, they can be cast in virtually any desired size and shape relatively quickly and inexpensively. Unfortunately, as we mentioned in the introduction, no glasses have exhibited performance levels even close to those achieved by the better scintillator crystals. But then again, most crystals do not do so either. Indeed glasses have a long and distinguished luminescence history, and are capable of remarkably high quantum efficiency and light output, as anyone familiar with the work on inertial confinement fusion [154] can readily attest.

As already discussed, however, in scintillation the emission step is only the last of three, and it is in the first two where glasses, as a group, have been deficient. Most glasses have relatively low densities and stopping powers, including the silicates that show the highest light outputs ($\approx$25% of BGO). High density glasses do exist [5], but these encounter their own individual problems: Some, containing high concentrations of lead and/or bismuth, have been known for many years, but these tend to have absorptions extending well into the visible. Others, such as the fluorohafniates developed some years ago for laser and other optical device applications, are tricky to fabricate and particularly susceptible to damage by ionizing radiation. Since the only scintillator applications without a premium on light output are those involving high radiation fluxes, such a deficiency is serious indeed.

But it is in the transfer step where most of the difficulties arise. By its very nature, glass is a highly disordered medium, with an abundance of point defects and trapping centers. As we have seen, such entities can seriously reduce the transfer efficiency and ultimate light output. And even if they did not introduce any nonradiative losses, the presence of deep traps will significantly slow the temporal response. Indeed glasses present so many difficulties that some respected experts in the luminescence community have voiced doubts as to whether amorphous materials are even intrinsically capable of good scintillator performance (see Ref. 2, p. 192).

Nevertheless, it is probably too early to write them off completely. We are only too aware of instances where similar experts proclaimed the impossibility of glass lasers or amorphous semiconductors, until facts demonstrated the contrary. While they may seem particularly intractable in glasses, none of the problems are unique to that medium. If the transfer or damage mechanisms can be characterized better even in crystals, the answers may be applicable to glasses as well, making it possible for them to find their own niche in the scintillator field.

4.7. Standardization

Finally we are compelled to mention the anarchy which characterizes the evaluation of scintillators, primarily the light output. We have seen that even in the case of such an established material as BGO the reported light output can vary by a factor of two. Absolute measurements are notoriously difficult because they involve an estimate of the light collection efficiency (fraction of the light generated in the crystal and intercepted by the detector, PMT or diode) which in turn depends on the refractive index of the crystal (which may be anisotropic to boot) and the geometry of the crystal/detector setup. Finally if we want to convert the number of electrons producing the signal in a PMT, or the number of electron-hole pairs created in a diode to the number of photons, we have to know the spectral sensitivity of the detector. There are several approaches to remedy these sources of error and uncertainty but much more work has to be done. To correct this problem is essential if we are to place any value on the interaction of theory with experiment. As it stands today the value of a particular scintillator is determined empirically by the user who chooses the material which, in his particular setup, produces the largest and fastest signal at the least cost.

These unsolved fundamental problems conspire to make prediction very difficult and risky. This is not an uncommon situation in solid state physics, being more the rule than the exception. There is plenty of room for progress, whether by accident and guesswork, or through an increased understanding of the underlying mechanisms. Hopefully the latter will play a greater role than it has in the past.

5. Conclusion

In this chapter we have endeavored to present an overview of the theory and practice of inorganic crystalline scintillators. We have shown that for a variety of reasons these materials, after a long period of neglect, have recently assumed great importance. Advances in particle physics, medical diagnostics and environmental concerns, all share a dependence on detection of high energy radiation. By examining the physical processes which determine scintillation, we demonstrated that they fall into several branches of solid state physics which are only partially capable of dealing with the process of converting the energy of a high energy particle (or photon) into visible light.

Where current knowledge falls short is in the details of the energy loss of a particle traversing a solid. In general this loss can be viewed as pair formation (favorable to scintillation) and heat generation (deleterious to scintillation). The problem is dealt with semi-empirically but almost entirely neglects the partitioning of these losses as the energy of the secondaries decreases during the cascade.

Using a phenomenological approach we have put considerable emphasis on the problem of transfer of the energy contained in the thermalized electron-hole population to the emitting centers. Once these centers are excited the problem becomes a matter of luminescence efficiency, to which all the previously developed analytical tools become applicable. We have shown, however, that it may be misleading to equate the luminescence efficiency following optical excitation with that excited by ionizing radiation, since (as in CeF_3) the efficiency of the emitting center may well depend upon which of its higher-lying states receive the excitation. Pending development of a more rigorous treatment, the phenomenological approach is only a stop-gap measure, necessitated by our limited knowledge. Exact conditions that would assure efficient transfer are difficult to pin down, other than the obvious point that the presence of lattice defects or impurities may provide competing nonradiative recombination paths or transient absorption, and should be avoided. It is quite encouraging to note that very efficient transfer can occur through high mobility of charge carriers, a property not normally associated with insulators.

What therefore are the signposts to guide future research on scintillators? The two major constraints are the use of optically allowed transitions and the presence of high atomic number constituents to assure high stopping power. The first directs us to the few cations (such as Ce^{3+}) with interconfigurational transitions, while the second runs afoul of the natural radioactivity of heavy nuclei. One thing is almost certain; successful scintillators must belong to the class of "wide band-gap" materials to which this book is devoted. We strongly believe that this is where the major advances in the field will ultimately be achieved.

6. Acknowledgments

The authors wish to express their great appreciation to Dariusz Wisniewski and Eugeniusz Zych for their invaluable and tireless assistance in assembling much of the data presented in the foregoing chapter. We thank John C. Sutherland and his staff at the National Synchrotron Light Source, Brookhaven National Laboratory, for their help and cooperation in performing some of the optical measurements reported here, Lynn A. Boatner and his staff at Oak Ridge National Laboratory, for their efforts in crystal growth and materials analysis, and Ralph H. Bartram of the University of Connecticut for his invaluable assistance and discussions on various aspects of solid state theory. We gratefully acknowledge the major financial support of the US Department of Energy, under Grant No. DE-FG-02-90ER61033 and Contract No. DE-FG02-94ER81702; the Public Health Service, US Department of Health and Human Services, under Grant No. R01 CA62330-02; the US Department of Transportation, under Contract No. DTRS-57-94-C-00175; and Martin Marietta Energy Systems, Inc., under Subcontract No. 80Y-HHM41V. We express our additional gratitude to the European Economic Community for the financial support provided to one of us (A. J. W.), under Grant No. ERBCIPDCT-940037, and to the Center for Photonics Research at Boston University, for funding some of the early LuAP crystal growth experiments. Finally, we also thank Boston University, N. Copernicus University, and ALEM Associates for the additional support, both tangible and intangible, that they provided in various aspects of this work.

References

[1] Kröger F. A., 1948: *Some Aspects of the Luminescence of Solids,* Elsevier Publishing Company Inc., New York NY, USA; Pringsheim P., 1949: *Fluorescence and Phosphorescence,* Interscience Publishers Inc., New York NY, USA.

[2] Blasse G. and Grabmaier B. C., 1994: *Luminescent Materials,* Springer-Verlag, Berlin, Germany.

[3] Birks J. B., 1964: *The Theory and Practice of Scintillation Counting*, Pergamon Press, Oxford, England, UK.

[4] Knoll G. F., 1989: *Radiation Detection and Measurement*, John Wiley & Sons, New York NY, USA.

[5] Poulain M., 1995: "Multicomponent Fluoride Systems for Scintillator Glasses", presented at the *Conference of the "Crystal Clear" Collaboration*, (Geneva, Switzerland, May 1995); Dumbaugh W., 1986: "Heavy-Metal Oxide Glasses Containing Bi_2O_3", *Phys. Chem. Glasses* **27**, pp. 119 - 123.

[6] *New Grolier Multimedia Encyclopedia*, Release 6, Version 6.0, 1993. Grolier Inc., New York NY, USA.

[7] Bushong S. C., 1975: *Radiologic Science for Technologists,* C. V. Mosby, St. Louis MO, USA.

[8] Rutherford E., Chadwick J., and Ellis C. D., 1930: *Radiations from Radioactive Substances,* Cambridge University Press, Cambridge, England, UK.

[9] Hofstadter R., 1948: "Alkali Halide Scintillation Counters", *Phys. Rev.* **74**, pp. 100 - 104.

[10] Menefee J., Swinehart C. F., and O'Dell E. W., 1967: "Calcium Fluoride as an X-Ray and Charged Particle Detector", *IEEE Trans. Nucl. Sci.* **NS-14**, pp. 720 - 724.

[11] Farukhi M. R. and Swinehart C. F., 1971: "Barium Fluoride as a Gamma Ray and Charged Particle Detector", *IEEE Trans. Nucl. Sci.* **NS-18**, pp. 200 - 204.

[12] Weber M. J. and Monchamp R. R., 1973: "Luminescence of $Bi_4Ge_3O_{12}$: Spectral and Decay Properties", *J. Appl. Phys.* **44**, pp. 5495 - 5499.

[13] Cho Z. H. and Farukhi M. R., 1977: "Bismuth Germanate as a Potential Scintillator Detector in Positron Cameras", *J. Nucl. Med.* **18**, pp. 840 - 844.

[14] Adeva B. *et al.* (L3 Collaboration), 1990: "The Construction of the L3 Experiment", *Nucl. Instr. & Meth.* **A289**, pp. 35 - 102.

[15] Newman H., 1993: "Precision Crystal Electromagnetic Calorimeters at High Energy Accelerators and Colliders", *Heavy Scintillators for Scientific and Industrial Applications, Proc. Crystal 2000 Int'l. Workshop* (Chamonix, France, Sept. 1992). Editors: DeNotaristefani F., Lecoq P., and Schneegans M., Editions Frontieres, Gif-sur-Yvette, France, pp. 5 - 32; Lecoq P., 1994: "The High Energy Physics Demand for a New Generation of Scintillators", *J. Lumin.* **60/61**, pp. 948 - 955.

[16] Melcher C. L., 1989: "Scintillators for Well Logging Applications", *Nucl. Instr. & Meth.* **B40/41**, pp. 1214 - 1218.

[17] Goldberg M. B., 1993: "Applications of Scintillators in Security Inspection Systems", *Heavy Scintillators for Scientific & Industrial Applications, Proc. Crystal 2000 Int'l. Workshop* (Chamonix, France, Sept. 1992). Editors: DeNotaristefani F., Lecoq P., and Schneegans M., Editions Frontieres, Gif-sur-Yvette, France, pp. 137 - 150.

[18] Autrata E., Schauer P., Kvapil Ji., and Kvapil Jo., 1983: "A Single Crystal of $YAlO_3$:Ce as a Fast Scintillator in SEM", *Scanning* **5**, pp. 91 - 96; Baryshevsky V. G., Korzhik M. V., Moroz V. I., Pavlenko V. B., and Fyodorov A. A., 1991: "$YAlO_3$:Ce: Fast-Acting Scintillators for Detection of Ionizing Radiation", *Nucl. Instr. & Meth.* **B58**, pp. 291 - 293.

[19] Laval M., Moszynski M., Allemand R., Cormoreche E., Guiner P., Odru R., and Vacher J., 1983: "Barium Fluoride - Inorganic Scintillator for Subnanosecond Timing", *Nucl. Instr. & Meth.* **206**, pp. 169 - 175.

[20] van Eijk C. W. E., 1994: "Crossluminescence", *J. Lumin.* **60&61**, pp. 936 - 941.

[21] Takagi K. and Fukazawa T., 1983: "Cerium-Activated Gd_2SiO_5 Single Crystal Scintillator"; *Appl. Phys. Lett.* **42**, pp. 43 - 45; Ishibashi H., Shimizu K., Susa K., and Kubota S., 1989: "Cerium Doped GSO Scintillators and its Application to Position Sensitive Detectors", *IEEE Trans. Nucl. Sci.* **36**, pp. 170 - 172.

[22] Anderson D. F., 1989: "Properties of the High-Density Scintillator Cerium Fluoride", *IEEE Trans. Nucl. Sci.* **36**, pp. 137 - 139.

[23] Moses W. W. and Derenzo S. E., 1989: "Cerium Fluoride, A New, Fast, Heavy Scintillator", *IEEE Trans. Nucl. Sci.* **36**, pp. 173 - 176.

[24] Melcher C. L. and Schweitzer J. S., 1992: "Cerium-Doped Lutetium Oxyorthosilicate: A Fast Efficient New Scintillator", *IEEE Trans. Nucl. Sci.* **39**, pp. 502 - 505.

[25] Lempicki A., Berman E., Wojtowicz A. J., Balcerzyk M., and Boatner L. A., 1993: "Cerium-Doped Orthophosphates: New Promising Scintillators", *IEEE Trans. Nucl. Sci.* **40**, pp. 384 - 387.

[26] Lempicki A., Randles M. H., Wisniewski D., Balcerzyk M., Brecher C., and Wojtowicz A. J., 1995: "$LuAlO_3$:Ce and Other Aluminate Scintillators", *IEEE Trans. Nucl. Sci.* **42**, pp. 280 - 284.

[27] Lempicki A., Wojtowicz A. J., and Berman E., 1993: "Fundamental Limits of Scintillation Performance", *Nucl. Instr. & Meth.* **A333**, pp. 304 - 311.

[28] Shockley W., 1961: "Problems Related to *p-n* Junctions in Silicon". *Solid State Electronics* **2**, pp. 35 - 67.

[29] van Roosbroeck W., 1965: "Theory of the Yield and Fano Factor of Electron-Hole Pairs Generated in Semiconductors by High-Energy Particles" *Phys. Rev. A* **139**, pp. 1702 - 1716.

[30] Klein C. A., 1968: "Bandgap Dependence and Related Features of Radiation Ionization Energies in Semiconductors", *J. Appl. Phys.* **39**, pp. 2029 - 2038.

[31] Robbins D. J., 1980: "On Predicting the Maximum Efficiency of Phosphor Systems Excited by Ionizing Radiation", *J. Electrochem. Soc.* **127**, pp. 2694 - 2702.

[32] Rothwarf A., 1973: "Plasmon Theory of Electron-Hole Pair Production: Efficiency of Cathode Ray Phosphors", *J. Appl. Phys.* **44**, pp. 752 - 756.

[33] Rodnyi P. A., Dorenbos P., and van Eijk C. W. E., 1994: "Creation of Electron-Hole Pairs in Inorganic Scintillators", *Scintillator and Phosphor Materials, MRS Symp. Proc.* (San Francisco CA, USA, April 1994). Editors: Weber M. J., Lecoq P., Ruchti R. C., Woody C., Yen W. M., and Zhu R. Y. Materials Research Society, Pittsburgh PA, USA. Vol. 348, pp. 379 - 385.

[34] Landsberg P., 1991: *Recombination in Semiconductors*, Cambridge Univ. Press, Cambridge, England, UK.

[35] Bartram R. H. (U. Connecticut), 1995: Private communication, to be published.

[36] Alig R. C. and Bloom S., 1977: "Cathodoluminescence Efficiency", *J. Electrochem. Soc.* **124**, pp. 1136 - 1138.

[37] Kittel C., 1986: *Introduction to Solid State Physics*, John Wiley & Sons Inc., New York NY, USA, pp. 253 - 263, 310 - 312; Burns G., 1990. *Solid State Physics*, Academic Press Inc., Boston MA, USA, pp. 486 - 498.

[38] Kingsley J. D. and Ludwig G. W., 1970: "The Efficiency of Cathode Ray Phosphors, I. Measurement", *J. Electrochem. Soc.* **117**, pp. 348 - 353; "The Efficiency of Cathode Ray Phosphors, II. Correlation with Other Properties", *ibid.* pp. 353 - 359.

[39] Vasil'ev A. N., 1995: "Final Stages of Inelastic Electron Scattering: Influence on Scintillator Efficiency", *SCINT95 Int'l Conf. on Inorganic Scintillators and their Applications*, Delft, Netherlands, Aug. 1995. Proceedings to be published.

[40] Holstein T., 1947: "Imprisonment of Resonance Radiation in Gases", *Phys. Rev.* **72**, pp. 1212 - 1233; Holstein T., 1951: *ibid.* **83**, pp. 1159 - 1168.

[41] Batchelder J. S., Zewail A. H., and Cole T., 1986: "Luminescent Solar Concentrators 2: Theoretical Analysis of their Possible Efficiencies", *Applied Optics* **20**, pp. 3733 - 3754.

[42] Wojtowicz A. J., Balcerzyk M., Wisniewski D., and Lempicki A., 1994: "Scintillation Light Trapping and Radiation Damage in CeF_3", *IEEE Trans. Nucl. Sci.* **41**, pp. 713 - 718.

[43] Sakai Y., Kawahigashi M., Minami M., Inoue T., and Hirayama S., 1989: "Deconvolution of Nonexponential Emission Decays Arising from Reabsorption of Light", *J. Lumin.* **42**, pp. 317 - 324.

[44] Henderson B. and Imbusch G. F., 1989: *Optical Spectroscopy of Inorganic Solids,* Oxford University Press, Oxford, England, UK.

[45] Visser R., Dorenbos P., van Eijk C. W. E., Meijerink A., Blasse G., and den Hartog H. W., 1993: "Energy Transfer Processes Involving Different Luminescent Centers in Cerium-Doped Barium Fluoride", *J. Phys.: Cond. Matter* **5**, pp. 1659 - 1680.

[46] Wanmaker W. L., Bril A., ter Vrgut J. W., and Broos J., 1966: "Luminescent Properties of Eu-Activated Phosphors of the Type $A^{iii}B^{v}O_4$", *Philips Res. Rep.* **21**, pp. 270 - 282.

[47] Davidov A. S., 1968: "The Radiationless Transfer of Energy of Electronic Excitation Between Impurity Molecules in Crystals", *Phys. Stat. Sol.* **30**, pp. 357 - 366.

[48] Nettle S. and Lempicki A., 1979: "Resonant Energy Transfer in Stoichiometric Rare Earth Compounds", *Opt. Comm.* **30**, pp. 387 - 390.

[49] Williams F., 1966: *Luminescence in Inorganic Solids*. Editor: Goldberg P., Academic Press, Inc., New York NY, USA.

[50] Wojtowicz A. J., 1995: "Scintillation Mechanism: The Significance of Variable Valence and Electron-Lattice Coupling in R.E.-Activated Scintillators", *SCINT95 Int'l Conf. on Inorganic Scintillators and their Applications*, Delft, Netherlands, Aug. 1995. Proceedings to be published.

[51] Murray R. B., 1975: "Energy Transfer in Alkali Halide Scintillators by Electron-Hole Diffusion and Capture", *IEEE Trans. Nucl. Sci.* **22**, pp. 54 - 57.

[52] Kaufman R. G., Hadley W. B., and Hersh H. N., 1970: "The Scintillation Mechanism in Thallium-Doped Alkali Halides", *IEEE Trans. Nucl. Sci.* **17**, pp. 82 - 88.

[53] Kaufman R. G. and Hadley W. B., 1966: "Luminescence Processes Generated by Ionizing Radiation in KI and KI:Tl", *J. Chem. Phys.* **44**, pp. 1311 - 1317.

[54] Kaufman R. G. and Hadley W. B., 1967: "Excitation of Luminescence from KI:Tl at 5 K", *J. Chem. Phys.* **46**, pp. 1598 - 1601.

[55] Mayer A. and Murray R. B., 1960: "Effect of Energetic Secondary Electrons on the Scintillation Process in Alkali Halide Crystals", *Phys. Rev.* **128**, pp. 98 - 105.

[56] Dietrich H. B., Purdy A. E., Murray R. B., and Williams R. T., 1973: "Kinetics of Self-Trapped Holes in Alkali Halide Crystals: Experiments in NaI(Tl) and KI(Tl)", *Phys. Rev. B* **8**, pp. 5894 - 5901.

[57] Hadley W. B., Polick S., Kaufman R. G., and Hersh H. N., 1966: "Energy Storage and Luminescence in KI:Tl at Low Temperatures", *J. Chem. Phys.* **45**, pp. 2040 - 2048.

[58] van Sciver W. J., 1960: "Fluorescence and Reflection Spectra of NaI Single Crystals", *Phys. Rev.* **120**, pp. 1193 - 1205.

[59] Song K. S. and Williams R. T., 1993: *Self-Trapped Excitons*, Springer-Verlag, Berlin, Germany.

[60] Schotanus P., Stam G., Gerritse E., Utts B., and Briaux B., 1992: *Scintillation Detectors*, Harshaw QS Catalog.

[61] Dietrich H. B. and Murray R. B., 1972: "Kinetics of the Diffusion of Self-Trapped Holes in Alkali Halide Scintillators", *J. Lumin.* **5**, pp. 155 - 170.

[62] Delbecq C. J., Gibson A., and Yuster P. H., 1966: "Trapping and Annihilation of Electrons and Positive Holes in KCl-TlCl", *Phys. Rev.* **151**, pp. 599 - 609.

[63] Creuzburg M. and Raether H., 1964: "Behavior of Nuclear Energy Losses of Electrons in Alkali Halides", *Solid State Comm.* **2**, pp. 175 - 177.

[64] Koepke Cz., Lempicki A., and Wojtowicz A. J., 1993: "Luminescence Quenching in Strongly Coupled Systems", *Phys. Stat. Sol. (b)* **179**, pp. 233 - 240.

[65] Tirsell K. G., Tripp G. R., Lent E. M., Lerche R. A., Cheng J. C., Harker L. P., and Lyons P. B., 1977: "Subnanosecond Plastic Scintillator Time Response Studies Using Laser-Produced X-Ray Pulsed Excitation", *IEEE Trans. Nucl. Sci.* **NS-24**, pp. 250 - 254.

[66] Blasse G., 1993: "Luminescence and Scintillation Mechanisms in Inorganic Scintillators", *Heavy Scintillators for Scientific and Industrial Applications, Proc. Crystal 2000 Int'l. Workshop* (Chamonix, France, Sept. 1992). Editors: DeNotaristefani F., Lecoq P., and Schneegans M., Editions Frontieres, Gif-sur-Yvette, France. pp. 85 - 97.

[67] Lempicki A. and Wojtowicz A. J., 1993: "New Trends in Scintillator Research", *Conference Record, IEEE Nucl. Sci. Symp. & Med. Imag. Conf.* (San Francisco CA, USA, Nov. 1993). Editor: Klaisner L. A., IEEE, Piscataway NJ, USA, pp. 71 - 75.

[68] Lempicki A. and Wojtowicz A. J., 1994: "Fundamental Limitations of Scintillators", *J. Lumin.* **60&61**, pp. 942 - 947.

[69] Wojtowicz A. J., Balcerzyk M., Berman E., and Lempicki A., 1994: "Optical Spectroscopy and Scintillation Mechanisms of $Ce_xLa_{1-x}F_3$", *Phys. Rev. B* **49**, pp.14880 - 14894.

[70] Wojtowicz A. J., Berman E., and Lempicki A., 1992: "Stoichiometric Cerium Compounds as Scintillators, II. CeP_5O_{14}", *IEEE Trans. Nucl. Sci.* **39**, pp. 1542 - 1548.

[71] Ishii M. and Kobayashi M.,1992: "Single Crystals for Radiation Detectors", *Prog. Cryst. Growth & Char. Mat'ls.* **23**, pp. 245 - 311.

[72] Holl I., Lorenz E., and Mageras G., 1988: "A Measurement of the Light Output of Common Inorganic Scintillators", *IEEE Trans. Nucl. Sci.* **35**, pp. 105 - 109.

[73] Melcher C. L. (Schlumberger Corp.), 1994: Private communication.

[74] Sakai E., 1987: "Recent Measurements on Scintillator-Photodiode Systems", *IEEE Trans. Nucl. Sci.* **34**, pp. 418 - 422.

[75] Butler K., 1980: *Fluorescent Lamp Phosphors*, Pennsylvania State Univ. Press, University Park PA, USA.

[76] Blasse G., 1980: "The Luminescence of Closed-Shell Transition-Metal Complexes: New Developments", *Structure and Bonding* **42**, pp. 1 - 41.

[77] Burnham C. A., Kaufman D. E., Chesler D. A., Chen Y., Gregoire M., and Brownell G. L., 1991: "MGH Cylindrical PET Detector Characteristics", *Conference Record, IEEE Nucl. Sci. Symp. & Med. Imag. Conf.* (Santa Fe NM, USA, Nov. 1991), pp. 1644 - 1647.

[78] Ivanov V. Y., Kruzhalov A. V., Pustovarov V. A., and Petrov V. L., 1987: "Electron Excitation and Luminescence in Bismuth Germanium Oxide ($Bi_4Ge_3O_{12}$) and Bismuth Silicon Oxide ($Bi_4Si_3O_{12}$) Crystals", *Nucl. Instr. & Meth.* **A261**, pp. 150 - 152.

[79] Hahn B., 1953: "Scintillations Produced by α-Particles in CsI", *Phys. Rev.* **91**, pp. 772 - 773.

[80] Utts B. K. and Spagno S. E., 1990: "An Investigation on the Characteristics of Pure CsI Crystals", *IEEE Trans. Nucl. Sci.* **37**, pp. 134 - 138.

[81] Woody C. L., Levy P. W., Kierstead J. A., Skwarnicki T., Sobolewski Z., Goldberg M., Horwitz N., Sounder P., and Anderson D. F., 1990: "Readout Techniques and Radiation Damage of Undoped Cesium Iodide", *IEEE Trans. Nucl. Sci.* **37**, pp. 492 - 499.

[82] Kirk F. W. K., 1990: "A Comparative Study of the Scintillation Response of Natural Crystals of Anglesite ($PbSO_4$) And Cerussite ($PbCO_3$)", *Nucl. Instr. & Meth.* **A297**, pp. 532 - 533.

[83] Moses W. W., Derenzo S. E., and Shlichta P. J., 1992: "Scintillation Properties of Lead Sulfate"; *IEEE Trans. Nucl. Sci.* **39**, pp. 1190 - 1194.

[84] Zhang J. C., Lund J. C., Cirignano L., Shah K. S., Squillante M. R., and Moses W. W., 1993: "Lead Sulfate Scintillator Crystal Growth for PET Applications" ; *Conference Record, IEEE Nucl. Sci. Symp. & Med. Imag. Conf.* (San Francisco CA, USA, Nov. 1993). Editor: Klaisner L. A., IEEE, Piscataway NJ, USA. pp. 76 - 80.

[85] Blasse G., 1991: "Search for New Inorganic Scintillators", *IEEE Trans. Nucl. Sci.* **38**, pp. 30 - 31.

[86] Dafinei L., Auffray E., Lecoq P., and Schneegans M., 1992: "Lead Tungstate for High Energy Calorimetry", *Heavy Scintillators for Scientific and Industrial Applications, Proc. Crystal 2000 Int'l. Workshop* (Chamonix, France, Sept. 1992). Editors: DeNotaristefani F., Lecoq P., and Schneegans M., Editions Frontieres, Gif-sur-Yvette, France. pp. 99 - 104; Korzhik M. V., Pavlenko V. B., Katchanov V. A., Peigneux J.-P., and Poulet M., 1992: "The Scintillation Mechanism in $PbWO_4$", *ibid.*, pp. 285 - 290.

[87] Nagornaya L. and Ryzhikov V., 1992: "Fast Scintillators Based on Large "Heavy" Tungstate Single Crystals", *Heavy Scintillators for Scientific and Industrial Applications, Proc. Crystal 2000 Int'l. Workshop* (Chamonix, France, Sept. 1992). Editors: DeNotaristefani F., Lecoq P., and Schneegans M., Editions Frontieres, Gif-sur-Yvette, France. pp. 367 - 374; Kobayashi M., Ishii M., Usuki Y., and Yahagi H., 1993: "$PbWO_4$ Scintillator at Room Temperature", *ibid.* pp. 375 - 379; Miassoedov L. V., Selivanov V., Sinitsin I., Torokhov V. D., Nagornaya L. L., Vostresov Y. Ia., and Tupitsina I. A., 1993: "$PbWO_4$: A Heavy, Fast and Radiation Resistant Scintillator for EM Calorimetry", *ibid.*, pp. 438 - 442.

[88] Ershov N. N., Zakharov N. G., and Rodnyi P., 1982: "Spectral-Kinetic Study of Intrinsic Luminescence Characteristics of Fluorite-Type Crystals"; *Opt. Spectrosc.* **53**, pp. 51 - 54.

[89] Valbis J. A., Rachko Z. A., and Yanson Ya. L., 1988: "Crossluminescence of KF and Related Compounds", *Solid State Comm.* **67**, pp. 183 - 185.

[90] Rodnyi P. A., 1994: "Choice of Compounds with Fast Core-Valence Transitions", *Scintillator and Phosphor Materials, MRS Symp.* (San Francisco CA, USA, April 1994). Editors: Weber M. J., Lecoq P., Ruchti R. C., Woody C., Yen W. M., and Zhu R. Y. Materials Research Society, Pittsburgh PA, USA. Vol. 348, pp. 77 - 88.

[91] Rodnyi P. A., 1992: "Core-Valence Band Transitions in Wide-Gap Ionic Crystals", *Sov. Phys. Sol. State* **34**, pp. 1053 - 1066.

[92] Woody C. L., 1991: "A Report on Radiation Damage in Barium Fluoride to the GEM Collaboration", Physics Dept., Brookhaven National Laboratory, Upton, NY, USA.

[93] Visser R., Dorenbos P., van Eijk C. W. E, and den Hartog H. W., 1992: "Energy Transfer Processes Observed in The Scintillation Decay of Lanthanum-Doped Barium Fluoride", *J. Phys.: Cond. Matter* **4**, pp. 8801 - 8812.

[94] Williams G. M., Edelstein N., Boatner L. A., and Abraham M. M., 1989: "Intensities of Electronic Raman Scattering Between Crystal Field Levels of Ce^{3+} in $LuPO_4$: Nonresonant and Near-Resonant Excitation", *Phys. Rev. B* **40**, pp. 4143 - 4152.

[95] Brewer L., 1972: "Energies of Electronic Configurations of Singly, Doubly, and Triply Ionized Lanthanides and Actinides", *J. Opt. Soc. Am.* **61**, pp. 1666 - 1686.

[96] Dieke G. H., 1968: *Spectra and Energy Levels of Rare Earth Ions in Solids*, Interscience Publishers, Inc., New York NY, USA.

[97] Weber H. P., 1975: "Neodymium Pentaphosphate Lasers", *Optical and Quantum Electronics* **7**, 431 - 442.

[98] Bimberg D., Robbins D. J., Wight D. R., and Jesser J. P., 1975: "Cerium Phosphate (CeP_5O_{14}): A New Ultrafast Scintillator", *Appl. Phys. Lett.* **27**, pp. 66 - 68.

[99] Wojtowicz A. J., Lempicki A., and Berman E., 1993: "Scintillation Mechanisms in Stoichiometric Cerium Compounds", *Heavy Scintillators for Scientific and Industrial Applications, Proc. Crystal 2000 Int'l. Workshop* (Chamonix, France, Sept. 1992*)*. Editors: DeNotaristefani F., Lecoq P., and Schneegans M., Editions Frontieres, Gif-sur-Yvette, France. pp. 179 - 183.

[100] Wojtowicz A. J., Berman E., Koepke Cz., and Lempicki A., 1991: "Stoichiometric Cerium Compounds as Scintillators, I: CeF_3", *IEEE Trans. Nucl. Sci.* **39**, pp. 494 - 501.

[101] Pedrini C., Moine B., Gacon J. C., and Jacquier B., 1992: "One- and Two-Photon Spectroscopy of Ce^{3+} Ions in LaF_3-CeF_3", *J. Phys.: Cond. Matter* **4**, pp. 5461 - 5470.

[102] Lecoq P., 1992: "Results on New Scintillating Crystals from the "Crystal Clear" Collaboration", *Conference Record, IEEE Nucl. Sci. Symp. & Med. Imag. Conf.* (Orlando FL, USA, Oct. 1992). Editor: Kirsten F. A., IEEE Inc., Piscataway, 1992, pp. 248 - 250; Dafinei I., Auffray E., Dezillie B., Lecoq P., Schussler M., Ferrere D., and Schneegans M., 1993: "Radiation Damage Studies on CeF_3 Crystals by the "Crystal Clear" Collaboration", *Heavy Scintillators for Scientific and Industrial Applications, Proc. Crystal 2000 Int'l. Workshop* (Chamonix, France, Sept. 1992*)*. Editors: DeNotaristefani F., Lecoq P., and Schneegans M., Editions Frontieres, Gif-sur-Yvette, France. pp. 459 - 466.

[103] Optovac Corp., North Brookfield MA, USA.

[104] Roos A., Buijs M., Wappenar K. E. D., and Schoonman J., 1985: "Dielectric Relaxation Properties of Tysonite-Type Solid Solutions $La_{1-x}Ba_xF_{3-x}$", *J. Phys. Chem. Sol.* **46**, pp. 655 - 664.

[105] Moses W. W., Derenzo S. E., Weber M. J., Ray-Chaudhuri A. K., and Cerrina F., 1994: "Scintillation Mechanisms in Cerium Fluoride", *J. Lumin.* **59**, pp. 89 - 100.

[106] McClure D. S. and Kiss Z., 1963: "Survey of the Spectra of the Divalent Rare Earth Ions in Cubic Crystals", *J. Chem. Phys.* **39**, pp. 3251 - 3257.

[107] Pedrini C., Belsky A. N., Vasil'ev A. N., Bouttet D., Dujardin C., Moine B., Martin P., and Weber M. J., 1994: "Fluorescence Properties of CeF_3 and of Some Other Cerium-Doped Crystals and Glasses under VUV and X-Ray Synchrotron Excitation", *Scintillator and Phosphor Materials, MRS Symp. Proc.*, (San Francisco CA, USA, April 1994). Editors: Weber M. J., Lecoq P., Ruchti R. C., Woody C., Yen W. M., and Zhu R. Y. Materials Research Society, Pittsburgh PA, USA. Vol. 348, pp. 225 - 234.

[108] Wojtowicz A. J., Lempicki A., Wisniewski D., Balcerzyk M., and Brecher C., 1995: "The Carrier Capture and Recombination Processes in Ln^{3+}-Activated Scintillators", *IEEE Trans. Nucl. Sci.* - in press.

[109] Williams R. T., Itoh C., and Tanimura K., 1995: "Time-Resolved Optical Absorption and Luminescence in LaF_3 and CeF_3, Irradiated by 1-MeV, 20-ns Electron Pulses", *SCINT95 Int'l Conf. on Inorganic Scintillators and their Applications*, Delft, Netherlands, Aug. 1995. Proceedings to be published.

[110] Wojtowicz A. J., Lempicki A., Wisniewski D., Brecher C., Bartram R. H., Woody C., Levy P., Stoll S., Kierstead J., Pedrini C., Bouttet D., and Koepke Cz., "Scintillation Mechanism and Radiation Damage in $Ce_xLa_{1-x}F_3$ Crystals", *Scintillator and Phosphor Materials, MRS Symp. Proc.* (San Francisco CA, USA, April 1994). Editors: Weber M. J., Lecoq P., Ruchti R. C., Woody C., Yen W. M., and Zhu R. Y. Materials Research Society, Pittsburgh PA, USA. Vol. 348, pp. 455 - 461.

[111] Blasse G., 1983: "Energy Transfer from Cerium (3+) to Europium (3+) in Yttrium Gadolinium Fluoride [$(Y,Gd)F_3$]", *Phys. Stat. Sol. (a)* **75**, pp. K41 - K43.

[112] Wojtowicz A. J., Lempicki A., Wisniewski D., and Boatner L. A., 1994: "Cerium-Doped Orthophosphate Scintillators", *Scintillator and Phosphor Materials, MRS Symp. Proc.* (San Francisco CA, USA, April 1994). Editors: Weber M. J., Lecoq P., Ruchti R. C., Woody C., Yen W. M., and Zhu R. Y. Materials Research Society, Pittsburgh PA, USA. Vol. 348, pp. 123 - 129.

[113] Hitachi Chemical Co., Shimodate, Japan.

[114] Melcher C. L., Schweitzer J. S., Utsu T., and Akiyama S., 1990: "Scintillation Properties of GSO", *IEEE. Trans. Nucl. Sci.* **37**, pp. 161 - 164.

[115] Suzuki H., Tombrello T. A., Melcher C. L., and Schweitzer J. S., 1992: "UV and Gamma-Ray Excited Luminescence of Cerium-Doped Rare-Earth Oxyorthosilicates", *Nucl. Instr. & Meth.* **A320**, pp. 263 - 272.

[116] Suzuki H., Tombrello T. A., Melcher C. L., and Schweitzer J. S., 1994. "Energy Transfer Mechanism in $Gd_2(SiO_4)O$:Ce Scintillators", *J. Lumin.* **60&61**, pp. 963 - 966.

[117] Lammers M. J. J. and Blasse G., 1987: "Luminescence of Tb^{3+}- and Ce^{3+}-Activated Rare Earth Silicates", *J. Electrochem. Soc.* **134**, pp. 2068 - 2072.

[118] Brandle C. D., Valentino A. J., and Berkstresser G. W., 1986: "Czochralski Growth of Rare-Earth Orthosilicates (Ln_2SiO_5)", *J. Cryst. Growth* **78**, pp. 308 - 315.

[119] Kobayashi M., Takamatsu K., Ide S., Mori K., Sugimoto S., Takaki H., Yuasa M., and Ishii M., 1991: "A Beam Test on a Fast EM-Calorimeter of Gadolinium Silicate, GSO(Ce)", *Nucl. Instr. & Meth.* **A306**, pp. 139 - 144.

[120] Melcher C. L. and Schweitzer J. S., 1992: "A Promising New Scintillator: Cerium-Doped Lutetium Oxyorthosilicate", *Nucl. Instr. & Meth.* **A314**, pp. 212 - 214.

[121] Ludziejewski T., Moszynska K., Moszynski M., and Wolski D., 1994. "Advantages and Limitations of LSO Scintillator in Nuclear Physics Experiments" *Conference Record, IEEE Nucl. Sci. Symp. & Med. Imag. Conf.* (Norfolk VA, USA, Oct. 1994), pp. 299 - 303.

[122] Dorenbos P., van Eijk C. W. E., Bos A. J. J., and Melcher C. L., 1994: "Scintillation and Thermoluminescence Properties of Lu_2SiO_5:Ce Fast Scintillation Crystals", *J. Lumin.* **60&61**, pp. 979 - 982; "Afterglow and Thermoluminescence Properties of Lu_2SiO_5:Ce Scintillation Crystals", *J. Phys.: Cond. Matter* **6**, pp. 4167-4180.

[123] Melcher C. L. (Schlumberger Corp.), 1994: Private communication

[124] Dorenbos P., de Hass J. T. M., van Eijk C. W. E., Melcher C. L., and Schweitzer J. S., 1993: "Nonlinear Response of the Scintillation Yield of $Lu_2SiO_5:Ce^{3+}$", *Conference Record, IEEE Nucl. Sci. Symp. & Med. Imag. Conf.* (San Francisco CA, USA, Nov. 1993). Editor: Klaisner L. A., IEEE, Piscataway NJ, USA. pp. 714 - 716.

[125] Suzuki H. (Schlumberger Corp.), 1994: Private communication.

[126] Nakazawa E. and Shiga F., 1977: "Vacuum Ultraviolet Luminescence-Excitation Spectra of $RPO_4:Eu^{3+}$ (R = Y, La, Gd, and Lu), *J. Lumin.* **15**, pp. 255 - 259.

[127] Green T. E., Riley M. E., Richards P. M., Loubriel G. M., Jennison D. R., and Williams R. T., 1989: "Auger-Induced Valence Processes in Alkali Halides Following Alkali Core-Hole Creation", *Phys. Rev. B* **39**, pp. 5407 - 5413.

[128] Kaminskii A. A., 1990: *Laser Crystals*, Springer-Verlag, Berlin, Germany.

[129] Robbins D. J., 1979: "The Effects of Crystal Field and Temperature on the Photoluminescence Excitation Efficiency of Ce^{3+} in YAG", *J. Electrochem. Soc.* **126**, pp. 1550 - 1555.

[130] Rotman S. R., 1992: "Defect-Property Correlations in Garnet Crystals. VI. The Electrical Conductivity, Defect Structure, and Optical Properties of Luminescent Calcium and Cerium-Doped Yttrium Aluminum Garnet", *J. Appl. Phys.* **71**, pp. 1209 - 1214.

[131] Robbins D. J., Cockayne B., Glasper J. L., and Lent B., 1979: "The Temperature Dependence of Rare-Earth Activated Garnet Phosphors": *J. Electrochem. Soc.* **126**, 1221 - 1228.

[132] Robbins D. J., Cockayne B., Lent B., Duckworth C. N., and Glasper J. L., 1979: "Investigation of Competing Recombination Processes in Rare-Earth Activated Garnet Phosphors", *Phys. Rev. B* **19**, pp. 1254 - 1269.

[133] Aizenberg G. E. and Rotman S. R., 1991: "A Nonradiative Energy Transfer Model for Cerium-Doped Garnet (Ce:YAG)", *Phys. Stat. Sol.* **126**, pp. 263 - 273.

[134] Moszynski M., Ludziejewski T., Wolski D., Klamra W., and Norlin L. O., 1994: "Properties of the YAG:Ce Scintillator", *Nucl. Instr. & Meth.* **A345**, pp. 461 - 467.

[135] Andreissen J., Dorenbos P., and van Eijk, C. W. E., 1994: "Calculation of Energy Levels of Cerium in Inorganic Scintillator Crystals", *Scintillator and Phosphor Materials, MRS Symp. Proc.*, (San Francisco CA, USA, April 1994). Editors: Weber M. J., Lecoq P., Ruchti R. C., Woody C., Yen W. M., and Zhu R. Y. Materials Research Society, Pittsburgh PA, USA. Vol. 348, pp. 355 - 365.

[136] Weber M. J., 1973: "Optical Spectra of Ce^{3+} and Ce^{3+}-Sensitized Fluorescence in $YAlO_3$", *J. Appl. Phys.* **44**, pp. 3205 - 3208.

[137] Ziegler S. I., Rogers J. G., Selivanov V., and Sinitzin I., 1991: "Characteristics of the New $YAlO_3$ Compared to BGO and GSO", *Conference Record, IEEE Nucl. Sci. Symp. & Med. Imag. Conf.* (Santa Fe NM, USA, Nov. 1991), pp. 156 - 161.

[138] Korzhik M. V., Misevich O. V., and Fyodorov A. A., 1992: "$YAlO_3$:Ce Scintillators: Application for X-Ray and Soft γ-Ray Detection", *Nucl. Instr. & Meth.* **B72**, pp. 499 - 501.

[139] Lempicki A., Brecher C., Wisniewski D., and Zych E., 1995: "Cerium-Doped Aluminate Scintillators", *SCINT95 Int'l Conf. on Inorganic Scintillators and their Applications*, Delft, Netherlands, Aug. 1995. Proceedings to be published.

[140] Trower W. P., 1994: "Cerium-Doped Yttrium Aluminum Perovskite (YAP): Properties of Commercial Crystals", *Scintillator and Phosphor Materials, MRS Symp. Proc.* (San Francisco CA, USA, April 1994). Editors: Weber M. J., Lecoq P., Ruchti R. C., Woody C., Yen W. M., and Zhu R. Y. Materials Research Society, Pittsburgh PA, USA. Vol. 348, pp. 131 - 136.

[141] Trower W. P., 1993: "The Nitrogen Camera and the Detection of Concealed Explosives", *Nucl. Instr. & Meth.* **B79**, pp. 589 - 592.

[142] Kaminskii A. A., Ivanov A. O., Sarkisov S. E., Mochalov I. V., Fedorov V. A., and Li L., 1976: "Comprehensive Investigations of the Spectral and Lasing Characteristics Of $LuAlO_3$ Crystal Doped with Nd^{3+}", *Sov. Phys. JETP* **44**, 516 - 524; Kaminskii A. A. and Petrosyan A. G., 1992: "Growth and Luminescence of $LuAlO_3$:Ti^{3+} Crystals", *Sov. Phys. Dokl.* **37**, pp. 422 - 423.

[143] Litton Airtron Corp., Charlotte NC, USA.

[144] Moses W. W., Derenzo S. E., Korzhik M., Gektin A., Minkov B., and Aslanov V., 1995: "$LuAlO_3$:Ce – A High Density, High Speed Scintillator for Gamma Detection", *IEEE Trans Nucl. Sci.* **42**, pp. 275 - 279.

[145] Bondar I. A., Shirvinskaya A. K., Popova V. F., Mochalov I. V., and Ivanov A. O., 1979: "Thermal Stability of Orthoaluminates of Rare Earth Elements of the Yttrium Subgroup", *Sov. Phys. Dokl.* **24**, pp. 289 - 292.

[146] Gumanskaya E. G., Egorcheva O. A., Korzhik M. V., Smirnova S. A., and Pavlenko E. B., 1992: "Spectroscopic Properties and Scintillation Efficiency of Ce-Doped $YAlO_3$ Single Crystals", *Opt. Spectrosc.* **72**, pp. 215 - 217.

[147] Baryshevsky V. G., Korzhik M. V., Minkov B. I., Smirnova S. A., Fyodorov A. A., Dorenbos P., and van Eijk C. W. E., 1993: "Spectra and Scintillation Properties of Ce-Doped $YAlO_3$ Single Crystals", *J. Phys.: Cond. Matter* **5**, pp. 7893 - 7902.

[148] Sausville J. (Sausville Chemical Co.), 1994: Private communication.

[149] Lee W., Bendahan J., Ryge P., and Gozani T., 1991: "Detector Selection for Nuclear-Based Explosive Detection System", *Proc. 1st Int'l. Symp. Explosive Detection Technol.* (Atlantic City, NJ, Nov. 1991), pp. 225 - 235.

[150] Bearden J. A. and Burr A. F., 1967: "Reevaluation of X-Ray Atomic Energy Levels", *Rev. Mod. Phys.* **39**, pp. 125 - 142.

[151] Armbruster A., 1976: "Infrared Reflection Studies on the Phosphates, Arsenates, and Vanadates of Lutetium and Yttrium", *Phys. Chem. Solids* **37**, pp. 321 - 327.

[152] Shannon R. D., 1993: "Dielectric Polarizabilities of Ions in Oxides and Fluorides", *J. Appl. Phys.* **73**, pp. 348 - 366.

[153] Halliburton L. E., and Edwards G. E., 1984: "Radiation Damage Mechanisms in Scintillator Materials: Applications to BaF_2 and CeF_3", *Scintillator and Phosphor Materials, MRS Symp. Proc.* (San Francisco CA, USA, April 1994). Editors: Weber M. J., Lecoq P., Ruchti R. C., Woody C., Yen W. M., and Zhu R. Y. Materials Research Society, Pittsburgh PA, USA. Vol. 348, pp. 423 - 434.

[154] Stokowski S. E., Saroyan R. A., and Weber M. J., 1978: *Nd-Doped Laser Glass: Spectroscopic and Physical Properties*, Lawrence Livermore Laboratory, Livermore CA, USA.

RARE EARTH-DOPED FIBER LASERS AND AMPLIFIERS

Bernard Jacquier

Laboratoire de Physico-Chimie des Matériaux Luminescents
UMR n°5620, CNRS-Université Claude Bernard Lyon 1
69622 Villeurbanne France

Abstract

Several technical revolutions over the last ten years have led to the increasing use of rare earth-doped fiber lasers and amplifiers as a key component of optical communication systems. In this chapter, we review the methods of producing these fibers and the effect of the host composition and the rare earth doping on the quality of the final product. We then review the physics of the dopants and the hosts involved which makes these doped fibers so valuable for optical communications. In the last section, the laser performance itself is summarized; new trends for the future are presented.

RARE EARTH-DOPED FIBER LASERS AND AMPLIFIERS

Bernard Jacquier

Contents

Rare Earth-Doped Fiber Lasers and Amplifiers

Bernard Jacquier

Laboratoire de Physico-Chimie des Matériaux Luminescents
*UMR n°*5620, CNRS-Université Claude Bernard Lyon 1
69622 Villeurbanne France

1. Introduction

The concept of fiber lasers and amplifiers was first demonstrated by Koester and Snitzer as early as 1964 [1], three years after the first laser operation of Nd^{3+} doped crystalline materials and glasses. The next step had to wait around six years before achieving laser action in a core/cladding waveguide configuration with end pumping by Stone and Burrus [2] who took advantage of the technological breakthrough in the quality of the fiber, especially its transparency. Again, it was necessary to wait until 1985 for the first demonstration of a single-mode rare earth doped fiber lasers by D.N. Payne & al. who pointed out the very low laser threshold [3]. This was made possible because of the development of chemical vapor deposition techniques in the process of fabricating low-loss single-mode fiber for long distance optical transmission systems. The advent of such low-loss single-mode fibers can be considered as the first technical revolution in optical telecommunications, followed rapidly by another one: the Erbium Doped Fiber Amplifier (EDFA) in 1987 demonstrated by several groups almost simultaneously [4-9]. This fiber amplifier operates in the third telecommunication window at 1.55 µm, producing outstanding gains with a low pump power, a high saturation power, a low noise feature and low interchannel cross-talk [10,11]. Now, large line broadening of the 1.55 µm emission and subsequent flatness of the gain, closely related to chemical composition (silica, multicomponent glass, fluorides...) or high doping level should shortly improve significantly the performance of the high bite rate transmissions through Wavelength-Division-Multiplexing techniques (WDM) [12]. However, soliton propagation should compete with WDM for long haul distance transmission at the dawn of the next century [13].

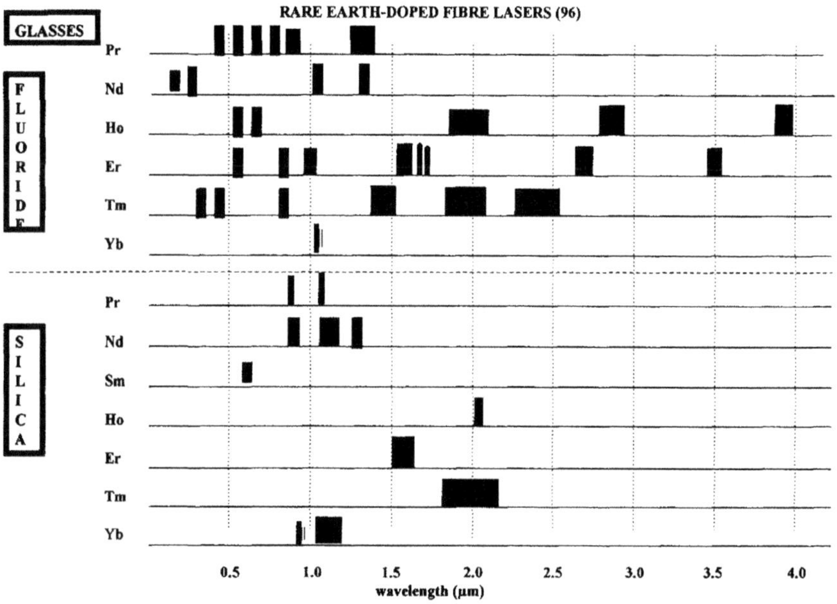

Figure 1. Fiber laser or amplifier operation achieved in the two main silica and fluoride materials (see the text for the different pumping schemes)

During the last decade, much effort has been concentrated on other types of glasses, such as fluorides, because of their potential in exhibiting losses lower than silica in the near infrared range. It turns out that they were much more developed for rare earth doped systems; the main composition (ZBLAN) was particularly suitable for incorporating high concentrations of rare earth ions [14]. Moreover, the low phonon energy of fluoride glasses is well known to increase the number of rare earth metastable states from which they can radiate in the whole infrared or visible or UV range [15]. Thus, laser action has been reported at many more wavelengths in fluoride fibers than in silica fibers. A technical challenge is to achieve UV or blue upconversion pumped lasers with red or infrared pump photons which may be provided by an assembly of standard materials, with respect to other more sophisticated devices involving non-linear properties or blue emitting semiconductor laser diodes [16]. Another reason for the development of rare earth doped fluoride fibers is the discovery of the Praseodymium Doped Fluoride Fiber Amplifier (PDFFA), in 1990, which operates in the second telecommunication window at 1.3 μm [17-19]. Though the impact was not comparable to the EDFA breakthrough, its potentialities for terrestrial communication applications may still be very important, and much effort is devoted to studies of new glass composition with low phonon energy to improve the gain versus power.

The third revolution which is now underway will bring new functions in the fiber such as narrow-band reflection filters, wavelength selective mirrors or even addressable properties... This is possible because of the photosensitivity of species entering in the chemical composition of the fibers as dopants. As early as in 1978, K. O. Hill et al. reported permanent index grating inscription into the core of germanium doped silica monomode fiber obtained by prolonged exposure to highly intense visible laser radiation launched into the fiber [20]. Now, gratings are written by holographic techniques to produce a relative modulation of the refractive index of the order of 10^{-5} up to 10^{-3} [21]. Some rare earth ions such as cerium or terbium have been recently used in fluoride materials where photoionization or oxydo-reduction processes can induce efficient gratings with the possibility of reversibility [22].

This chapter is aimed at providing an overview of the specific spectroscopic properties which are emphasized in rare earth doped fiber lasers or amplifiers with respect to bulk materials already mentioned in other chapters of this book. In the first part, the materials and elaboration methods are presented showing the compatibility of host composition and rare earth doping with pulling techniques giving rise to a limited number of best low-loss single mode fibers. In the second section, rare earth spectroscopic properties are reviewed; emphasis is put on multiple mechanisms which are greatly enhanced as a result of confinement of the laser excitation. Then, amplification properties are discussed within the classical three and four level systems as exemplified by neodymium, erbium or praseodymium ions. Particular attention will be drawn to the limitations of the gain performances versus pump power or versus signal wavelength. In the last section, rare earth doped fiber lasers are reviewed keeping in mind the new extensive possibility to achieve a laser cavity with Bragg gratings and with a special emphasis on upconversion effects. Finally, new trends are schematically presented for the future.

2. Materials and Elaboration Methods

2.1 INTRODUCTION

The waveguide configuration is designed, in general, to produce the best overlap between the pump beam, the signal and the active ion distribution in the core of the fiber. In order to fulfill these conditions, single mode propagation of the electromagnetic waves (pump and signal) must be confined to a restricted section on the order of a few micron square depending on the refractive index difference between core and cladding. The condition for a single-mode operation is expressed by the relation $V < 2.405$ where V, the normalized frequency, establishes the relationship between the core radius a, the index difference Δ and the wavelength as follows:

$$V = \frac{2\pi n a}{\lambda}\sqrt{2\Delta} \qquad (1)$$

For example, for a single-mode operation at 900nm, the core diameter of the fiber should be 7.3 µm if n=1.5 and Δ=0.002. It is important to note that this condition is sometimes difficult to achieve because of the wavelength difference between pump beam and signal, for example in upconversion pumped lasers or in the EDFA with the pump beam setting at 980 nm. Overlap of the TM_{00} mode of the pump beam with the active ion distribution gives the best compromise for the laser threshold P_{th}, which is proportional to:

$$P_{th} \propto \frac{A}{\sigma\tau P} \tag{2}$$

where σ and τ are the stimulated cross section and the lifetime of the metastable state respectively, P the absorbed power and A the core area.

Since the evanescent part of the pump wave can extend into the clad, other waveguide configurations have been considered, such as a multimode pump guide using ray-propagation into the clad with a variety of circular or elliptical cores. This design may provide excellent output laser power with a diode pump laser.

As noticed from equations 1 and 2, mode confinement as well as rare earth ion distribution concentrated at the center of the core contribute to lower the pump threshold. It is clearly required to optimize elaboration methods with appropriate materials to fabricate single-mode fibers. As a matter of fact, this requirement is also necessary for high bit rate transmission.

2.2 MATERIALS

Silica based glass composition is the most popular one to draw single mode fiber though it may be not the best host for rare earth ions. Indeed, an extensive compilation of optical properties of rare earth doped glasses shows that multicomponent glass, phosphate glass or fluoride glass exhibit higher absorption and emission cross-sections, parameters relevant to quantify laser potential [23]. Furthermore, they may accept rare earth concentrations higher than silica with a subsequent line broadening in favor of the Wavelength Division Multiplexing (WDM) application [12]. From a historical point of view, very weak background attenuation was first reached in the silica host using MCVD techniques. In many applications based on fluorescence quantum efficiency, it is evident that increasing the concentration of active ions is a worthwhile approach as long as ion-ion interactions are negligibly small. This is an important consideration because we already mention the necessity to concentrate the active ion at the center of the core of the fiber to achieve the best pump conditions. The relation between fluorescence efficiency and concentration has to deal with rare earth solubility for which the glass net must adopt the highly coordinated environment, between 8 and 12, required for rare earth ions [24]. If the network is compact, as the tetrahedra of silica is, rare earth ions are encased, and it is easily assumed that clusters rapidly occur even at concentration as low as ppm and a fortiori with increasing concentration. As a general consequence of increasing concentration, ion-ion interactions are favored, inducing energy transfer processes which contribute to losses or to optimized conditions. However, clustering effects are difficult to really characterize; unless they are directly detected by imaging the real structure (from electron microscopies or others), indirect measurements through theoretical models are required. This will be illustrated by the spectroscopic studies of several examples concerning the 1.5 μm or 1.3 μm amplifier as well as upconversion fiber lasers.

As far as oxides are concerned and from a structural point of view, several classes of glasses can be considered. The most important one for fiber application is the silica based glass for which the tetrahedral net can be slightly modified by germanium oxide (few percent) or phosphorus oxide (less than a percent) to achieve waveguide conditions. However, such compositions are not well designed for high rare earth concentrations because of clustering. Arai et al. have nicely shown clustering effect in neodymium doped silica glass, looking at fluorescence dynamics of the well known 1.06 μm emission [25]. Indeed, a complex decay is observed, including a fast component due to pairs or higher order aggregates (fluorescence dynamics will be detailed later); this behavior can be removed when Al_2O_3 enters in the host compositions to facilitate RE ion dispersion. It is known that Al_2O_3 is soluble in SiO_2 while RE_2O_3 is not and RE_2O_3 is soluble in Al_2O_3. Furthermore, alumina has six-fold coordinated aluminum ions intermediate to the fourfold coordination of silica and 8/9fold coordination of rare earth oxides. It is generally admitted that an RE concentration of one thousand ppm is a good compromise between absorption or scattering losses and the performance of the optical amplifier. But it is important to note that concentration quenching effects can be very different from one rare earth ion to another. Other parameters such as fluorescence line shape or excited state absorption or upconversion mechanisms will be considered in a later section. The addition of modifier ions (Li, Na, K or Be, Mg, Ca...) has produced a wide range of multicomponent glasses for which neodymium laser action has been reported and classified [23].

Though they have not be considered for fiber applications, other RE doped oxide glasses (germanate, borate...) have been extensively studied for their laser properties. Among them, phosphate and fluorophosphate glasses exhibit remarkably high stimulated emission cross sections; their chain-like structures contribute to accommodate high coordination number required for RE ions as well as to allow screening effects between RE ions which tend to limit concentration quenching as compared to the closed compact structure of other glasses. Despite these qualities, only a few results have been reported for phosphate or fluorophosphate glass fiber lasers [78] due to the presence of OH^- ions which both limit the transmission of the fiber and reduce the quantum efficiency of rare earth ions due to coupling with the high phonon energy of the OH group.

Fluoride glasses and more specifically Heavy Metal Fluoride Glasses, discovered by a French group at the University of Rennes [26], represent a new trend in the domain of optical waveguide application for the major reason that a much wider transparency than silica is expected because of low phonon energy. However, experimental losses due to absorption and scattering are still much higher than theoretical predictions (0.01dB/km between 2 μm and 3 μm) and close to the silica value without considering the reliability of optical quality or optogeometrical parameters or mechanical strengths [28].

Contrary to silica fiber, fluoride fibers are obtained from bulk materials; they are fabricated by melting the high purity fluoride starting material (or obtained by fluoridation) at high temperatures depending on the composition (850°C for ZBLAN, 580°C for BIG) in platinum or vitreous carbon crucibles under an inert atmosphere. Small quantities of oxidizing gas, either oxygen or HF, are used in the flow and ammonium bifluoride is generally added to burn off any residual organics or oxide impurities of the starting materials. It is important to note that any kind of small size particles arising from impurities can induce strong scattering effects which reduce the transparency.

Table 1: Non-radiative relaxation parameters of rare earth-doped glasses.

Host	C ($10^9 s^{-1}$)	$\alpha(10^{-3} cm^{-1})$	$\hbar\omega$ (cm^{-1})
Borate*	2900	3.8	1400
Phosphate*	5400	4.7	1200
Silicate*	1400	4.7	1100
Germanate*	34	4.9	900
Tellurite*	63	4.7	700
ZBLAN**	8.4	5.2	500
BIG**	9.4	5.3	450
Chlorofluoride**	0.44	4.7	375
Sulphide*	10^{-3}	2.9	350
Chalcogenenide	NA	NA	250-350

* from [23] and ** from [27]

At the present, fluoride fibers are not challenging silica for long haul distance telecommunications. But, because of the low phonon energy and the capacity to accept high rare earth concentration, RE doped fluoride fibers have created an upsurge of interest, from 1990, in the field of solid state lasers (upconversion pumped lasers for example) and optical amplifiers (see figure I.1). The first composition was ZBLA (Zr, Ba, La, Al) which is greatly improved by adding NaF for fiber application; the typical ZBLAN composition refers to mole percentages: ZrF_4 53%, BaF_2 20%, LaF3 4%, AlF3 3%, NaF 20%, Slight variations of the chemical composition is generally used to match index refraction differences between clad and core or by the addition of PbF_2 into the core or partially replacing Zr by Hf into the clad [28]. The glass formation and stability with respect to crystallization limits the extent of fluoride glass composition to only a few other examples such as fluoroaluminate or fluorozirconoaluminate. Special attention will be drawn to indium based HMF glass because they offer new possibilities for optically active waveguides due to low phonon energy, higher stability with heavier rare earth forming material and high RE doping levels compared to ZBLAN glass [29].

Finally, within the trend to find new glass compositions exhibiting low phonon energy, mixed sulfide glasses (with lanthanum and gallium) have been recently considered with some promising features, especially as a host for optically active RE ions [30]. Other compositions including chlorofluorides [31] and other chalcogenides [32] either are thermodynamically unstable or are hygroscopic or do not host RE ions easily.

2.3 FABRICATION OF FIBERS

As mentioned before, low loss communication fibers e.g. silica based optical fibers, were readily available with the use of the well known Modified Chemical Vapor Deposition techniques MCVD. The situation is very different for HMF glasses which all start from melting techniques and crucibles.

2.3.1 Silica Fibers

There are two main categories of fabrication of silica preforms depending upon the method of the reaction of the halide components ($SiCl_4$, $GeCl_4$...) to produce mixed oxides on a mandrel or in a tube. In the first case, the reaction occurs in a hydrogen flame and oxides formed in the soot are deposited on a boule; in the second more used technique, chlorides are simultaneously reacting and deposited inside of a substrate tube using either a torch plasma or microwaves. Index changing (either raising or lowering) doping ions are introduced into the reaction stream as germanium or phosphorus or aluminum halide vapors carried by oxygen just above room temperature. Then, by heating one end around 2000°C, the hollow tube collapses to create a preform the dimensions of which are of the order of a millimeter for the core and more than a centimeter for the cladding depending on the desired optogeometrical parameters of the fiber.

Rare earth doping requires specific techniques because of the high temperature vapor pressure of rare earth halides. In the MCVD configuration, several methods have been successfully used: heated sources of chloride (heated frit, or injector...), aerosol (with ultrasonic nebulizer) or organic compound (chelate) delivery. Special care is taken for keeping the low OH content in the range of ppm. Because of the vapor pressure of rare earth chloride, more important techniques are the solution doping techniques which have been used for VAD, OVD and MCVD. Aqueous or, even better, ethyl alcohol or other organic solvents solutions impregnate a porous SiO_2 layer deposition; then, drying is performed under chlorine and oxygen gas flow before the preform collapses, in the case of MCVD. More recently, a sol/gel dip coating process has been used to form, inside the substrate tube, a micron-size gel layer whose thickness is controlled by the viscosity of the gel.

For more detailed information, the reader may refer to several general reviews already published [33].

2.3.2 Fluoride Fibers

Fabrication of HMF glass fibers is only based on glass melting techniques. Many attempts have been done to achieve vapor deposition without success, until now. This is because all the elementary components of HMF exhibit very different vapor pressures versus temperature (see for example the ZBLAN composition). The only composition, to my knowledge, fulfilling the thermodynamic conditions is PZG (PbF_2, ZnF_2, GaF_3) which has been recently demonstrated to produce RE doped planar waveguides [34]. Four main techniques have been used to prepare fluoride fibers: built-in casting, rotational casting, suction casting and the double-crucible drawing. They mostly consist of preparing the cladding first and removing the center of a filled mold during the cooling process by different approaches: inverting the mold or rotating the glass melt. In the double-crucible technique, the core glass melt is flowed through the cladding glass melt in the drawing procedure which offers the potential of fabricating long lengths of single-mode fibers. The achievement of low loss fibers has been limited by internal defects such as microcrystals and bubbles in the core or at the interface core/cladding. The best optical transparency lies around 1dB/km at 2.55 μm, recorded on a hundred meter long ZBLAN fiber [28]. 1.3 μm amplification using praseodymium ion has renewed the interest to prepare fluoride glasses with lower phonon energy than ZBLAN because it increases the fluorescence quantum yield of the laser transition. This is the case of the indium/gallium based fluoride composition for which the praseodymium doped fiber has been drawn with encouraging attenuation and amplification properties [35].

As mentioned in the bulk material section, the optical properties of praseodymium doped low-phonon energy glasses are extensively studied, having found some promising decay times at 1.3 μm, more than three times of the ZBLAN:Pr value. Unfortunately, no single mode fiber of such composition has been drawn until now.

3. Rare Earth Spectroscopy of Glasses

Rare earth doped materials have all the needed criteria of optical properties to achieve laser operation or amplification. Indeed, they possess strong absorption bands and sharp fluorescence lines in the whole ultraviolet (UV), visible (VIS) and near-infrared (IR) range. They generally exhibit numerous metastable states depending upon the host and a reasonably high quantum efficiency for the fluorescent transitions of interest. This is because of the well shielded 4f electrons responsible for these transitions. All these statements are particularly verified for RE doped crystals in which well-defined sites are occupied by RE ions as a result of cation substitution. On the contrary, optical transitions in glasses are between ten and hundred times broader than in crystals. This inhomogeneous broadening results from the disordered structure of the glass which involves a distribution of RE sites arising with different crystal field strength. These considerations are very important because they allow us to explain limited stimulated emission cross-sections, on one hand, but easy diode pump capability and wide tuning range of amplified stimulated emission, on the other hand. In that respect, special attention will be paid to broadening and relaxation processes as well as the non-linear population dependent mechanism which are greatly enhanced in the confined structure of a waveguide. In the following, the general optical properties of trivalent rare earth ions are featured with an emphasis on the determination of parameters relevant to laser applications.

3.1 ELECTRONIC STRUCTURE

In condensed matter, rare earth ions appear mostly as trivalent cations (with some remarkable exceptions: Sm^{2+}, Eu^{2+},Yb^{2+} and some tetravalent Ln^{4+} ions) which implies that the optical transitions occur within the $4f^n$ electronic configuration (see [36] and Chapter 2 of this book for detailed developments). The electronic structure and the maximum degeneracy of each trivalent ionic configuration depend specifically on the number of 4f electrons when the three external electrons (2*6s and 1*4f or 1*d) are removed. The shielding of the 4f electrons with respect to ligand field is responsible for the small energy shift of the electronic levels when the rare earth ions sit in different anionic environment. In other words, the ligand field causes only a small perturbation to the one electron atomic wavefunctions, solutions of the schrodinger equation for the free ion hamiltonian:

$$H = \text{Hconfiguration} + \text{Helectrostatique} + \text{Hspin-orbit} + \text{Hstatic ligand field} \quad (3)$$

The examples of the lifting of degeneracy for Pr^{3+} and Er^{3+} ions are shown in figures 2 and 3. Within the $4f^n$ configuration, the term is represented by ^{2S+1}L notation as a result of diagonalising the two first terms of H. Considering Russel Saunders coupling of the angular momentum and the spin momentum, spin orbit interaction lifts the degeneracy into J states. Then, the total angular momentum J is no longer a good quantum number considering the effective crystal field potential of the lattice. In crystals, it is generally assumed that the rare earth ion sustains the static potential of the substitution site. This electrostatic potential can be expanded in terms of spherical harmonics involving the well known phenomenological $B^k{}_q$ parameters.

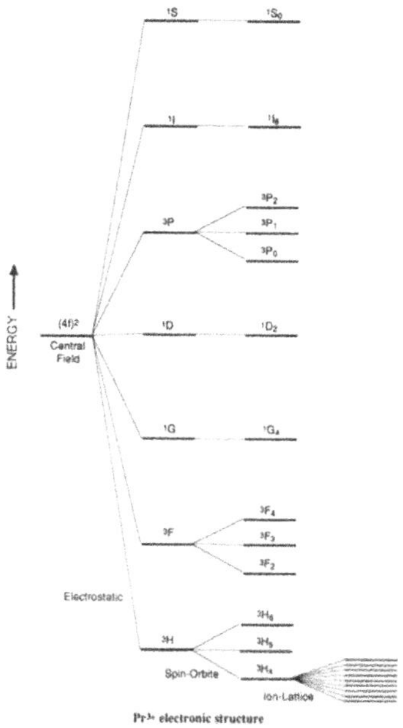

Figure 2: Electronic structure of trivalent praseodymium.

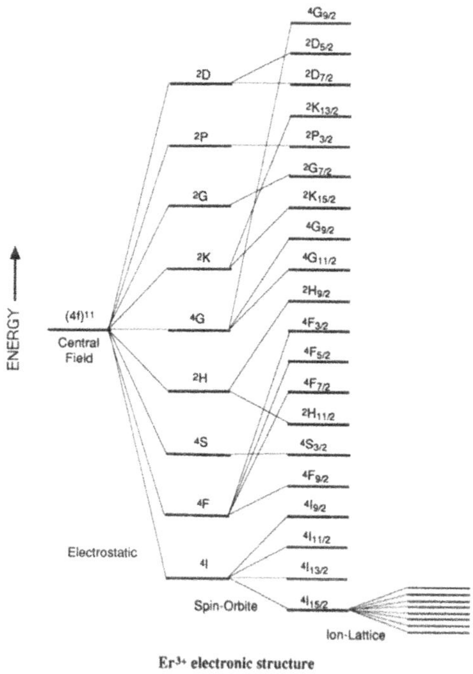

Figure 3: Electronic structure of trivalent erbium.

Only the even-k term in the expansion lifts the J-degeneracy according to the symmetry of the site, assuming that the crystal field hamiltonian is invariant within the point group symmetry at the rare earth site. Then, in the absence of (or with very low) symmetry, the crystal field potential can completely remove the degeneracy of each J-state into Γ crystal field states, except for Kramers doublets appearing for an odd number of electrons in the 4f ionic configuration. Crystal field splitting is generally observed on the order of 10-1000 cm^{-1}.This behavior is also encountered in amorphous solids, including a broadening of the $\Gamma \rightarrow \Gamma'$ transitions due to the distribution of the local crystal field environments of the rare earth ions. It is to be noted that such a distribution behaves roughly as the spread of the energy shift of the transitions in various crystalline environments. If the number of crystal field components is generally well accounted for by symmetry considerations of the crystal field hamiltonian in the point charge approximation, energy shifts are much more difficult to reproduce. Experimental observations and phenomenological approaches have pointed out the character of covalent bonds between a rare earth ion and its first neighbors which is emphasized in the so-called nephelauxetic effect. These empirical considerations are of importance to explain some general trends e.g., covalent glasses (such as silicates) absorb or emit generally at longer wavelengths than ionic glasses (fluorides). As usual, such empirical rules have to be used very cautiously with respect to the nature of the electronic states involved.

3.2 RADIATIVE PROBABILITIES

Interaction of the activator ion with the electromagnetic field gives rise to absorption and emission of photons through electric and magnetic dipole interaction operators:

$$\boldsymbol{D} = \sum e\, \mathrm{r_i} \quad \text{and } \boldsymbol{D} = \sum (\mathrm{l_i} + 2\mathrm{s_i})e/2\mathrm{m} \quad \text{respectively.}$$

In a simple two-level system, the total downward transition rate is the sum of the induced and spontaneous contributions following Einstein [37]:

$$W'_{21} = B_{21}\,\rho(\nu) + A \qquad (4)$$

where $\rho(\nu)$ is the energy density per unit frequency.

At thermal equilibrium, the number of upward and downward transitions are equal; thus, $N_2[\,B_{21}r(n) + A\,] = N_1\,B_{12}\rho(\nu)$, and the ratio N_2/N_1 is given by the Boltzmann factor including the degeneracy dependence g_2/g_1.

Taking into account the expression of the radiation density $\rho(\nu)$ [37], it is required that

$$B_{12} = B_{21}\,g_2/g_1 \text{ and } A/B_{21} = 8\pi n^3 h\nu^3/c^3$$

Then the induced transition rate can be written as:

$$W_{21} = B_{21}(\nu)\rho(\nu) = Ac^3\rho(\nu)/8\pi n^3 h\nu^3 \qquad (5)$$

It is noted that the spontaneous rate A is just the inverse of the radiative lifetime τ.

On the other hand, the probability of spontaneous emission can be expressed as a function of the line strength of the transition S_{21} through the general equation:

$$A = A_{21} = \frac{1}{4\pi\varepsilon_0} * \frac{64\pi^4 nc^2}{2\hbar(2J+1)\overline{\lambda}^3}\left(\frac{E_{loc}}{E}\right)^2 S_{21} \qquad (6)$$

where $1/(2J+1)$ takes account of the degeneracy of the initial state and $(E_{loc}/E)^2$ is a local field correction due to the polarisability of the medium. This factor is equal to $(n^2+2)^2/9$ and n^2 for electric dipole or magnetic dipole transitions, respectively.

S_{21} is directly related to the absorption oscillator strength by :

$$A_{21} = \frac{1}{4\pi\varepsilon_0} * \frac{8\pi^2 ne^2}{mc} \frac{1}{\overline{\lambda^2}} (\frac{E_{loc}}{E})^2 f \qquad (7)$$

Finally, f is known by measuring the absorption k_ν (m^{-1}), using Beer's law:

$$f = 4\pi\varepsilon_0 \frac{mcn}{\pi e^2} (\frac{E_{loc}}{E})^2 \int \sigma_{21}(\nu) d\nu$$

and then,

$$f = 4\pi\varepsilon_0 \frac{mcn}{\pi e^2} (\frac{E_{loc}}{E})^2 \int \frac{k_\nu}{N} d\nu \qquad (8)$$

where $\sigma_{12}(\nu)$ is the absorption cross section (m^{-2}) and N the density of absorbing centers (m^3). f is a dimensionless number which is close to 1 for fully allowed transitions such as the interconfigurational transition (e.g. $f^n \rightarrow f^{n-1}d$ for Ce^{3+}) and 10^{-6} to 10^{-8} for forced electric dipole or magnetic dipole intracongurational transitions in most of the trivalent rare-earth. Indeed, Laporte's selection rule on parity can be removed in a non-centrosymetric environment because the odd parity term of the crystal field hamiltonian can mix ground and excited configurations of opposite parity. That is the reason such f-f transitions are often called forced electric dipole transitions.

Considering the first order perturbation term in the development of the transition matrix element:

$$M \propto \left| -\sum_i [E_{nf}^{-1} \langle g | E * D | i \rangle \langle i | H_{cf}^{odd} | f \rangle + E_{ng}^{-1} \langle g | E * D | i \rangle \langle i | H_{cf}^{odd} | f \rangle] \right|^2 \qquad (9)$$

Judd [38] and Ofelt [39] have separately proposed the so-called closure relation over the i intermediate states as $\sum | i \rangle\langle i | = 1$ because of the closure approximation in which all the energies of the states of the opposite parity configuration are regarded as constants and equal to an average value. It greatly simplifies the mathematics, and the final result of line strengths is written in terms of reduced matrix elements $U^{(\lambda)}$ and the well known Judd Ofelt parameters Ω_λ:

$$S = e^2 \sum \Omega_\lambda \, |<| \, U^{(\lambda)} \, | >|^2 \tag{10}$$

Considering electric dipole and magnetic dipole transitions, we end up with the useful formula:

$$S^{meas}(J \to J') = \frac{9n}{(n^2+2)^2}\left[\frac{3hc}{8\pi^3 e^2}\frac{(2J+1)}{N\bar{\lambda}}\int k(\lambda)d\lambda - S_{md}\right] \tag{11}$$

in which

$$S_{md}(J \to J') = \frac{3h(2J+1)\bar{\lambda}}{8\pi^2 mc} F_{md} \tag{12}$$

where the magnetic dipole oscillator strength has been already calculated and depends on the host.

As usual, room temperature absorption spectra provide a set of integrated absorption coefficients which allow us to estimate the Ω's parameters using a mean square procedure. Knowing the line strength, the individual transition probabilities A are calculated which give the branching ratios β of each of the transitions:

$$\beta = \frac{A(J \to J')}{\sum_{J'} A(J \to J')} \tag{13}$$

and the total lifetime of each excited level is then:

$$\tau_{rad} = \frac{1}{\sum_{J'} A(J \rightarrow J')} \tag{14}$$

with $$A(J \rightarrow J') = A_{ed} + A_{md}$$

Finally, it is straightforward to obtain quantum fluorescence spectra (cross sections in m^{-2}) using the room temperature absorption spectra and applying the Judd Ofelt approximation to compute transition probabilities. Besides the appropriateness of the Judd Ofelt approximation which will be discussed later in the concrete case of Pr^{3+} ions, two complications may arise: the case of an insufficient number of absorption bands and the problem of equal populations in the Stark components of the ground multiplet which is required for the validity of the Judd Ofelt approach. The first complication can be circumvented by using alternative methods [40]. The reciprocity method which is based on the relationship between cross sections $\sigma_{ij} = \sigma_{ji}$ can be easily applied to the $^4I_{15/2} \leftarrow\rightarrow {}^4I_{13/2}$ transitions of erbium ions, taking into account the Boltzmann factor between Stark components of each multiplet [41]. The second method uses the relation between Einstein A and B coefficients which gives rise to the Fütchbauer-Ladenburg equation [42]:

$$\sigma_{em}(\nu) = \frac{c^3}{8\pi n^2 g(\nu)\nu^4} * \frac{1}{\tau_{rad}} \tag{15}$$

where g(ν) is the normalized lineshape function. This should be the simplest method as far as τ_{rad} is known. However, even though it is easy to measure lifetime, a variety of mechanism may affect its value either shortening (by energy transfer) or lengthening it (by trapping effects). A third category takes advantage of the fiber configuration by direct measurements of small signal gain or saturated fluorescence [43].

3.3 NON-RADIATIVE TRANSITIONS

In one-center processes or concentration independent processes, relaxations of excited states can occur via emission of photons or nonradiatively by means of the emission of phonons. It is generally formulated in terms of probability: the emission probability (the inverse of the measured lifetime) is the sum of the radiative probability (determined by the Judd-Ofelt approximation, for example) and of the nonradiative probability. Two basic regimes can be considered depending on the energy gap between the excited state and the next lower states as compared to phonon energies of the material. If more than one or two phonons are necessary to fill the energy gap, there is a competition between light emission and nonradiative relaxation; otherwise very fast nonradiative relaxation occurs predominantly as a result of the electron-phonon interaction. Furthermore, this is because most of the $4f^n$ electronic states exhibit radiative lifetimes longer than a few μs. When more than two phonons are involved, higher order perturbation theory is required. Multiphonon processes are then probable because of the strength of the electron-phonon coupling and of the high density of phonon states, many orders of magnitude greater than the photon states. Besides the energy gap, the number of phonons participating in the radiationless transition and the temperature are the most important parameters. In that respect, multiphonon nonradiative relaxations are strongly material dependent but not as rare earth ion dependent. The empirical theoretical dependence of multiphonon relaxation probability versus the energy gap is known as the "energy gap law" which was firstly applied to crystalline materials [44] and more recently to glasses [45]. The energy gap law express the exponential dependence of the nonradiative rate W_{nr} versus ΔE as follows:

$$W_{nr}\ (\Delta E, T=0\ K) = W_0 \exp(-\alpha\, \Delta E) \qquad (16)$$

where W_0 and α are experimentally determined parameters [46].

Very often W_0 is expressed as $C[n(T) + 1]^p$, where p is the number of phonons matching the energy gap and n(T) is the Bose-Einstein occupation number for the effective phonon mode with ω, generally taken as the cut-off frequency of the material (see table 1). Then,

$$n(T) = 1/\,[\exp(h\omega/kT) - 1]$$

In practice, the temperature dependence of the nonradiative rate taken as the inverse of the measured lifetime allows us to determine C and α parameters which are material dependent and the number of phonons required to fill the energy gap with respect to the phonon cut-off frequency of the material. It is important to note that the energy gap law, depicted in figure 4, is independent of the rare earth ion. A systematic study of nonradiative rates of one rare earth ion embedded in a given material permits us to predict the same information for other rare earth ions. One can see that the non-radiative rates increase with the phonon energy for a given gap, the slope (α) is rather similar for oxides and fluorides but different for chloride and sulfide materials. Ionicity or covalency effects cannot completely explain such behavior. However, it is interesting to note that chlorofluorides behave much more like chloride than fluoride materials. As a matter of fact, fluorophosphates would be very interesting to look at as a function of fluoride concentration; this has not been yet reported.

From figure 4, it is evident that the 1.06 µm and the 1.5 µm emissions of neodymium and erbium ions exhibit almost negligible nonradiative rates. This is not the case for the 1.3 µm emission of Pr^{3+} or Dy^{3+} which are candidates for optical waveguide amplifiers [17,47]. Considering the energy gaps below the metastable states $^1G_4$ and $^6H_{9/2}$, roughly 2700cm^{-1} and 1800cm^{-1} respectively, low phonon energy materials are required, such as fluorides or, even better, chalcogenides or sulphides (especially with Dy^{3+} [47]).

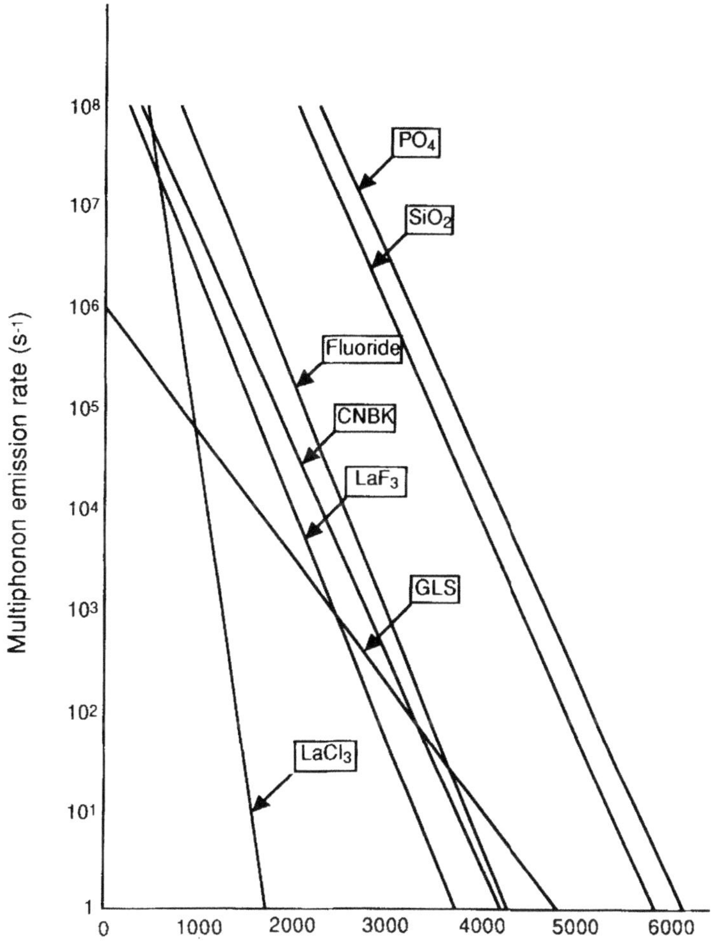

Figure 4: Energy gap law for a number of glasses, two crystalline materials have been included for comparison.

3.4 INHOMOGENEOUS AND HOMOGENEOUS LINE BROADENING

An interesting feature of RE-doped glass spectroscopy appears in the observation of absorption and fluorescence lines ten or hundred times broader than the ones measured in crystals [48]. In crystalline hosts, RE ions generally substitute cations embedded in quasi identical crystal field environments; then, the inhomogeneous broadening always present in real crystals is much less than the crystal field splitting of the ground and excited states involved in the transitions. At room temperature, Raman processes (with a temperature dependence of T^2 at intermediate and high temperature) which are the dominant line broadening mechanisms, result in homogeneous line broadening. On the other hand, the disordered structure of the glass creates a wide range of crystal field environments at the rare earth site which, in turns, induces a broad spectral distribution of transition energies. As a consequence, inhomogeneous line broadening may be comparable to the crystal field splitting or even larger; this is remarkably exemplified by comparing the fluorescence linewidths of the 1.5 μm and 1.06 μm laser transitions of erbium and neodymium-doped silica fibres, respectively. Fluorescence line narrowing experiments at low temperature have allowed us to measure the crystal field splitting of the ground state in the range of 20 to 50 cm^{-1} for erbium and more than 60 cm^{-1} for neodymium while site distributions on the order of 40-50 cm^{-1} have been estimated [49]. This question of homogeneous/inhomogeneous linewidth should be addressed carefully because it can have a drastic influence on gain tunability and laser stability. Indeed, in an inhomogeneous line profile, amplification or laser action taking place in a class of sites (that is to say at a specific wavelength) are readily related to stimulated emission cross sections and excited state lifetimes. They may differ from one class of sites to another; also, their different saturation properties may cause instability in the laser cavity. A clear advantage of large homogeneous line broadening for optical amplifiers is their potential to transmit and then to amplify signals at different wavelengths. Furthermore, many efforts are concentrated in keeping the gain constant over a broad range of wavelengths for WDM applications and very promising results of the eight or sixteen individual signals simultaneously amplified within the same EDFA have been recently obtained by several groups [12]. In turn, it raises the interesting fundamental problem of the relationship between the homogeneous line broadening and the chemical composition of the glass. Surprisingly, the best result of flat gain over 30 nm has been obtained in an erbium-doped ZBLAN fluoride fiber amplifier [50] while 13 nm is measured in a silica based EDFA with a high concentration of aluminum. At the present, there is no clear explanation but the crystal field splitting of the $^4I_{13/2}$ and $^4I_{15/2}$ states of Er^{3+} may differ notably between Al/Ge silica glass or Ge silica glass and fluorozirconate glass.

3.5 EXCITED STATE DYNAMICS AND ION-ION INTERACTIONS

As a first simple idea, one may guess that increasing pump power or active ion concentration should result in a maximum fluorescence quantum efficiency, but it is known that some specific electronic transitions suffer from competing processes due to transition energy coincidences (between excited states: ESA or ground state: GSA) involving single or two centers.

As a best illustration, ESA affects the $^4F_{3/2} \rightarrow ^4I_{13/2}$ transition of neodymium ion around 1.3 μm. Indeed, that transition energy almost coincides with the $^4F_{3/2} \rightarrow ^4G_{7/2}$ ESA transition but with a smaller absorption cross section than the 1.3 μm emission cross section which results in a weak but measurable gain or laser action in a number of crystalline or glassy systems. Unfortunately, the best gain obtained, mainly in fluorozirconate materials, is limited to roughly 10dB and its maximum occurs at wavelengths longer than 1.32 μm (between 1.34 and 1.36 μm) which is unusable for the second telecommunication window [50]. An impressive number of glassy materials have been investigated to improve the performance and shift the gain maximum without success [23]. The main reason is that ESA is well centered on the high energy side of the $^4F_{3/2} \rightarrow ^4I_{13/2}$ spectral transition of neodymium ions, as well as the subsequent line broadening, whatever the material; also, this emission suffers from an unfavorable branching ratio as compared to the $^4F_{3/2} \rightarrow ^4I_{9/2}$ and $^4I_{11/2}$ transitions. The Pr^{3+} 1.3 μm emission also exhibits ESA and GSA which perturb the long wavelength side of the IR emission, fortunately.

Another example is ESA of the 800nm pump of the EDFA. In that process, pump photons are used both for GSA or for an efficient two-step excitation due to a good match between $^4I_{15/2} \rightarrow ^4I_{9/2}$ and $^4I_{13/2} \rightarrow ^4F_{3/2}$ and $^2H_{11/2}$ electronic transition energies of the erbium ion [23]. This is one of the reasons that the 800nm pump wavelength is less efficient than 980nm or 1.48 μm for EDFA and does not compete for commercially available systems, though powerful laser diodes are more reliable and less expensive at 800nm.

The other parameter which plays an important role in phosphors in general is the so-called concentration quenching effect which is directly related to ion-ion interaction. For that purpose, several conditions have been extensively considered: the energy mismatch (as above for ESA) between initial and final states of both ions, the interaction length which depends on the rare earth concentration, the nature of the interaction (either electric dipole (long-distance range type) or electric multipole or exchange (short-distance range type)) and the nature of the ligands (acting as bridging species between two active ions) [51].

Dynamics of the population of the $^4F_{3/2}$ state of Nd^{3+} is dominated by a non-resonant cross relaxation mechanism ($^4F_{3/2}\rightarrow{}^4I_{15/2}$ on one ion and $^4I_{9/2}\rightarrow{}^4I_{15/2}$ on another) as the Nd^{3+} ion concentration increases. However, the concentration quenching range may vary from a percent (YAG crystal, silicate glasses...) to more than ten percent (phosphate crystals...) of neodymium ions depending not only on distances but on the geometry and ionicity of the ligands. Moreover, the non-homogenous rare earth distribution is especially notable in the rare earth-doped silica-type fiber which has the effect of increasing concentration quenching to concentrations smaller than one percent. However, the cross relaxation effect may contribute to improve the fluorescence efficiency. This is nicely exemplified by the case of the 2 μm Tm^{3+} laser emission in oxide crystals for which a thulium ion is pumped into the $^3H_4$ level around 800nm providing two ions in the $^3F_4$ state by cross relaxation energy transfer, known as a 1 to 2 process (the photonic quantum rate can be ideally equal to two!). On the other hand, the reverse process, that is to say: two to one, has been proposed to explain the absence of a saturation effect of the 1.5 μm amplified emission of EDFA (silica fibers) as the pump power increases [52]. Indeed, at high excitation density providing two erbium ions in the $^4I_{13/2}$ excited state, ion-ion interaction may promote one ion to a higher excited state and the other to the ground state; the process does not reach a saturation regime because the upper state relaxes either radiatively to the ground state or non-radiatively to the $^4I_{13/2}$ state from which it emits a photon. From this and from the ESA mechanism, it is evident that optimization of such upconversion processes provides other pump channels for solid state lasers; this is not a new idea, but has been usefully demonstrated only recently.

In the following figure, all the main mechanisms discussed above have been schematically drawn in figure 5. The second part of the graph shows some less probable mechanisms called cooperative energy transfers (which is often inappropriately used for all upconversion energy transfers). They involve two excited ions and a "virtual" state (cooperative sensitization) or two "virtual" excited ions and a real excited state (cooperative absorption). Since they deal with a "virtual" (non-equilibrium) state, these processes shall be considered for some specific energy level diagram (terbium ions pumped in the infrared range or visible Yb^{3+} luminescence...) or when the excitation density is high in the range of the optical non-linearity of the material.

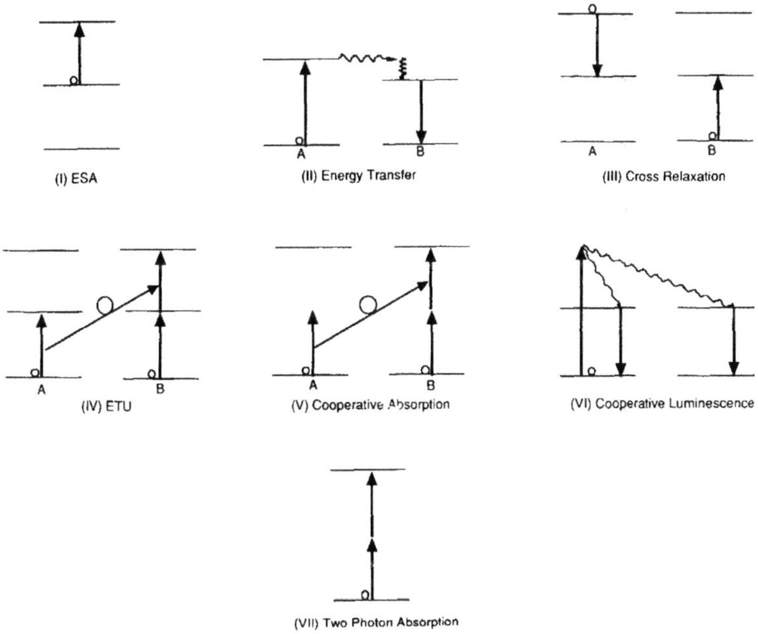

Figure 5: Schematic diagrams for different upconversion pump mechanisms; they are less and less effective from the top to the bottom.

It is important to distinguish between all these processes; concentration dependencies can provide a good tool to separate one and two ion processes. The power dependence of the fluorescence intensity is generally not sufficient and, very often, it is better to analyze fluorescence dynamics under short pulsed excitation. The simplest example is to compare ESA and two ion upconversion energy transfer (APTE); both upconverted fluorescences are quadratically dependent on pump power but they differ in concentration dependence (not always available) or from their decay profiles, one exponential law in the case of ESA (with a time constant equal to the upper excited state lifetime) and a risetime followed by a decay profile in the energy transfer process case. Often, an intermediate situation is observed and a branching ratio of the two processes can be extracted from decay data. Fluorescence dynamics of cooperative luminescence is somewhat more complex and the rate equation model developed and justified in the case of dilute or fully concentrated materials may not be adapted.

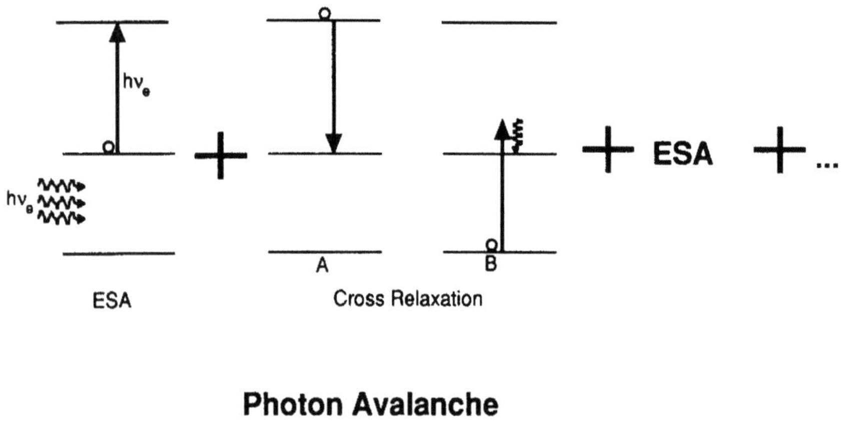

Figure 6: Schematic diagram for photon avalanche: it is a combination of very weak GSA, a strong resonant ESA followed by a cross relaxation, then again ESA + cross relaxation resulting in two, four...excited ions in the intermediate state.

Finally, a relatively new mechanism reported for the first time in 1979 introduces a novel channel for up-conversion lasers which was firstly demonstrated in 1989 [53,54]. Since that time, the number of laser systems has grown and photon avalanche is now observed in rare earth-doped waveguides [55,56], but not yet in laser operation. There are two main features for characterizing the so-called up-conversion avalanche process: (i) a very weak GSA and simultaneously a strong ESA (ii) an efficient cross relaxation which depletes the upper excited state to create a population reservoir (see figure 6). Then, a single laser beam resonant with a strong ESA is weakly absorbed as a first step, ESA occurs followed by a cross relaxation which creates two excited ions into the reservoir level, then the system becomes self-fed (GSA is no longer necessary) until the reservoir is almost filled, favoring the radiative recombination from the upper excited state [57]. Experimentally, absorption and upconverted luminescence intensities have non-linear responses with a clear power threshold and a steep power dependence above the threshold (see figure 7). A typical temporal behavior is also observed for each of the regimes (below and above threshold) which are nicely reproduced by a three level rate equation model (see figure 8).

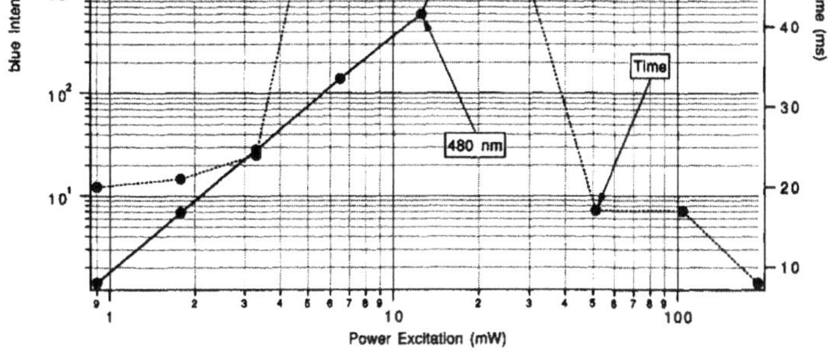

Figure 7: Graphs of 480 nm fluorescence intensity (left) and fluorescence rise time (right) versus 633 nm pump power in Tm:ZBLAN fiber [56].

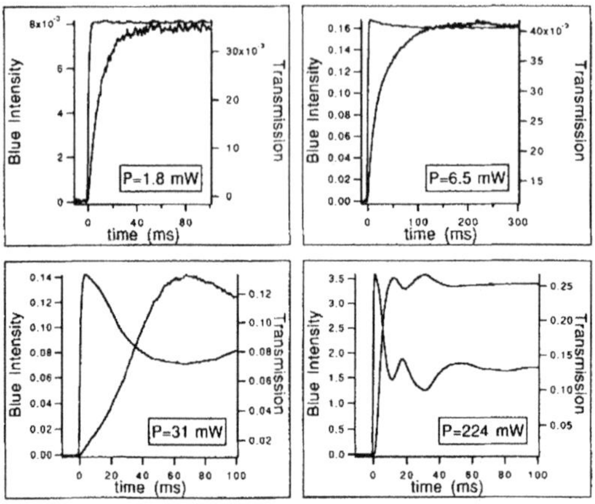

Figure 8: Examples of fluorescence intensity (left) and pump depletion (right) as a function of time and for different pump power in Tm:ZBLAN fiber [56]

The weak GSA or the small cross section of the first step of the avalanche process is well adapted to waveguide configurations as well as the easily achieved high excitation density, but concentration effects and radiative reabsorption versus the length of the waveguide should be optimized. Finally, we have pointed out the negative effect of GSA for the avalanche; this is seen in glassy materials where broad band absorption exists due to the disordered structure [58].

3.6 PHOTOSENSITIVY AND NON-LINEAR EFFECTS

Glass photosensitivity results in a permanent change of the refractive index of the glass with light irradiation. Recent work has been devoted to the underlying mechanisms of the photosensitivity of germanosilicate fibers and other related compositions [21]. This is particularly helpful for developing new methods for writing gratings for very narrow band filters and highly monochromatic laser sources for WDM applications, for example. In germanosilicate fiber, the role of germanium has been pointed out by considering the germanium oxygen defect center absorption band near 243 nm. However, refractive index changes as large as 10^{-3}-10^{-2}, obtained under low temperature high pressure H_2 loading, cannot be fully explained by using a Kramer-Kronig relationship applied to induced absorption [59]. Several models have considered local compactness effect [60] or stress release [61] and corrugations have been recently observed at the surface of GeO_2 and SiO_2 preforms after Bragg grating inscriptions which cannot fully account for refractive index change assuming plastic and elastic strains [62].

Other photosensitivity-inducing dopants have been considered because of the need for a better knowledge of the fundamental processes involved and because incorporation of a high rare earth concentration requires the presence of alumina in the fiber core, as discussed above, and the development of germanium-free phosphosilicate based waveguides. As a result, the photosensitivity of rare earth ions in optical fibers has been recently observed, though it was discovered more than ten years ago in multicomponent silicate glasses. Models involving either a structural change of the glass or a change of glass polarisability via charge transfer processes are currently investigated. It turns out that refractive index changes have been measured for most of the rare earth ions incorporated into aluminosilicate fibers ; again, H_2 loading enhances the photosensitivity effect [63].

Finally, it has been very recently reported that rare earth-doped heavy metal fluoride fibers [64] and waveguides [65] exhibit photosensitivity effects with refractive index changes as high as in H_2 loaded germanosilicate fibers. Again, absorption change due to photoionisation species involving rare earth ions do not explain refractive index changes of the order of 10^{-3}-10^{-2}. These exciting results and the tentative description of the mechanisms make this field very attractive for applied and fundamental research.

Non-linear properties are also enhanced in optical fibers as a result of the confinement of laser light over a long range interaction with media. The most well known effect is the stimulated vibrational Raman shift due to second order susceptibility which has been observed a long time ago in silica and fluoride fibers providing a quasi continuum of light at wavelengths longer than the laser excitation. Brillouin scattering as well as the stimulated Raman effect combined with Bragg gratings offer a nice opportunity to achieve amplification with high gain in a long fiber. Electronic Raman shifting in cerium-doped ZBLAN fiber has been measured, but no stimulated effect has yet been reported [66]. $\chi^{(2)}$

and $\chi^{(3)}$ effects have been identified in rare earth-doped silica fibers with non-negligible doubling frequency efficiency in neodymium-doped silica fiber or generating sum or difference of frequencies as a result of parametric interaction or four-wave-mixing. Much effort is now addressed to the investigation of glass composition achieving a high effective non-linear coefficient d_{eff} of the non-linear susceptibility.

A more fundamental breakthrough appears with the optical soliton in fibers as a result of a solitary wave for which the non-linear propagation of light (Kerr effect) compensates the dispersion of the group velocity (bringing opposite contributions to the non-linear schrodinger-type equation of motion). It is related more to the amplitude modulation of a signal then to pulse propagation. Now, it is understandable that pulses as short as hundred of femtoseconds may propagate without spectro-temporal modifications allowing very long distance transmission if the signal is amplified periodically. 20,000 km transmission at 1.5 μm in a regenerating loop with a cascade of EDFAs has been reported showing the potential of virtually unlimited transmission distances with error and penalty-free propagation at 10 GBits [67].

4. Fiber Amplifiers

In this section, we present the basic concept of a fiber amplifier with theoretical expressions for optical gain. A discussion is devoted to a review of 1.5 μm and 1.3 μm fiber amplifiers which represent typical examples of a three-level laser system and a four-level laser system (figure 9). Instead of developing the whole mathematical formalism of gain curves, saturation effects, noise figures of merit, etc... which are reported in recent excellent text books [23, 67], we report the qualitative procedures and discuss the major spectroscopic approximations.

4.1 INTRODUCTION

Optical gain is generally achieved by coupling a pump beam of wavelength λ_p into a core fiber for which the radial rare earth distribution is assumed constant and does not vary along the length l of the fiber. In general, the optical gain of a laser material with population inversion can be modeled by the well-known rate equations which describe the relationship between the pump and signal photon densities and the population densities. Pump and signal beam intensities are functions of both the radial and longitudinal positions in the fiber; mode profiles and propagation (forward and backward) of the beams must be accounted for. The calculations can be very complicated, and numerical resolutions are generally required for the exact solutions. Two categories of approximations have been applied depending on the waveguide and spectroscopic characteristics. Furthermore, pump power and signal power considerations cause several regimes to exist: the small signal gain and saturation regimes.

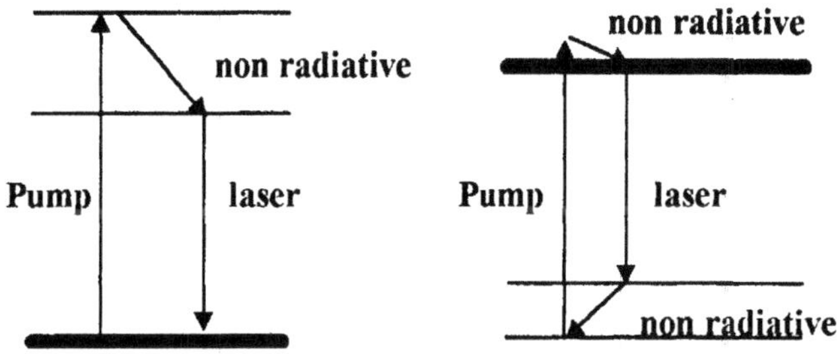

Figure 9: Schematic diagram of the three and four level laser systems.

The interaction between light and atom in the waveguide is a local effect of the field rather than an average of the propagating power mode. In the case of weakly step-index fibers, the LP_{01} mode profile is a good approximation of the fundamental mode as defined by Bessel functions. The gaussian approximation of the LP_{01} mode based on power coupling efficiency has been widely used to provide analytical solutions of the optical gain. However, this LP_{01} gaussian approximation does not reproduce correctly the exact Bessel functions and it is better to use the LP_{01} gaussian envelope approximation which is defined by equating the 1/e power mode radius and the power radius of the Bessel function.

The second category of approximations concern the atomic system interaction with the exciting light. Ideally speaking, three-level or four level systems are generally considered good models of an EDFA (980nm or 1480nm pump wavelength) and a neodymium amplifier at 1060nm (Figure 9). However, in many cases, these models fail to reproduce laser or amplification characteristics because of the complexity of the atomic system. This is particularly true when resonance occurs between the excited state energy gaps of the activator and pump and / or signal photon energies, or when ion-ion interactions are strong enough to induce losses in the fluorescence quantum yield.

4.2 THEORETICAL BACKGROUND

Let us consider the case of a "pure" three-level or four-level system. The steady state populations involved in the laser transition, called N_1 and N_2 for the final and initial levels of the laser transition, are only functions of the absorption W_{abs} and emission W_{em} transition rates and the lifetime of the metastable state and the pump rate W_p. Assuming that the total population $N_0=N_1+N_2$, the solution of the rate equations provides the expressions of the population densities N_1 and N_2:

$$\frac{N_1}{N_0}=\frac{W_{em}+1/\tau_2}{W_{abs}+W_e+\frac{1}{\tau_2}+W_p}$$

$$\frac{N_2}{N_0}=\frac{W_e+W_p}{W_{abs}+W_{em}+\frac{1}{\tau_2}+W_p} \qquad (17)$$

with $W_p=\sigma_p I_p/h\nu_p$, $W_{abs}=\sigma_s I_s/h\nu_s$ and $W_{em}=\sigma_{se} I_s/h\nu_s$ where σ_p, σ_s and σ_{se} are the pump and signal absorption cross sections and the stimulated emission cross section, respectively.

The pump intensity evolution $I(r,\phi,z)$ along the fiber is described by:

$$\int_0^\infty \int_0^{2\pi} \frac{dI_p}{dz} r dr d\phi = -\int_0^\infty \int_0^{2\pi} \sigma_p N_1 I_p r dr d\phi \tag{18a}$$

using cylindrical coordinates.

Similarly, the evolution of the signal can be written as follows:

$$\int_0^\infty \int_0^{2\pi} \frac{dI_s}{dz} r dr d\phi = \int_0^\infty \int_0^{2\pi} (\sigma_e N_2 - \sigma_a N_1) I_s r dr d\phi \tag{18b}$$

Assuming a small-signal regime, i.e., $W_s \cong W_e$, the exact solutions of the previous equations have been derived:

- For the pump power evolution:

$$\frac{dp}{dz} = -\frac{2\pi N_0 h\nu_p}{\tau\, P_p(0)} \int_0^a B_n(r,z) r dr \tag{19}$$

$$B_n(r,z) = 1 - \frac{1}{1 + \dfrac{\sigma_p \tau_2 P_p(0)}{h\nu_p} p(z) r_n(r)}$$

for n=0, and

$$B_n(r,z) = 1 - \frac{1}{(1+\frac{\sigma_p \tau_2 P_p(0)}{h\nu_p} p(z) r_n(r))^{1/2}}$$

for n>0

in which $r_n(r)$ is the radial dependence of the pump power in the mode n and $P_p(0)$ is the input power.

- For the signal evolution, the gain, defined as $g_0(z)=\ln(s(z))$, has the form:

$$\frac{dg_0}{dz} = -\sigma_a N_0 \eta_s + \sigma_e N_1 (1+\frac{\sigma_a}{\sigma_e}) 2\pi \int_0^a S_0(r) B_n(r,z) r dr \qquad (20)$$

where η_s is the fractional signal power in the core, and $s_0(r)$ is the normalized power density in the signal mode.

We note that the ratio σ_s/σ_{se} is equal to zero in the four-level system but one in the three-level system. In the EDFA case, this ratio is rather close to zero because of the signal wavelength dependence of the cross sections.

4.3 INFLUENCE OF PUMP MODE

Only numerical solutions of the equations above are available but approximate solutions can be obtained accounting for the spread of the pump mode. For instance, it has been proposed to consider a uniform pump distribution $R_p/A\eta_p$ into the core and zero outside where A is the core area and η_p the fractional pump power in the core.

Then. the approximate solution for the small signal gain at a length l is:

$$g_0(l) = -\sigma_a N_0 \eta_s l + (1 + \frac{\sigma_a}{\sigma_e}) \frac{\sigma_e \tau_2}{h\nu_p} \frac{P_{abs}}{A} \frac{F_n}{\eta_p} \xi \quad (21)$$

where P_{abs} is the absorbed pump power at l, F_n is the normalized overlap integral between the pump and the signal in the core of the fiber, and the parameter ξ accounts for the effect of ground state depletion.

In the case of the four-level system, $g_0(l)$ is again formally reduced by setting σ_a=0. Finally one of the key parameter is $\frac{F_n}{\eta_p}\xi$ which contains the overlap between the propagation characteristics of the waveguide and the doping radial profile in the active media. As mentioned before, other approximations such as a gaussian envelope allow us to develop analytical solutions in the small-signal regime and unsaturated conditions. It corresponds to a number of practical applications or to the behavior of the laser oscillator at threshold.

A great amount of work has been devoted to the development of analytical models useful for the investigation of numerous parameters involved in the fiber amplifier and fiber laser, such as absorbed pump power (backward and forward), gain dependence on the concentration of dopants, as well as mode profiles, noise figures, etc... From these works, it has been possible to fabricate erbium-doped fiber amplifiers which correspond to the best solutions for practical applications of telecommunication devices.

4.4 INFLUENCE OF FLUORESCENCE DYNAMICS

Now, we shall discuss the approximations used for the description of the atomic system. Because of the richness of the RE energy level diagram and of the broadening of the electronic transitions in the glass media, pump and signal wavelengths may be resonant with several transitions of the ion incorporated into the fiber.

4.4.1 Excited State Absorption

The best illustration of the effect known as Excited State Absorption is the EDFA when the pump wavelength is set around 800nm (figure 10) and NDFA (Neodymium Doped Fiber Amplifier) working at 1.3 µm (figure 11).

In the former case, the absorbed pump power for the amplification is reduced by an amount of $\sigma'_p N_2$ in equation 19, where σ'_p is the ESA cross section. The whole development stays the same with a more complex pump power evolution than equation 19:

$$\frac{dp}{dz} = -\sigma_p N_0 [\eta_p \frac{\sigma'_p}{\sigma_p} p(z) - (1 - \frac{\sigma'_p}{\sigma_p}) \frac{2\pi h\nu_p}{\sigma_p \tau_2 P_p(0)} \int_0^a B_n(r,z) r dr] \tag{22}$$

and a small-signal gain given by the equation 20.

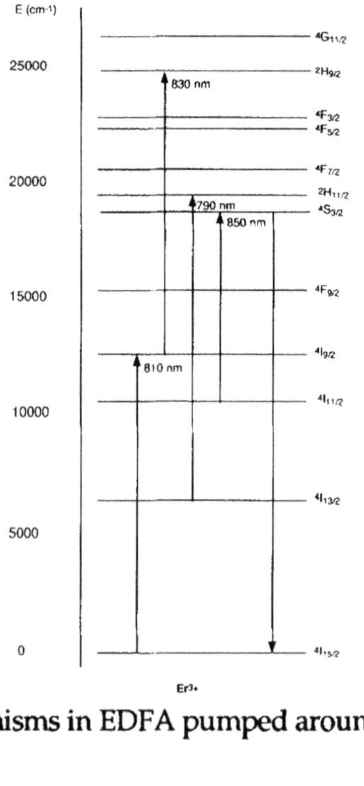

Figure 10: ESA mechanisms in EDFA pumped around 800 nm.

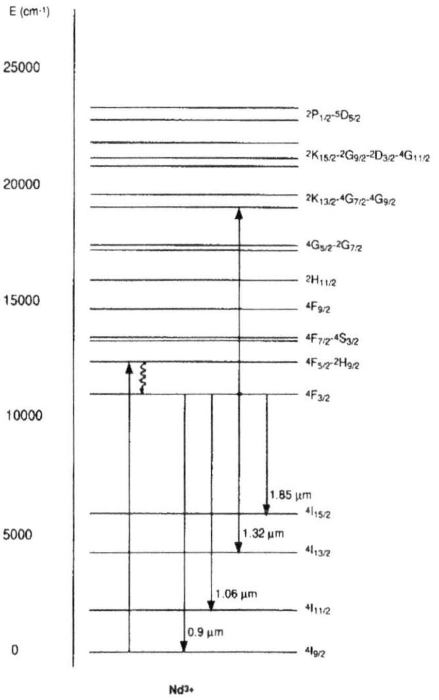

Figure 11: ESA mechanism of the signal at 1.3 µm in NDFA.

In the presence of ESA of the signal, the pump power evolution is not affected, but the gain is reduced by an amount $\sigma_a^{'} N_2$ which should be added to the GSA of the signal. In the NDFA case working at 1.3 µm, GSA is negligible, while this is not true for the PDFFA.

From an experimental point of view, pump ESA occurs in the general case but it can be greatly reduced; by choosing a pump wavelength ESA such as 980nm or 1480nm for EDFA, it can then be considered a second order effect compared to the ESA of signal for NDFA, NDFFA or PDFFA.

4.4.2 Energy Transfer Mechanisms:

To optimise the performance of fiber amplifiers or fiber lasers, co-doping (for instance: Yb,Er or Yb,Pr) requires efficient energy transfer between a strongly absorbing center called a sensitizer (or donor) and an emitting center called an acceptor. If the two species are identical, they may reduce the quantum yield by the concentration quenching effect. These issues are still open because of the lack of knowledge of parameters, such as the energy level diagram, the distribution of active species in the glass and the strength of the ion-ion interaction. Furthermore, two main types of energy transfer can be distinguished: that which provides population in higher excited states than the metastable state (the laser level) and that which depletes the population of the metastable state through cross relaxation, multiphonon relaxation or migration to traps. Since ion-ion interactions are strongly dependent upon the distance between two interacting ions, $1/r^6$ for electric dipole coupling or e^{-r} for exchange interaction, the chemical composition and the structure of the glass play a major role in determining the distribution of active centers. As mentioned before, the silica with tetrahedral coordination does not easily accept rare earth ions for which the number of nearest ligands is between 6 and 12 [24]. The condition of minimum energy is possibly reached with a non-homogeneous distribution or cluster effects. Pair-induced effects have been considered in the interpretation of fast and slow fluorescence decays of neodymium-doped silica glass without alumina but never seen in EDFA.

Delevaque & al have modeled the unsaturated residual absorption by a pair-induced quenching effect in which an upconversion energy transfer takes place between two excited erbium ions in the metastable level. An APTE (Addition de Photons par Transfert d'Energie) process occurs to produce one ion in a higher excited state which relaxes to the metastable level via multiphonon relaxation while the other ion returns to the ground state. It results in a net loss of one excited ion which is now available for absorption [52]. In their model, these authors assume the same lifetime for isolated and pair centers which is in contradiction with their findings of high transfer rate. This discrepancy can be removed if the two centers are differentiated in the rate equation description. Then, isolated and cluster centers contribute on a different time scale to the fluorescence decay. However, no evidence of fast decay has been observed yet (except at very high doping level of a few percent); this may be due to the rather small proportion of expected clusters. Another illustration of the cross relaxation effect is the behavior of the NDFFA working at 1.06 µm which involves a non-resonant energy transfer between transitions $^4F_{3/2} \rightarrow ^4I_{15/2}$ on one ion and $^4I_{9/2} \rightarrow ^4I_{15/2}$ on another neodymium ion nearby. This effect has been already well studied in laser bulk solid state lasers and the waveguide structure does not influence this 4-level system to first order.

As compared to the silica EDFA, the Erbium Doped Fluoride Fiber Amplifier (EDFFA) exhibits a concentration quenching by upconversion energy transfer, but the difference between EDFA and EDFFA occurs in the weak non-radiative rate for EDFFA favoring direct visible fluorescence from highly excited states. EDFFA works with a pump wavelength at 1480 nm because of the bottleneck created by the long lifetimes of the $^4I_{9/2}$ (1ms) and $^4I_{11/2}$ (~7ms) upper states. It is not likely to invoke pair-induced effects to explain losses of 1.5 μm emission intensity because multistep (up to five absorbed photons) upconversion mechanisms have been identified and used to explain laser action. Moreover, the ZBLAN composition includes 4% of LaF_3 which can be substituted by a similar amount of ErF_3, but this is still a subject of discussion since some authors have detected two categories of RE sites in the ZBLAN glass.

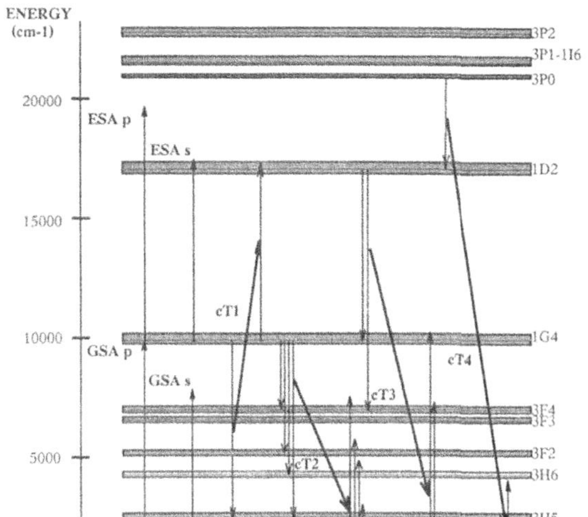

Figure 12: Main pump and relaxation mechanisms in PDFFA.

The last example (the more complex one) is the case of PDFFA where most of the mechanisms discussed above have been pointed out, except that the ESA is less significant. Because of the great number of energy level coincidences, the 1.3 µm emission due to the $^1G_4 \rightarrow ^3H_5$ transition of Pr^{3+} ions, that is already strongly non-radiative (the quantum yield is about 1.5%), suffers from cross relaxation, upconversion energy transfer and both the GSA and ESA of the signal.

As a first approach, a very simple model with an analytical gain formula:

$$G(dB/mW) \propto \frac{1}{A_{eff} h\nu_p} \sigma_{se} \tau_{\exp} \tag{23}$$

has been used where spectroscopic parameters such as transfer rates or ESA cross sections have been introduced as a single parameter in the fitting procedure. As compared to EDFA, it is suspected that the confinement of the pump and the signal is more crucial to optimise the gain. In that respect, a 30dB gain has been achieved with a 100 mW power launched into a 2.3 µm core size of high NA fiber to maintain a cutoff wavelength around 1.3 µm [68]. In that experiment, the ZBLAN fluoride fiber was doped with 2000 ppm of Pr^{3+} ions. Demonstration of 52mW output (17.2dBm) has been obtained with M-MOPA laser diodes showing a great improvement in the small packaging [69]. In order to be more useful, models must use more accurate parameters which have been recently measured by conventional spectroscopy.

4.4.3 Line Broadening Mechanisms

Considerable interest in the spectrally broad gain tunability of EDFA has been recently enhanced by the potential use for WDM transmission in which a number of channels lie in a series of equally spaced signal wavelengths. The limitation of the number of channels depends on the capability to discriminate each wavelength (narrow filters) and on the spectral response of the EDFA for long haul transmission. For an inhomogeneously broadened transition, amplifier power efficiency as well as the line width of the laser can be affected and can distort the output signal after several passes in the amplifiers. This raises a current fundamental question of line broadening of electronic transitions in amorphous media, as mentioned before. In the development of theoretical models for amplifiers, it is generally considered that the lines are homogeneously broadened which allows for an easier calculation of the gain. Formally, it should be possible to improve such calculations with the knowledge of the site distribution and a subsequent estimation of the homogeneous line broadening of individual species. Such an approach has been discussed by Desurvire [67]. As far as WDM transmissions are concerned, homogeneous line broadening and flatness of the spectral gain are required for high bit rate transmissions. As we saw before, the chemical composition of the fiber contributes largely to the assessment of both conditions. This is one of the reasons why aluminum doped silica fiber has been preferred to germanosilicate but is much less competitive than fluoride fiber for which performances as high as 16*10 Gbit/s spanning a wavelength of 24 nm over more than 531 km have been obtained [12].

4.5 THE EXAMPLE OF PDFFA

In this section, we have chosen to focus our attention on the spectroscopic investigation of the Praseodymium Doped Fluoride Fiber Amplifier for two reasons. It is still a controversial subject to consider PDFFA as a candidate for amplification in the 1.3 μm telecommunication window and almost all the mechanisms described above have been recently pointed out, while the physics and applications of NDFA or EDFA are nicely handled and understood.

Pr^{3+} is a unique ion which, embedded in a solid state, develops the three basic colors for TV as well as important laser emissions in the visible and in the infrared range. In this section, attention is mainly drawn to the 1.3 μm emission resulting from the relaxation $^1G_4 \rightarrow {}^3H_5$, but relaxation from the upper excited states $^1D_2$, $^3P_{0,1,2}$, and $^1I_6$ is considered too. The energy level diagram of Pr^{3+} ions in ZBLAN glass is presented in figure 12; All the mechanisms giving rise to possible losses limiting the performances of the PDFFA are included. In current models, only three parameters have been considered: the ESA of the pump, the ESA of the signal and energy transfer by upconversion; they are adjusted by fitting gain curves that limit their physical meaning [70]. Our approach was to measure the most intense radiative emissions of all the levels likely to store energy in the ZBLAN bulk excited at 1017 nm, allowing an estimation of the ESA cross section of the pump and energy transfer by an upconversion coefficient, called cT_1 [71].

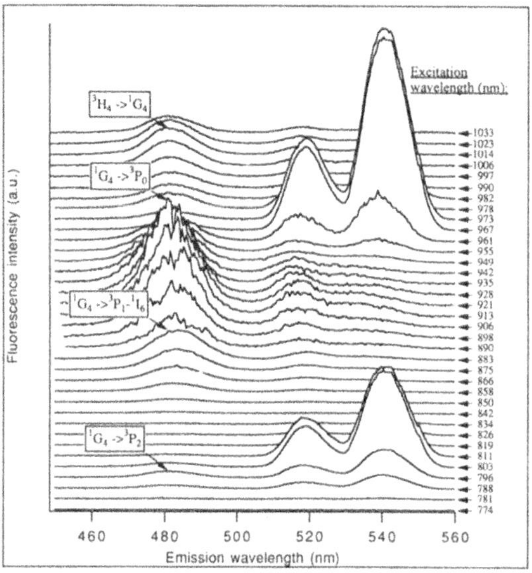

Figure 13: Upconversion pump fluorescence spectra at the output of a praseodymium-doped fluoride fiber as a function of pump wavelength [72].

Before doing this, all the mechanisms involved have been identified by the laser spectroscopy of selectively excited states, namely $^1G_4$, $^1D_2$, $^3P_{0,1,2}$ and $^1I_6$. When one uses the entire spectral range of a cw Ti-sapphire laser, the visible fluorescence spectra in the Pr^{3+} doped ZBLAN multimode fiber have been recorded with a constant power of 60mW measured at the output of the fiber (figure 13) [72]. One recognizes two types of luminescent centers: a Pr^{3+} center responsible for both the blue (480 nm) emission and the weak contribution around 520 nm assigned to the $^3P_1 \rightarrow {}^3H_5$ transition and a quite strong green fluorescence due to erbium ions which are impurities often present in rare earth raw materials. This well-known erbium fluorescence due to the relaxation of $^2H_{11/2}$ and $^4S_{3/2}$ states is pumped by a resonant ESA process via the $^4I_{9/2}$ state at 810 nm or via the $^4I_{11/2}$ state at 978 nm. For the blue fluorescence wavelength, one successively identifies the various ESA transitions as indicated on the figure.

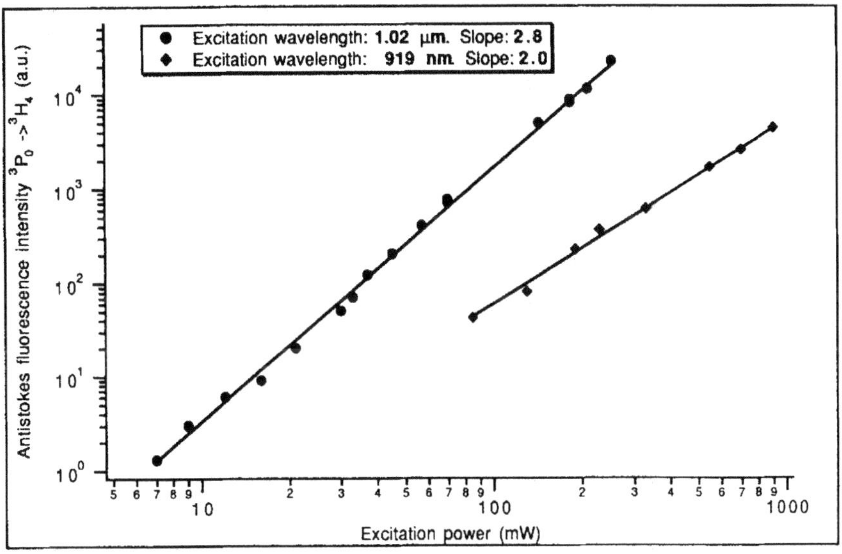

Figure 14: Power dependence of the upconverted blue fluorescence pumped at two wavelengths [72].

Moreover, the power dependence of the main excitation maximum reported on figure 14 indicates a three-photon mechanism when GSA (at 1.02 µm) is present and a two-photon process for the resonant ESA. In the latter case, the first non-resonant step (GSA) seems to be the dominant process because the upconverted fluorescence intensity decreases with wavelength. Indeed, two laser beam ESA experiments in ZBLAN: Pr^{3+} fiber or one laser beam ESA experiment of co-doped Yb-Pr ZBLAN bulk and fiber indicate more intense peaks at the $^1G_4 \rightarrow {}^3P_{0,1,2}, {}^1I_6$ wavelengths because of a larger population of the $^1G_4$ state achieved either by direct absorption or by energy transfer through ytterbium ions respectively (see figure 15).

Another particular experimental behavior is the strong line shifts with temperature involving all the infrared Pr^{3+} levels ($^3H_4$, $^3H_5$,....$^1G_4$). With the aid of IR pulsed laser excitation at 1.02 µm, the fluorescence dynamics of the metastable states ($^1G_4$, $^1D_2$, $^3P_0$) were investigated to probe the population build-up mechanism.

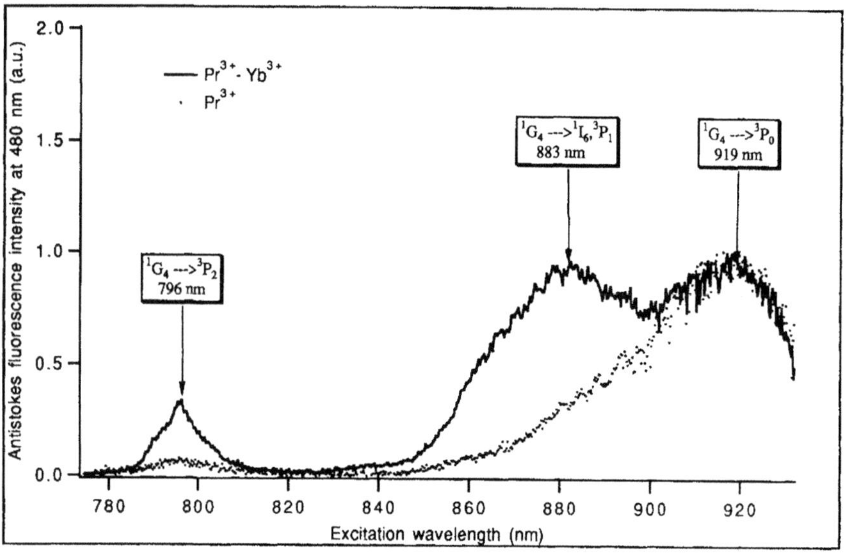

Figure 15: Excitation spectra of the blue fluorescence in Pr^{3+} and codoped Yb^{3+} fluoride fibers.

Fluorescence decays of the $^1D_2$ and the $^3P_0$ levels recorded for different Pr^{3+} concentrations are non-exponential with a rise time starting at a non-zero intensity for t=0, reaching a maximum and then decaying with time. Both the profile and its concentration dependence indicate that the dynamics of the upconversion processes involve more than one mechanism, ESA and ETU being the next likely. The red fluorescence of the $^1D_2$ state can result from ETU and ESA involving the $^3F_4$ intermediate level: $^3F_4 \rightarrow {}^1D_2$; both processes contribute to a clear quadratic dependence with pump power [72]. A simulation based on population equations is able to reproduce the shape of the experimental curves (figure 16). However, in the cw regime (the amplification regime, for instance), the ESA of the pump should not be considered because the excitation density is several orders of magnitude smaller than in our pulsed excitation measurements. Besides, in the amplification regime, the red $^1D_2$ fluorescence is greatly enhanced by efficient ESA of the 1.3 μm signal.

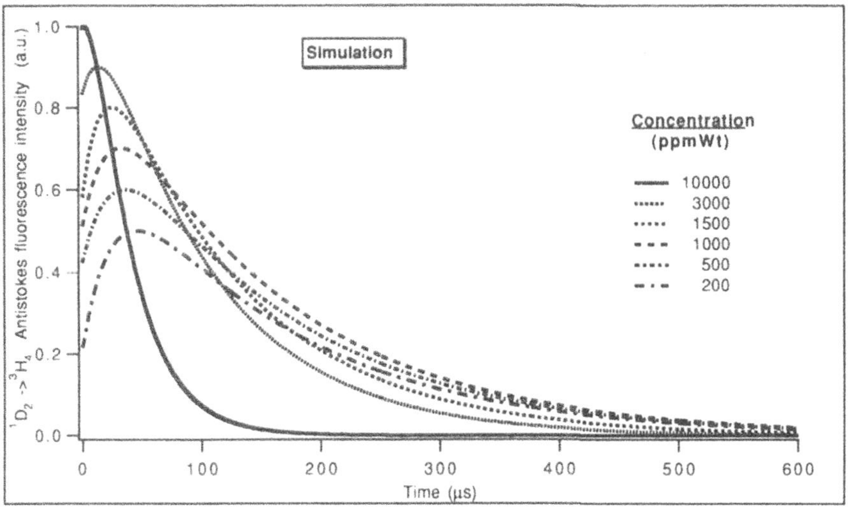

Figure 16: Simulation of the $^1D_2$ upconversion pump fluorescence decay curves as a function of praseodymium concentration [72].

Blue fluorescence dynamics were analyzed through resonant ESA at 796 nm, 883 nm and 919 nm according to a quasi-instantaneous build-up of the population of the ^3P_j states, while a rise time with a non-zero initial intensity is measured for GSA at 1.02 μm. A third power dependence of the blue fluorescence is found using CW laser excitation which indicates the role of the $^1D_2$ intermediate state and probes a three photon process (figure 14). Alternatively, in a pulsed laser experiment, the branching ratio between ESA and ETU is different because of the high photon density of the laser pulse, ESA is not negligible compared to ETU and the fluorescence dynamics are very similar to the one from the $^1D_2$ level.

5. Fiber Laser

As shown in figure 1, the wide range of laser wavelengths accessible with a rare earth-doped fiber is quite impressive and with the potential to increase because of the probable advent of low phonon energy materials. In that respect, the first laser operation has been recently reported in sulfide glasses (the La:Ga:S type) [73]. The most attractive aspects of fiber lasers is their integration with pigtailed diode laser pump sources and imprinted Bragg reflectors. However, the main drawback of rare earth-doped is their slow gain dynamics in the ms or μs range which removes any possibility of direct internal high speed modulation (GHz) and requires external electro-optic devices to achieve such a function. In this section, we emphasize more the pump mechanism scheme with various RE ions than on single frequency laser operation or short pulse laser output, though they are important for many purposes.

Nd^{3+}-doped fiber lasers have drawn the most attention because they offer an efficient four-level system and several ranges of wavelength depending upon the resonator, the chemical composition of the fiber and the pumping scheme. Besides the well known transition around 1.06 μm laser emission, a diode-pumped laser has been achieved both in the 1.34 μm band (with limited success because of the ESA of the signal), and, more interestingly in the 900nm range of the $^4F_{3/2} \rightarrow {}^4I_{9/2}$ transition. In that range, both cw and pulsed operations have been obtained earlier [74] with a quite broad tunability [75]. Moreover, picosecond pulses have been achieved with a fiber laser mode-locked with a linear external cavity [76]. Dynamic aspects of the laser involving mode competition are still a subject of theoretical work in terms of instability and chaos; this is also true for the erbium fiber laser [77]. 1.31 μm laser emission has been reported for a fluoride fiber while emission as short as 1.34 μm has only been achieved with silica based fiber which is quite far from the second telecommunication window. Furthermore, its threshold is generally high because of the competition with the main laser transition $^4F_{3/2} \rightarrow {}^4I_{11/2}$ for which most of the population inversion is used through amplified spontaneous emission [78].

As we discussed previously, fluoride fibers are best suited for upconversion pumped lasers. UV and violet laser lights have been obtained under red light excitation at 590 nm [79] in a 1000 ppm Nd^{3+}-doped fiber. A two-step excitation mechanism involving a resonant GSA is the most probable but it is to be noted that a resonant excitation wavelength between excited states in conjunction with a high doping concentration would favor an avalanche process as shown in [80].

For the case of erbium, a much greater number of laser channels as well as excitation routes have been identified resulting in visible and infrared coherent emissions. The importance of the matrix in which erbium ions are imbedded is crucial as shown in figure 1. In silica, only laser operation in the 1.55 µm band has been observed because of the small energy gaps and high phonon energy. On the other hand, several excitation wavelengths are available to pump this laser emission. Much effort has been put into achieving pulse operation as short as a few hundred of femtoseconds from a passively modelocked erbium-doped fiber laser [81, 82]. One may note single frequency operation in the range of MHz thanks to a very short cavity and Bragg grating inscriptions with at least 30 nm tunability and power as high as 60mW in a MOPA structure [83]. As we noticed before, the ZBLAN fluoride fiber allows more radiative channels as well as laser transitions; for instance a new upconversion pump (791 nm) cascade laser has been observed at 1.7 µm and 2.7 µm. The longest wavelength ever reached in a rare earth-doped fiber laser has been obtained with a Ho:ZBLAN fluoride fiber at 3.9 µm, working at room temperature [85]. A green upconversion laser has been achieved with Ho:ZBLAN with red Kr laser pump light [86], but a commercial prototype with Er:ZBLAN provides a few mW output from a standard laser diode pump [16].

Tm^{3+} exhibits a very similar situation to erbium with an early demonstration of laser action at many wavelengths in a fluorozirconate fiber [87]. In silica again only, the well known 2 µm laser emission is very efficient and provides a broad tunability around that wavelength and could be very useful for sensor applications or LIDAR techniques for pollution detection. It is best pumped around 0.8 µm in the $^3F_4$ level with a subsequent cross relaxation process; improvement has been achieved using a Bragg grating [88]. However, very efficient laser action has been recently achieved at 1.9 μm under an unusual 1.57 μm pump wavelength which is of importance for eye-safe laser applications [89]. A 1 W cw laser has been obtained with a 4 W pump at 1.47 μm making the thulium-doped fluoride fiber very attractive for pump EDFA's [90]. Of particular interest is the case of the 0.8 μm laser or an amplifier pumped around 790 nm for short distance operation in telecommunications [91]. A thulium-doped fluoride fiber is also a promising candidate for integrated blue laser devices with a three-photon pump process exhibiting a rather good efficiency but requiring unusual pump wavelengths around 1.12 μm [92]. New mechanisms such as photon avalanche may provide a better excitation route involving only one red pump wavelength and possible 450 nm laser operation [56]. Shorter wavelengths arising from $^1I_6$ or ^3P states could be reached since they have been already observed, but an easy excitation route must be found. However, the photodarkening effect has been observed in Tm-doped fluoride fibers under infrared radiation which may restrain the development of upconversion laser devices or benefit by understanding the controlling of the photosensitivity of the material [93].

The Pr^{3+} ion is a unique case (with erbium to a lesser extent) for which fiber technology has provided a tremendous of interest for both infrared emission and an upconversion pumped laser. The recent interest was mainly due to the use of the $^1G_4$ metastable level though, in most materials, its emission probability is highly quenched by multiphonon relaxation. However, its radiative lifetime is very long, generally in the ms range, as determined from absorption measurements with a Judd-Ofelt analysis [71]. Laser action with a pump at 1.017 μm, 8% slope efficiency and a maximum output of 50 mW has been obtained with butted coating dichroic mirrors on cleaved fiber endfaces [94]. For telecommunication applications, short pulse operation has been demonstrated for soliton propagation at 1.3 μm taking advantage of the very weak dispersion of the silica fiber at that wavelength [95]. The energy level diagram of trivalent praseodymium shows that upconversion pumped fluorescence arising from the ^3P level is best achieved with two wavelengths because of the necessity to populate efficiently the weakly absorbing $^1G_4$ level. A lot of work has been devoted to the optimization of these GSA + ESA processes using a model [96] or experimentally with all-solid state devices [97]. An alternative requiring only one pump wavelength involves codoped ytterbium ZBLAN fluoride fiber demonstrated by Allain et al. [98] and achieved with semiconductor lasers [99]. However, only red laser operation has been measured with that pumping process.

Trivalent ytterbium has the special behavior to operate as three and four laser systems and with the best efficiency among the rare earth ions. It has been considered only recently because of its electronic structure with absorption bands efficiently pumped by diode lasers either at 975 nm or approximately 900 nm. Its simple electronic structure as one hole in the electronic configuration which gives rise to two states split by the crystal field is responsible for an efficiency close to unity. Indeed, quenching effects involving higher or lower energy states through cross relaxation or ETU (energy transfer by upconversion) can not be considered as for other rare earth ions. Only migration to defects or pair processes to a lesser extent may be invoked to explain any reduction in efficiency. The long lifetime of the $^2F_{5/2}$ excited state is favorable for storing energy, but the absorption strength can be high depending upon the wavelength. Thus, long fiber lasers are pumped out of resonant electronic transitions. Contrary to other rare earth fibers, silica fiber is a very favorable material to operate as a three level system at 975 nm and, even more interestingly, as a four level system in the broad range 1 μm to 1.15 μm [100]. Powerful laser action at 1.02 μm has been reported [101] which may represent a serious advantage for the upconversion pumped system or for femtosecond laser operation. Finally, it should be mentioned that the optical properties of ytterbium ions have been well considered for upconversion scheme through energy transfer and ESA in the ytterbium praseodymium fluoride fiber laser, for instance [99], or the thulium-doped blue upconversion fiber laser [102].

6. Perspective

As far as silica and fluoride materials are concerned, the major laser transitions have been observed successfully. The advent of Bragg gratings as well as new available pump sources may create conditions for new transitions or new pump routes including resonances between excited states. However, one should observe more results with chalcogenides and sulphides glasses where most of the RE emissions are shifted towards a long wavelength. Another major improvement to consider for the future is the optical non-linearities of the fiber glasses and the possibility of internal high speed modulation.

References

[1] Koester, C.J. and Snitzer, E. 1963. Amplification in a fibre laser, *Appl. Optics,* **3**, 1182.

[2] Stone, J. and Burrus, C.A. 1973. Nd^{3+} doped SiO_2 lasers in end pumped fibre geometry, *Appl. Phys. Lett..,* **23**, 388.

[3] Mears, R.J., Reekie, L., Poole, S.B. and Payne, D.N. 1985. Neodymium doped silica single-mode fibre lasers, *Electron. Lett.,* **85**, 740.

[4] Mears, R.J., Reekie, L., Poole, S.B. and Payne, D.N. 1986. Low threshold, tunable cw and Q-switched fibre laser operating at 1.55 μm, *Electron. Lett.,* **22(3)**, 159.

[5] Mears, R.J., Reekie, L., Jauncey, J.M. and Payne, D.N. 1987. High-gain rare earth-doped fibre amplifier at 55 μm, *Proc. Opt. Fib. Com. (OFC)* Reno (NV).

[6] Mears, R.J., Reekie, L., Jauncey, J.M. and Payne, D.N. 1987. Low-noise erbium-doped fibre amplifier operating at 1.54 μm, *Electron. Lett.* **23**, 1026.

[7] Desurvire, E., Simpson, J.R. and Becker, P.C. 1987. *Opt. Lett.* **12**, 888.

[8] Nakazawa, M., Kimura, Y. and Suzuki, K. 1989. Efficient Er^{3+}-doped optical fibre amplifier pumped by a 1.48 μm InGaAs laser diode, *Appl. Phys. Lett.,* **54**, 295.

[9] Fevrier, H., Augé, J., Parlier, V., Bousselet, P., Dursin, A., Marcerou, J.F. and Jacquier, B. 1989. Erbium-doped fibre amplifier with outstanding gain characteristics, in *Proc. ECOC'15th*, Gothenburg.

[10] Saito, S., Imai, T., Sugie, T., Okhawa, O., Ichihashi Y., and Itoh, T. 1990. An over 2200 km coherent transmission experiment at 2.5 Gbit/s using erbium-doped fibre amplifiers, *Proc. OFC'90*, San Francisco.

[11] TAT 12: Transmission link between Europe and US with EDFA's achieved at the end of 1995.

[12] Artigaud, S., Chbat, M., Nouchi, P., Chiquet, F., Bayart, D., Hamon, L., Pitel, A., Goudeseune, F., Bousselet, P. and Beylat, J.L. 1996. Transmission of 16*10 Gbit/s channels spanning 24 nm over 531 km of conventional single-mode fibre using 7 in-line fluoride based EDFAs, *Proc. OFC'96*, post deadline paper.

[13] Nakazawa, M., Suzuki, K.,Yamada, E. and Kimura, Y. 1990. Dynamical optical soliton communication, *IEEE J. Quantum Electron.*, 26, 295.

[14] Davey, S.T. and France, P.W. 1989. Rare earth doped fluorozirconate glasses for fibre devices, *British, Telecom. Technol. J.* 7, 58; France P.W., 1989. A review of fluoride glass fibre lasers, in *Proc. 6th Int. Symp. Halide Glasses*, Clausthal-Zellerfeld, 51.

[15] Yeh, D.C., Sibley, W.A., Schneider, I., Afzal, R.S. and Aggarwal, I. 1987. Intensity-dependent upconversion efficiencies of Er^{3+} ions in heavy-metal fluoride glass, *J. Appl. Phys.*, 69(3), 1648.

[16] Peilher, D. 1993. Upconversion process creates compact blue/green lasers, *Laser Focus World*, Nov. 1993, 95; Hanna, D., Tropper, A. 1995. Fibre lasers offer higher powers and shorter wavelengths, *Laser Focus World*, May 1995, 123.

[17] Ohishi, Y., Kanamori, T., Kitagawa, T., Takahashi, S., Snitzer, E. and Sigel, G.H. Jr., 1991, Pr^{3+}-doped fluoride fibre amplifier operating at 1.31 µm, *Proc. OFC'91*, San Diego.

[18] Durteste, Y., Monerie, M., Allain, J.Y., and Poignant, H. 1991. Amplification and lasing at 1.3 μm in praseodymium-doped fluorozirconate single mode fibres, *Electron. Lett.*, **27**, 626.

[19] Carter, S.P., Szebesta, D., Davey, S.T., Wyatt, R., Brierley, M.C. and France, P.W. 1991. Amplification at 1.3 μm in a Pr^{3+}-doped single mode fluorozirconate fibres, *Electron. Lett.*, **27**, 628.

[20] Hill, K.O., Fujii, Y., Johnson, D.C. and Kawasaki, B.S. 1978. Photosensitivity in optical fiber waveguides: Applications to réflection filter fabrication, *Appl. Phys. Lett.*, **32**, 647.

[21] Niay, P. 1994. Bragg grating inscription in photosensitive fibres and glasses, *Proc. R.E. Doped Optical Waveguides*, Les Houches, June 1994, **11**, 697.

[22] Xie, W.X., Bernage, P., Ramecourt, D., Douay, M., Taunay, T., Niay, P., Boulard, B., Gao, Y., Jacoboni, C., Da Costa, A., Poignant, H. and Monerie, M. 1996. UV induced permanent gratings in Ce^{3+}or Eu^{2+} doped PZG glass thin-film waveguides deposited on CaF_2 substrates, *Optics Com.* Submitted.

[23] Miniscalco, W.J. 1993. Optical and electronic properties of rare earth ions in glasses, in *Rare Earth Doped Fiber Lasers and Amplifiers* ed. by Digonnet, M.J.F., M. Dekker Inc., 19.

[24] Caro, P.E., Antic, E., Beaury, L., Beaury, O., Derouet, J., Faucher, M., Guttel, C., Moune, O.K. and Porcher, P. 1977. L'évolution des niveaux d'énergie des configurations $4f^n$ en phase solide en fonction des structures cristallines, in *Proc. Spectroscopie des Eléments de Transition et des éléments lourds dans les solides*, ed. Colloques internationaux C.N.R.S. n°255, 71.

[25] Arai, K., Namikawa, H., Kumata, K., Honda, T., Ishii, Y.and Handa, T. 1986. Aluminium or phosphorus co-doping effects on the fluorescence and structural properties of neodymium-doped silica glass, *J. Appl. Phys.*, **59(10)**, 3430.

[26] Poulain, M., Poulain, M. and Lucas, J. 1975. Verres fluorés au tétrafluorure de zirconium: Propriétés optiques d'un verre dopé au Nd^{3+} , *Mat. Res. Bull.*, **10**, 243.

[27] Adam, J.L., Matecki, M., L'Helgouach, H.and Jacquier, B. 1994. Non-radiative emissions in Er^{3+}-doped chloro-fluoride glasses, *Eur. J. Solid State Inorg. Chem.*, **31**, 337.

[28] France, P.W., Drexhage, M.G., Parker, J.M., Moore, M.W., Carter, S.F. and Wright, J.V. 1990. Fluoride Glass Optical fibers, *France P.W. ed.*, CRC Press Boca Raton (FL).

[29] Guery, C., Adam, J.L. and Lucas, J. 1988. Optical properties of Tm^{3+}ions in indium-based fluoride glasses, *J. Luminescence*, **42**, 181.

Adam J.L., Rigout, N., Denoue, E., Smektala, F. and Lucas, J., 1991. Optical properties of Ba-In-Ga-based fluoride glasses for amplification at 1.3 µm, *SPIE*, **1581**, 155.

[30] Becker, P.C., Broer, M.M., Lambrecht, V.G., Bruce, A.J. and Nykolak, G. 1992. Pr^{3+}: La-Ga-S: a promising material for 1.3 µm fiber amplifier, *Proc.OOA'92*, Santa Fé, 1992, PD16.

[31] Matecki, M. and Poulain, M. 1992. *J. Non-Cryst. Solids*, **140**, 82.

[32] Hewak, D.W, Deol, R.S., Wang, J., Wylangowski, G., Mederios, Neto, J.A., Samson, B.N., Laming, R.I., Brocklesby, W.S., Payne, D.N., Jha, A., Poulain, M., Otero, S., Surinach, S. and Baro, M.D. 1993. Low phonon-energy glasses for efficient 1.3 µm optical fibre amplifier, ***Electron. Lett.***, **29**, 237.

[33] Li, T. 1985. *Optical Fiber Communications Vol.1, Fibre Fabrication*, Academic Press, New York.

Miller S.E. and Kaminow I.P., 1988, *Optical Fiber Telecommunications Vol. II*, Academic Press, New York.

[34] Boulard, B. and Jacoboni, C. 1990. Preparation of fluoride glass films by evaporation, *Mat. Res. Bull.*, **25**, 671.

Perrot O., Boulard, B. and Jacoboni, C. 1993. Physical vapour deposition of rare earth fluorides, *J. Non-Cryst. Solids*, **161**, 23.

[35] Yanagita, H., Itoh, K., Ishikawa, E., Aoki, H. and Toratani, H. 1995. 26 dB amplification at 1.31μm in a novel Pr^{3+}-doped InF_3/GaF_3 based fibre, *Proc. OFC'95*, PD 2.

[36] Wybourne B.G. 1965. *Spectroscopic Properties of Rare Earths,* Wiley, New York.

Dieke G.H. 1968. *Spectra and Energy Levels of Rare Earth Ions in Crystals,* Wiley, New York.

Hüfner S., 1978, *Optical Spectra of Transparent Rare Earth Compounds,* Academic Press, New York.

[37] For a full treatment see for example: Yariv A. 1975. *Quantum Electronics 2nd Edition,* Wiley, New York, 158.

[38] Judd, B.R. 1962. Optical absorption intensities of rare earth ions, *Phys. Rev.,* **127**, 750.

[39] Ofelt, G.S. 1962. Intensities of crystal spectra of rare earth ions, *J. Chem. Phys.,* **37**, 511.

[40] Miniscalco, W.J. and Quimby, R.S. 1991. General procedure for the analysis of cross sections, *Optics Lett.,* **16(4)**, 258.

[41] McCumber, D.E. 1964. Theory of phonon-terminated optical masers, *Phys. Rev.,* **134(2A)**, 299.

[42] Edwards, J.G. 1968. Measurement of the cross-section for stimulated emission in neodymium-doped glass from the output of a free-running oscillator, *Br. J. Phys. D,* **1**, 449.

[43] Barnes, W.L., Laming, R.I., Tarbox, E.J. and Morkel, P.R. 1991. Absorption and Emission cross sections of Er^{3+}doped silica fibers, *IEEE J. Quantum Electron.,* **27(4)**, 1004.

[44] Riseberg, L.A. and H.W. Moos, 1968. Multiphonon orbit-lattice relaxation of rare earth ions in crystals, *Phys. Rev.,* **174**, 429.

[45] Reisfeld, R., in *Structure and bonding,* Dunity, J.D., Ed., Springer-Verlag, Berlin, 1975, 123.

[46] Reifeld, R. and Jorgensen, C.K. 1987. Excited State Phenomena in vitreous materials, Elsevier, Amsterdam.

[47] Samson, B.N., Medeiros Neto, J.A., Laming, R.I. and Hewak, D.W. 1994. Dysprosium doped Ga-La -S glass for an efficient optical fibre amplifier operating 1.3 µm, *Electron. Lett.*, **30**, 1617.

[48] Macfarlane, R.M. and Shelby, R.M. 1987. Homogeneous line broadening of optical transitions of ions and molecules in glasses, *J. Luminescence*, **36**, 179.

[49] Briançon, A.M., Jacquier, B., Gâcon, J.C., Le Sergent, C., and Marcerou, J.F. 1990. Inhomogeneous line broadening of optical transitions in Er^{3+} and Nd^{3+} doped preforms and fibres, in *Proc. SPIE Conf. On Fiber Laser Sources and Amplifiers 11*, **1373**, 9.

[50] Miyajima, Y., Komurai, T., Sugawa, T. and Yamamoto, T. 1994. Rare earth- doped fluoride amplifiers and fiber lasers, *Opt. Fiber Technol.*, **1**, 35.

[51] See for example: Auzel, F. 1980. Multiphonon processes, cross-relaxation and upconversion in ion-activated solids, exemplified in minilaser materials, in *Radiationless processes*, ed. By Di Bartolo, B., Nato series, Plenum New York, 213.

[52] Delevaque, E., Georges, T., Monerie, M., Lamouler, P. and Bayon, J.F. 1993. Modeling of pair-induced quenching in erbium-doped silicate fibers, *IEEE Phot. Technol. Lett.*, **5**, 73.

[53] Chivian, J.S., Case, W.E. and Eden, D.D. 1979. The photon avalanche: A new phenomenon in Pr^{3+}-based infrared quantum counters, *Appl. Phys. Lett.*, **35**, 124.

[54] Tong, F., Macfarlane, R.M. and Lenth, W. 1989. Laser emission at 413 and 730 nm in upconversion-pumped $YLiF_4$:Nd^{3+}, *Proc. QELS' 89*, paper THKK4.

Lenth, W. and Macfarlane, R.M. 1990. Excitation mechanisms for upconversion lasers, *J. Luminescence*, **45**, 346.

[55] Auzel, F., Chen, Y. and Meichenin, D. 1993. Room temperature photon avalanche up-conversion in Er doped ZBLAN glass, *J. Luminescence*, **60&61**, 692.

[56] Guy, S., Shepherd, D.P., Joubert, M.F., Jacquier, B. and Poignant, H. 1996. Blue avalanche upconversion in Tm: ZBLAN fibre, *J. Opt. Soc. Am. B.*

[57] Joubert, M.F., Guy, S., Jacquier, B. and Linarès, C. 1994. The photon-avalanche effect: review, model and application, *Opt. Mat.*, **4**, 43.

[58] Joubert, M.F., Guy, S., Linarès, C., Jacquier, B. and Adam, J.L. 1995. Avalanche upconversion in Tm^{3+}-doped BIGaZYTZr glass, *J. Non-Cryst. Solids*, **184**, 98.

[59] Lemaire, P.J., Atkins, R.M., Mizrahi, V. and Reed, W.A. 1993. High pressure H_2 loading as a technique for achieving ultrahigh UV photosensitivity and thermal sensitivity in GeO_2 doped optical fibres, *Electron. Lett.*, **29**, 1191.

[60] Russel, P.St.J., Hand, D.P., Chow, Y.T. and Poyntz-Wright, L.J. 1991. Optically induced creation, transformation and organisation of defects and color-centres in optical fibres, in *Int. Workshop on Photoinduced Self-Organisation Effect in Optical Fibre*, Ouellette, F., Ed. Proc. Soc. Photo-opt. Instrum. Eng., **1516**, 47.

[61] Sceats, M.G. and Poole, S.B. 1991. Stress-relief the mechanism of photorefractive index control in fibre cores, in *Proc. Of the 16th Australian Conf. On Opt. Fobre Technol.*, Adelaide, 1991, 302.

[62] Poumellec, B., Niay, P., Douai, M. and Bayon, J.F. 1995. UV induced densification during Bragg grating writing, in *Technical Digest 22 OSA*, 1995, Photosensitivity and quadratic non linearity in glass waveguides: fundamentals and applications, Portland, Paper SaD4, 54.

[63] Taunay, T., Bernage, P., Douay, M., Xié, W.X., Martinelli, G., Niay, P., Bayon, J.F., Delevaque, E. and Poignant, H. 1996. UV enhanced photosensitivity in cerium doped alumino-silicate fibres and glasses through high pressure H_2 loading, *J.O.S.A:B*, submitted.

Taunay, T., Bernage, P., Martinelli, G., Douay, M., Niay, P., Bayon, J.F. and Poignant, H. 1996. Photosensitisation of terbium doped alumino-silicate fibres through high pressure H_2 loading, *Optics Com.*, submitted.

[64] Poignant, H., Boj, S., Delevaque, E., Monerie, M., Taunay, T., Niay, P., Bernage, P. and Xié, W.X. 1994. Efficiency and thermal behaviuor of cerium- doped fluorozirconate glass fibre Bragg gratings, *Electron. Lett.*, **30**, 1339.

[65] Xié, W.X., Bernage, P., Ramecourt, D., Douay, M., Taunay, T., Niay, P., Boulard, B., Gao, Y., Jacoboni, C., da Costa, A., Poignant, H. and Monerie, M. 1996. UV induced permanent gratings in Ce^{3+} or Eu^{2+} doped PZG glass thin-film waveguides deposited on CaF_2 substrates, submitted.

[66] Bouazaoui, M., Jacquier, B. and Poignant, H. 1995. Electronic Raman scattering in cerium-doped fluorozirconate fibres, *Opt. Mat.*, **4**, 553.

[67] Desurvire, E. 1994. Erbium-doped fiber amplifiers, ed. *J. Wiley & Sons*, N.Y.

[68] Miyajima, Y., Sugawa, T. and Fukasaku, Y. 1991. 38.2 dB amplification at 1.31 μm and possibility of 0.98 μm pumping in Pr^{3+}-doped fluoride fibre, *Electron. Lett.*, **27**, 1706.

[69] Sanders, S., Dzurko, K., Parke, R., O'Brien, S., Welch, D.F., Grubb, S.G., Nykolak, G. and Becker, P.C. 1996. Praseodymium doped fibre amplifiers (PDFAs) pumped by monolithic master oscillator power amplifier (M-MOPA) laser diodes, *Electron. Lett.*, **32**, 343.

[70] Ohishi, Y., Kanamori, T., Terunuma, Y., Shimizu, M., Yamada, M. and Sudo, S. 1994. Investigation of efficient pump scheme for Pr^{3+}-doped fluoride fiber amplifier, *IEEE Photonics Technol. Lett.*, 1994, **6**, 195.

Pedersen, B., Miniscalco, W.J. and Quimby, R.S. 1992, Optimization of Pr^{3+}: ZBLAN fiber amplifier, *IEEE Photonics Technol. Lett.*, **4**, 446.

[71] Remillieux, A., Jacquier, B., Linarès, C., Lesergnet, C., Artigaud, S., Bayart, D., Hamon, L. and Beylat, J.L. 1996. Upconversion mechanisms of a praseodymium- doped fluoride fibre amplifier, *J. Phys. D: Appl. Phys.*, **26**, 1.

[72] Remillieux, A. and Jacquier, B. 1996. IR-to-visible conversion mechanisms in Pr^{3+}-doped ZBLAN fluoride glasses and fibres, *J. of Lumin.*, accepted.

[73] Schweiser, T., D.W. Hewak, D.N. Payne, T. Jensen and G. Huber, 1996, Rare earth doped chalcogenide glass laser, *Electron. Lett.*, **32**, 666.

[74] Jauncey, I.M., J.T. Lin, L. Reekie and R.J. Mears, 1986, Efficient diode pumped CW and Q-Switched single-mode fibre laser, *Electron. Lett.*, **22**, 199.

[75] Alcock, I.P., Ferguson, A.I., Hanna, D.C. and Tropper, A.C. 1986. Tunable continuous-wave neodymium-doped monomode-fibre laser operating at 0.9-0.945 and 1.070-1.135 μm, *Opt. Lett.*, **11**, 709.

[76] Sargsjan, G., Stamm, U., Unger, C. and Muller, M. 1991. Picosecond pulses from a neodymium-doped fibre laser mode-locked with a linear external cavity, *SPIE, Fiber Laser Sources and Amplifiers*, Boston, MA, Sept 1991.

[77] Bielawski, S., Derozier, D. and Glorieux, P. 1992. Antiphase dynamics and polarisation in the Nd-doped fibre laser, *Phys. Rev. A*, **46**, 2811.

[78] Digonnet, M.J.F. and Snitzer, E. 1993. Nd^{3+} and Er^{3+}-doped silica fibre lasers in *Rare Earth Doped Fibre Lasers and Amplifiers*, Digonnet, M.J.F., ed. Dekker Inc, N.Y.

[79] Funk, D.S., Carlson, J.W. and J.G. Eden, 1994. Ultraviolet (381 nm), room temperature laser in neodymium-doped fluorozirconate fibre, *Electron. Lett.*, **30**, 1859.

[80] Jacquier, B., Remillieux, A., Joubert, M.F., Christensen, P. and Poignant, H. 1993. Upconversion and relaxation of high lying states in rare earth-doped ZBLAN bulk and fibres, *J. Non-Crystalline Solids*, **161**, 241.

[81] Tamura, K., Ippen, E.P. and Haus, H.A. 1994. Optimisation of filtering in soliton fibre lasers, *IEEE Photonics Technol. Lett.*, **6**, 1433.

[82] Tamura, K., Kimura, Y. and Nakazawa, M. 1995. Femtosecond pulse generation over 82 nm wavelength span from passively modelocked erbium- doped fibre laser, *Electron. Lett.*, **31**, 1062.

[83] Ball, G.A., Holton, C.E., Hull-Allen, G. and Morey, W.W. 1994. 60mW 1.5 μm single-frequency low-noise fibre laser MOPA, *IEEE Photon. Technol. Lett.*, **6**, 192.

[84] Ghisler, C., Pollnau, M., Bunea, G., Bunea, M., Lüthy, W. and Weber, H.P. 1995. Upconversion cascade laser at 1.7 μm with simultaneous 2.7 μm lasing in erbium ZBLAN fibre, *Electron. Lett.*, **31**, 373.

[85] Schneider, J. 1995. Fluoride fibre laser operating at 3.9 μm, *Electron. Lett.*, **31**, 1250.

[86] Allain, J.Y., Monerie, M. and Poignant, H. 1990. Room temperature cw tunable green upconversion holmium fibre laser, *Electron. Lett.*, **26**, 261.

[87] Allain, J.Y., Monerie, M. and Poignant, H. 1989. Tunable CW lasing around 0.82, 1.48, 1.88 and 2.35 μm in thulium-doped fluorozirconate fibre, *Electron. Lett.*, **25**, 1660.

[88] Delevaque, E., Boj, S., Allain, J.Y., Bayon, J.F., Niay, P., Bernage, P., Saccoccio, M. and Aubry, Y. 1994. Lasers à fibre à haut rendement pompés par diode àcavité photoinscrite accordée dans la bande 1.9 à 2.1 μm, *JNOG'94*, Proc. JNOG 94, Besançon, 124.

[89] Yamamoto, T., Miyajima, Y. and Komukai, T. 1994. 1.9 μm Tm-doped silica fibre laser pumped at 1.57 μm, *Electron. Lett.*, **30**, 220.

[90] Miyajima, Y., Komukai, T. and Sugawa, T. 1993. 1-W cw Tm-doped fluoride fibre laser at 1.47, *Electron. Lett.*, **29**, 660.

[91] Carter, N.J., Smart, R.G., Hanna, D.C. and Tropper, A.C. 1990. Lasing and amplification in the 0.8 μm region in thulium doped fluorozirconate fibres, *Electron. Lett.*, **26**, 1759.

[92] Grubb, S.G., Bennett, K.W., Cannon, R.S. and Humer, W.F. 1992. Blue upconversion fibre laser, *Electron. Lett.*, **28**, 1243.

[93] Barber, P.R., Paschotta, R., Tropper, A.C. and Hanna, D.C. 1995. Infrared- induced photdarkening in Tm-doped fluotide fibres, *Optics Lett.*, **20**, 2195.

[94] Döring, H., Peupelmann, J. and Wenzel, F. 1995. Pr^{3+}-doped 1.3 μm fibre laser using direct coated dichroic mirrors, *Electron. Lett.*, **31**, 1068.

[95] Peter, D.S., Onishchukov, G., Hodel, W. and Weber, H.P. 1994. 570 fs pulses from a Pr^{3+}-doped fibre laser modelocked by pump pulse induced cross-phase modulation, *Electron. Lett.*, **30**, 1595.

[96] Tropper, A.C., Carter, J.N., Lauder, R.D.T., Hanna, D.C., Davey, S.T. and Szebesta, D. 1994. Analysis of blue and red laser performance of the infrared- pumped praseodymium-doped fluoride fibre laser, *J. Opt. Soc. Am. B*, **11**, 886.

[97] Zhao, Y. and Fleming, S. 1996. All-solid state and all-fibre blue upconversion laser, *Electron. Lett.*, **32**, 1199.

[98] Allain, J.Y., Monerie, M. and Poignant, H. 1991. Red upconversion Yb- sensitised Pr fluoride fibre laser pumped in 0.8 μm region, *Electron. Lett.*, **27**, 1156.

[99] Baney, D.M., Yang, L., Ratcliff, J. and Kok Wai Chang, 1995. Red and orange Pr^{3+}/Yb^{3+} doped ZBLAN fibre upconversions lasers, *Electron. Lett.*, **31**, 1842.

[100] Hanna, D.C., Percival, R.M., Perry, I.R., Smart, R.G., Suni, P.J. and Tropper, A.C. 1987. Yb-doped monomode fibre laser: Broadly tunable operation from 1.010 μm to 1.162 μm and 3-level operation at 974 nm, *J. of Modern Optics*, **37**, 329.

[101] Allain, J.Y., Bayon, J.F., Monerie, M., Bernage, P. and Niay, P. 1993. Ytterbium-doped silica fibre laser with intracore Braggs gratings operating at 1.2 μm, *Electron. Lett.*, **29**, 309.

[102] Mackenzie, C.J., Barnes, W.L., Carman, R.J., Townsend, J.E. and Hanna, D.C. 1993. High power ytterbium doped fibre laser operating at around 1.2 μm, *Proc. Advanced Solid- State Lasers*, Ed. A. Pinto, New Orleans, **15**, 192.

INDEX

GPSR Compliance

The European Union's (EU) General Product Safety Regulation (GPSR) is a set of rules that requires consumer products to be safe and our obligations to ensure this.

If you have any concerns about our products, you can contact us on ProductSafety@springernature.com

In case Publisher is established outside the EU, the EU authorized representative is:

Springer Nature Customer Service Center GmbH
Europaplatz 3
69115 Heidelberg, Germany

Batch number: 10755284

Printed by Printforce, the Netherlands